Classic Queen Rearing Compendium

by G.M. Doolittle,
C.C. Miller,
Henry Alley,
Jay Smith,
Isaac Hopkins
and Frank Pellett

Overview by Michael Bush

Classic Queen Rearing Compendium

ISBN 978-161476-059-7

Introduction

When I first got interested in Queen Rearing these were the books I worked so hard to find. In my opinion they are the best books available on queen rearing and I wanted to make them available to the public. I put most of these on my web site after I finally found them, but wanted to make them available in a book as well. I think these men laid the groundwork and worked out the details of practical systems of queen rearing. Pellet did a lot of distilling down of the various systems, which is very valuable, Smith perfected a lot of what Doolittle started. Miller and Hopkins were the queen rearing examples for the beekeeper who just wanted a few good queens for themselves.

If you want to raise queens, these books are the place to start.

For the reader's convenience some small sections are repeated e.g. the Hopkins method from Pellet and from Hopkins are put together in the Hopkins section but Pellet's version is also in *Practical Queen Rearing* where it occurs in the original. The Hopkins queen rearing chapter from 1886 and 1911 are both included because there are some significant differences and both are very useful.

I wrote the overview because while there are a lot of queen rearing books with a lot of good information, some of the basic concepts are often not clearly delineated. I wanted to clarify some of the principles so that these classic books might make more sense to someone who is new to queen rearing. Also I wanted to point out reasons and goals today for queen rearing, which have changed somewhat from the time when these were written.

Contents

An Overview Of Queen Rearing

by Michael Bush

An Overview of Queen Rearing

X-Star Publishing Company
Founded 1961

Table of Contents

Genetics

The Need for Genetic Diversity

In any species that uses sexual reproduction, genetic diversity is essential for the overall success and health of the species. A lack of it leaves the population vulnerable to any new pest, disease or problem that comes along. A lot of it greatly improves the odds of having the necessary traits to survive such things. This need seems at odds with the concept of selective breeding, and to some extent it is. Selective breeding is just that—selective. Meaning you breed out traits you don't like. Of course this narrows the gene pool, hopefully in a positive way, but still it limits the variety as you keep selecting from fewer and fewer ancestors. Whether you believe in a Creator or evolution as the origin of nature, sexual reproduction has as its obvious goal, diversity. The queen mates, not just with one drone, but several, the hives make many drones to keep their genes out there, and even a hive doomed to die from queenlessness will put drones out there to try to preserve those genes in the pool. Every disease narrows the pool to only those that can survive that disease, and every pest narrows the pool to only those that can survive that pest. We beekeepers keep limiting that pool even more by selecting one queen and raising thousands of queens from her, something that never happens in nature, and by buying queens from only a few breeders, who do the same and who share stock with each other, we narrow it even more. The more we narrow the gene pool the less likely it is that the remaining genes will be sufficient to survive the next onslaught of diseases and pests. This is a scary prospect. And all of this is ignoring the built in control over this with the bees' method of gender control being sex alleles that limit the success of inbred bees. An inbred line of bees has many diploid (fertilized) drone eggs (because similar sex alleles line up) that will not be allowed by the bees to develop.

Feral Bees Have Maintained This

The depth of the gene pool, for many years, has been maintained by the large pool of feral bees. In recent years, however this pool has shrunk significantly from the influx of diseases and pests not to mention loss of habitat, use of pesticides and fear of AHB.

What can we do?

We cannot propagate bees with a limited gene pool and expect them to survive, let alone thrive. So what can we do to promote genetic diversity and still improve the breed of bees we raise? We can change our view from picking only the one best queen we have for the mother and the next best for the drone mother and start thinking in terms, instead of only breeding out the worst. In other words, if a queen has bad traits we don't want, such as bad tempered workers, then we cull those out. But if they have good traits we don't try to replace them with only the genetics of our best queen, but rather try to keep that line going by doing splits, or raising queens, or using the drones from those other lines. Don't use the same mother for every batch of queens. Don't requeen feral colonies that you remove or feral swarms that you catch. If a hive is hot but has other good traits, try raising a daughter and see if you can lose that trait instead of just wiping out that queen's line. Raise your own bees from the local survivors instead of buying queens. Raise your own bees even from the commercial queens you have so they will mate with the feral survivors. Support small local queen breeders so they can keep more genetic lines going. Do more splits and let them raise their own queens rather than buying queens, so that each colony can continue their line.

Feral Bees

There is much talk that the feral bees died. In my observation there was a serious shift in what I found when catching feral bees. I used to find "leather" colored Italian looking bees. Now I'm finding more black bees with a little brown mixed in. I'm breeding these survivor bees for myself and for sale

Typically I'm asked how I know these are feral survivors instead of recent escapees. First, they act differently than any of the domestics. Just little things, mostly, but also they overwinter in very small clusters and are very frugal. They are also very variable in aspects that are usually bred for, like propolis or being runny. Also they are typically smaller when you find them being from natural sized comb.

Swarms

...are the easiest way to get feral bees. But a lot of swarms are, and a lot of swarms aren't, feral bees. I'd take them either way, but if you're looking for feral survivor bees to raise queens then look for the smaller bees. Swarms with small bees are probably feral survivors. Swarms with larger bees are probably swarms from someone's hive. To get swarms, notify the local police and rescue people and the county agricultural extension office. If you want to do a lot of them run a yellow pages ad for swarm removal.

Capturing a swarm

Much has been written and each situation is both similar and unique. A swarm is a bunch of homeless bees with a queen. They may have already decided where they think they want to go, or they may still have scouts out looking. Swarms usually happen in the morning and they usually leave by early afternoon, but they may swarm in the afternoon and they may leave in a few minutes or a few days. If you chase swarms you will often get there too late and often get there in time. Both will happen. It's best to have all your equipment with you all the time. If you have to go get your equipment, you will probably be too late. Have a box with a screened bottom attached. This can be attached by nail little squares of plywood into both the box and the bottom or with the 2" wide staples that are sold by bee suppliers for moving hives. You need a lid. I like a migratory cover because it's simple. Less moving parts. I like to have a #8 screen cut and bent to 90 degrees to block the door (but not attached yet). A stapler is nice for anchoring the screen to the door and the cover to the hive. The best are the ones labeled as light duty staplers instead of the heavy duty ones. They penetrate better and stay better. I don't know why. The ones that take the T50 staples are *not* the right ones, although if you already have one you can use it. The ones that take the J21 staples are easier to use. You need a veil minimum, but I like a jacket or a suit. Gloves and a brush are helpful. You can make or buy a rig with a 5 gallon bucket to knock them down with. The idea is that you add EMT (conduit) to it as a long handle and you slam it under the swarm to dislodge them into the bucket. Then you pull on the rope to put the lid on and lower the whole thing back down and dump them into a box. The main trick to swarms is to get the queen. If you can reach and see, try to find the queen. If you know you see her and can make sure she ends up in the box, close it up, brush off the stragglers and leave. If you're not sure, then let them settle in. It helps if the box smells like lemongrass essential oil. Either put some lemongrass essential oil in it (lasts longer) some swarm lure (costs more but works well) or actually spray some lemon pledge (cheap, easy to find, but doesn't last as long) in the box before you put the swarm in. If you pay attention when you buy a package or hive a swarm you'll notice it's what they smell like. Sometimes they will settle into the box. Sometimes you didn't get the queen, or she likes the branch she was on better, and they all start accumulating on the branch again. I just keep shaking them in until they stay. It usually works. In my observation, honey, brood etc. are no help in hiving a swarm although they may help anchor them once they have decided to move in. They are not look-ing for an occupied house, they are looking for an emp-ty or abandoned house. Old empty comb sometimes helps. Some brood might help anchor them so they don't leave though. It's also well worth having some Queen Mandibular Pheromone. You can either keep your old retired queens in a jar of alcohol (queen juice), or buy Bee Boost (last I checked available from Mann Lake).

Always wear protective equipment. Swarms don't usually get mean, but can be unpredictable. Also be careful of power lines and falling off of ladders. It sounds redundant, but when a lot of bees are buzzing you, and especially if one gets in your bonnet, it's hard to stay calm, but it is a requirement if you are on top of a ladder.

My current favorite method for getting a swarm is skip ladders altogether. Take enough boxes to make a good size (one deep, two mediums) and preferable ones that have been lived in. Some old comb if you have it. Some QMP (a quarter of a stick of Bee Boost or the end of a Q-tip dipped in queen juice) and some lemongrass essential oil. Dip the other end of the Q-tip in the lemongrass oil. Drop the Q-tip in the hive, put the lid on, put it near the swarm and come back after dark. They probably will have moved into the box. Staple screen over the entrance and take them home.

Removal

Sometimes called a "cut out". This is not the easiest way to get bees. It is exciting and fun, but sometimes requires some construction skills and lots of courage. The idea is to remove all of the bees and all of the combs from a tree, a house, or whatever they are living in. It often involves removing sections of walls and someone repairing them afterwards. It is not usually financially worth it unless you are being paid to remove them or you have a lot of free time.

Each removal is a separate situation. Sometimes they are in an old abandoned building and the owner doesn't care if you rip the wallboard off or tear the siding off. Usually it does matter and you can't go tearing it up, you have to put everything back when you are done or make it clear to the homeowner that they will need to hire a carpenter to do so. Ignoring, for the moment, the construction issues, if you get to the combs, whether they are in a house or a tree or whatever, you need to cut the brood to fit frames and tie around the frames to hold it in. This does not work well for honey, especially in new comb, because it's too heavy, so scrap the honey. Throw it in a five gallon bucket with a lid to keep out the bees trying to clean up the spill. Try to put the brood in an empty hive box and keep brushing or shaking the bees off into it. If you see the queen, then catch her with a hair clip queen catcher or put her in a cage and put her in the hive box. If you get some brood and the queen in the hive box the rest of the bees will eventually follow. If you don't see the queen, then just keep putting bees in the box and brood comb in frames in the box and honey in the bucket until the combs are all gone. Take the bucket and, if you can, leave for a few hours and let the bees figure out where the queen and the other bees are. The will all settle into the new box. At dark they should all be inside and you can close it up and take it home.

Cone Method

This method is used when it's impractical to tear into a hive and remove the comb or there are so many bees you don't want to face them all at once. This is a method where a screen wire cone is placed over the main entrance of the current home of the bees. All other entrances are blocked with screen wire stapled over them. Make the end of the cone so it has some frayed wires so that a bee can push the wires enough to get out (including drones and queens) but can't get back in. Aim it a bit up and it helps some on keeping them from finding the entrance. Now you put a hive that has just a frame of open brood, a couple of frames of emerging brood and some honey/pollen, right next to the hive. You may need to build a stand or something to get it close to where the returning foragers are clustered on the cone. Sometimes they will move into the box with the brood comb. Sometimes they just hang on the cone. The biggest problem I've had is that this causes many more bees to be looking for a way in and circling in the air and the homeowners often get antsy and spray the bees with insecticide because they are afraid of them. If you think this is likely, then *don't* put the box with the brood here, but rather at your beeyard, hopefully at least 2 miles away, and you vacuum or brush the bees off into a box every night and take them and dump them in the box with the brood, you will eventually depopulate the hive. If you keep it up until no substantial number of bees are in it anymore, you can use some sulfur in a smoker to kill the bees (sulfur smoke is fatal but does not leave a poisonous residue) or some bee quick to drive the rest of them out of the tree (or house or whatever). And if you use the BeeQuick you may even get the queen to come out. If you do, catch her with a hair clip queen catcher and put her in a box and let the bees move into the box. Since the cone is still on the entrance they can't get back in the old hive. I'd leave it like this for a few days and then bring a strong hive and put it close to the old hive. Remove the cone and put some honey on the entrance to entice the bees to rob it. This is most effective during a dearth. Mid summer and late fall being likely dearths. Once they start robbing it, they will rob the entire hive out. This is especially important if removing them from a house, so that the wax doesn't melt and honey go everywhere or the honey attract mice and other pests. Now you can seal it up as best you can. The expanding polyurethane foam you buy in a can at the hardware store is not too bad for sealing the opening. It will go in and expand and make a fairly good barrier. Joe Waggle came up with this option, if you can keep a good eye on it is when they swarm, put the cone on and then the virgin queen will leave to mate and not be able to get back in. Then you can get a swarm with a queen from the cone.

Bee Vacuum

I will preface this that I don't like Bee vacuums. They kill a lot of bees make it hard to find the queen and likely to kill her. I hardly ever use them. They are nice for cleaning up the last stragglers at a colony, but I prefer to use a spray bottle of water to keep them from flying so much and a bee brush or shaking to get them off. I think a Bee vacuum is often a replacement for finesse and skill. Since they are occasionally useful, let's talk about them.

Brushy Mt. Bee Farm makes these, but you can modify a cheap shop vac to do it. The most important issues are these:

If you have too much vacuum it will kill too many bees. If you are converting a shop vac, cut a hole in the top or use a hole saw and drill a hole. You'll have to adjust this to fit the way the vac is designed, but if there is room you could just drill a three inch hole. If not you could drill and saw to make a longer hole. The idea is that we will take a piece of wood or plastic and make a damper by putting a screw through it on one corner and pivoting the damper to make a larger or smaller hole. This hole is covered on the inside by hardware cloth or screen wire. I just glue it with epoxy on the inside. Now when you adjust the damper to be more open there is less vacuum. When you close it more, there is more vacuum.

If the bees hit the bottom of the vacuum too hard they will die or be injured. The solution to that is put a piece of foam rubber on the bottom. Or wad up some newspaper and put it on the bottom—anything to soften their landing so they don't hit the hard plastic bottom.

Bees get torn up hitting the corrugations of the tube. If you get a smooth hose there will be less of this. If you get smaller corrugations there will be less of this.

If you run the vacuum too long the bees inside get hot, regurgitate their honey and die. If this happens you will notice they are a sticky mess. Don't run the vacuum any longer than you have to.

Adjust the vacuum carefully. You want just enough vacuum to pick the bees off the comb and no more. Too much and you'll have a canister full of squashed bees.

This tool can be used for bee removal. Getting bees off of the combs and not in the air is very helpful. Be careful. I have used them with good luck and I have also killed a lot of bees when I didn't mean to.

Transplanting Bees

Moving bees from one "hive" to another. (trees, old hives or other homes of bees)

People often have bees in an old rotting hive that is crumbling to pieces and is so cross-combed they can't manipulate it. Or they have a hive in a log gum, a box hive (no frames), a skep, a piece of a tree that fell down or some oddball equipment that they want to retire or even that they want to move them from all deeps to all mediums etc. If you want bees to abandon some current abode that can be taken home and manipulated here are some methods that I've used, and some variations that I have not used, but should work.

I have used this on box hives and log gums. You want the bees to abandon their old home, but you don't want to sacrifice all the brood. You want to get most of the bees and the queen out of the old hive into a box that is connected to the old hive. In other words there needs to be some connection between the two. A piece of plywood that is as large as the largest dimension of either one in both directions can then had a hole cut in the middle of it that is as large as the smaller of either on in both directions. By putting this between the new hive body and the old hive you have connected the two.

The next decision is whether you want to use Bee Go, Bee Quick (similar but smells nicer) or smoke and drumming or just patience.

It helps if the new hive has some drawn comb and, better yet, a frame of brood.

If you want to use the fumes (Bee Go and Bee Quick) then you put the old hive on top and the new hive body on the bottom. Have a queen excluder handy. Use a soaked rag for fumes and put it as near the top of the old hive as you can. This will drive the bees down into the box. When the box seems pretty full and the old hive seems pretty empty put the excluder between. If you can easily do it, put the old hive so that the combs are upside down from what they used to be. That way the bees will be more likely to abandon it eventually because honey runs out of the cells and the combs are the wrong way for brood.

If you want to smoke and drum, then you put the old hive on the bottom and the new one on the top. Smoke the old hive heavily and tap on the side with a pocketknife or a stick. You don't have to do it hard like a bass drum, just a tap tap tap. Lots of smoke helps. Again, when it looks like most of the bees are in the top put in the excluder. It doesn't matter what the orientation of the combs is for driving the bees out, but it helps if it is upside down now. The queen should be in the top and they will finish the brood in the bottom and then rework it for honey or abandon it.

If you want to use patience, just put the new hive on the top and wait for the bees to move up. This may or may not work for some time because the queen wants to stay in the brood chamber.

Bait hives.

Bait hives are empty boxes that are set out to try to entice a swarm to move in. They will not entice a hive to swarm, but they may offer a nice place for a hive that wants to swarm. I use Lemongrass oil and sometimes queen pheromone. You can by QMP (Queen Mandibular Pheromone brand name Bee Boost). It is little tubular pieces of plastic that have the smell impregnated in them. When I use these for bait, I cut each of them into four equal pieces and use one piece and some lemongrass oil or some swarm lure. Swarm lure and QMP are available from bee supply places. You can get your own QMP by putting all your old queens when you requeen and any unused virgin queens in a jar of alcohol. Put a few drops of this in the bait hive. Old empty combs are nice too and using boxes that have had bees in them helps. I set out about seven of these last year and got one swarm. Not great odds, but I got some nice feral bees. There are things that have been researched to increase your odds such as the size of the box, the size of the opening and the height in the tree. There seem to be a lot of exceptions though. So far my best luck has been a box the size of a deep five frame nuc or a 8 frame medium with some kind of lure (homemade or otherwise), 12 feet or so up a tree, with about the equivalent of a 1" hole for an entrance. And foundationless frames (frames with a comb guide, see chapter *Foundationless Frames*). My problems have been wasps moving in, finches moving in and wax moths eating old combs and kids knocking them out of trees with rocks and destroying them. Try putting nails in the hole to make an "X" to make it hard for the finches or cover the hole with #4 hardware cloth. Paint them brown or "tree" colored to make them harder for the kids to see. Use starter strips or clean dry old comb so the wax moths don't move in or spray the old comb with Certan. Remember, this is like fishing. I would not count on it if you're trying to get started beekeeping. You might catch one the first year or you might not catch one for several years, or you might catch several. It's like fishing because you want fish for supper. You may be better off to buy some fish.

Queen Rearing

For a live presentation by the author of this try a search for videos on the web for "Michael Bush Queen Rearing".

Why rear your own queens?

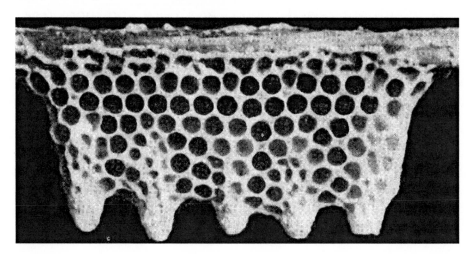

Cost

A typical queen costs the beekeeper $20 and up counting shipping and may cost considerably more.

Time

In an emergency you order a queen and it takes several days to make arrangements and get the queen. Often you need a queen yesterday. If you have some in mating nucs, on hand, then you already have a queen.

Availability

Often when you need a queen there are none available from suppliers. Again, if you have one on hand availability is not a problem.

AHB

Southern raised queens are more and more from Africanized Honey Bee areas. In order to keep AHB out of the North we should stop importing queens from those areas.

Acclimatized bees

It's unreasonable to expect bees bred in the deep South to winter well in the far North. Local feral stock is acclimatized to our local climate. Even breeding from commercial stock, you can breed from the ones that winter well here.

Mite and disease resistance

Tracheal mite resistance is an easy trait to breed for. Just don't treat and you'll get resistant bees. Hygienic behavior, which is helpful to avoid AFB (American Foulbrood) and other brood diseases as well as Varroa mite problems. And yet most queen breeders are treating their bees and not selecting, either on purpose or by default for these traits. The genetics of our queens if far too important to be left to people who don't have a stake in their success. People selling queens and bees actually make more money selling replacement queens and bees when the bees fail. Now I'm not saying they are purposely trying to raise queens that fail, but I am saying they have no financial incentive to produce queens that don't. This is not to say that some respon-sible queen breeders aren't doing the right thing here, but most are not. Basically to cash in on the benefits of not treating, you need to be rearing your own queens.

Quality

Nothing is more important to success in beekeep-ing than the queen. The quality of your queens can often surpass that of a queen breeder. You have the time to spend to do things that a commercial breeder cannot afford to do. For instance, research has shown that a queen that is allowed to lay up until it is 21 days old will be a better queen with better developed ovari-oles than one that is banked sooner. A longer wait will help even more, but that first 21 days is much more critical. A commercial queen producer typically looks for eggs at two weeks and if there are any it is banked and eventually shipped. You can let yours develop better by spending more time.

Concepts of Queen Rearing

Reasons to rear queens

Bees rear queens because of one of four condi-tions:

Emergency

There is suddenly no queen so a new queen is made from some existing worker larvae.

Supersedure

The bees think the queen is failing so they rear a new one.

Reproductive swarming

The bees decide there are enough bees, and enough stores and enough of the season left to cast a swarm that has a good chance of building up enough to survive the winter without endangering the survival of the colony.

Overcrowding swarm

The bees decide that there are too many bees and not enough room or not enough stores to continue under the current conditions, so they cast an over-crowding swarm as population control. This swarm doesn't have the best chance of survival but the colony believes it improves the colony's chances of survival.

We get the most cells and the best feeding for the queens if we simulate both Emergency and Overcrowd-ing.

A beekeeper can easily get a queen simply by making a queenless split with the appropriate aged larvae. So why would we want to do queen rearing?

Most Queens with Least Resources

The underlying concept of queen rearing is to get the most number of queens from the least resources from the genetics chosen for the traits you want.

To illustrate the resource issue let's examine the extremes. If we make a strong hive queenless. They could have, during that 24 days of having no laying queen, reared a full turnover of brood. The queen could have been laying several thousand eggs a day and a strong hive could easily rear those several thousand brood. Then we have lost the potential for about 30,000 or more workers by making this hive queenless and resulted in only one queen. And, actually, this hive made many queen cells, but they were all destroyed by the first queen out.

If we made a small nuc we would only have a couple of thousand queenless bees rearing several queen cells and those couple of thousand bees could only have reared a few hundred workers in that time. But again they made several queen cells and the results were only one queen.

In most queen rearing scenarios we are making the least number of bees queenless for the least amount of time and resulting in the most number of laying queens when we are done.

Where queens come from

A queen is made from a fertilized egg, exactly the same as a worker. It's the feeding that is different and that is only different from the fourth day on. So if you take a newly hatched worker egg, and put it in a queen cell (or in something that fools the bees into thinking it's a queen cell) in a hive that needs a queen (swarming or queenless) they will make those into queens.

Methods of getting larvae into "queen cups"

There are many methods. You can find the original books for many of these here:
    http://bushfarms.com/beesoldbooks.htm

Here are a few of them:

The Doolittle Method

Originally published by G.M. Doolittle, is to graft the appropriate aged larvae into some homemade wax cups. This requires a bit of dexterity and good eyesight, but is the most popular method used. Today plastic cups are often used in place of wax. The queen is sometimes confined to get the right aged larvae all in one place for easy selection. #5 hardware cloth works well for this as the workers can pass through it but the queen cannot. This is usually put on old dark brood comb to make the larvae easier to see and to make the cell bottom more sturdy for grafting. Once you have a good eye for the right age larvae this is less critical and one can do this by simply finding the right age larvae. On day 14 these are usually put in mating nucs.

The Jenter method

Several variations of this are on the market under various names. The concept is that the queen lays the eggs in a confinement box that looks like worker cells. Every other cell bottom of every other row has a plug in the bottom. When the eggs hatch the plug is removed and placed in the top of a cup. This accomplished the same thing as the Doolittle method without the need for so much dexterity and eyesight. On day 14 these are usually put in mating nucs.

Jenter Box Front

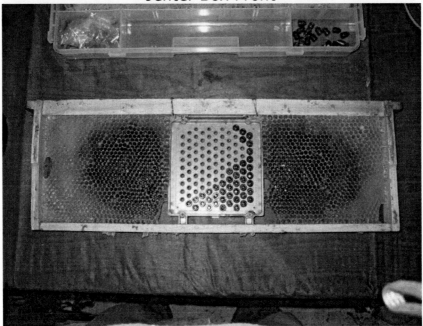

Jenter Box Back

Jenter Box Top

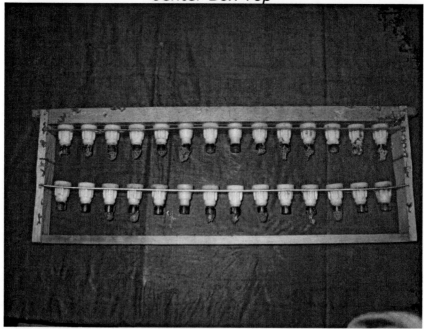

Missed Queen Cell

Jenter queen rearing system pictures. Front, back and top of the queen box and then a picture of a cell bar where I missed a queen cell the bees built in the cell builder. 17 queens dead.

Advantages to Jenter

- If you are a newbee you get to see exactly what the right age larvae looks like as you know when they were laid.
- If your eyesight isn't so good you don't have to be able to see the larvae (mine isn't the greatest)
- If you are not very coordinated (and I'm not) you don't have to be able to pick up something very tiny and down inside a cell without damaging it. You just move the plugs.

Advantages to grafting

- If the queen didn't lay in the Nicot cage and I'm on a schedule, I don't have any larvae the right age unless I go find some and graft (or do the Better Queens method).
- If I was too busy to confine the queen four days ago, I can just graft.
- If the queen mother is in an outyard, I don't have to make two trips, one to confine her, and another to transfer larvae.
- I don't have to buy a queen rearing kit.

The Hopkins Method

#5 hardware cloth queen confinement cage

In my variation, the queen is confined with #5 hardware cloth to get her to lay in the new comb and so we know the age of the larvae (as the Doolittle method but on new comb empty instead of old comb). This should be wax, preferably with no wires so you can cut the cells out without wires in the way, although Hopkins says you should used wired comb so it doesn't sag. If you use wired comb, be sure to work around the wires when leaving larvae so the wires won't interfere. Release the queen the next day. You can also just put the new comb in the middle of the brood nest and check every day to see if the queen has laid in it yet, to judge the age of the larvae.

On the fourth day (from when the queen was confined or she laid in the comb) the larvae will be hatched. In every other row of cells *all* the larvae are destroyed by poking them with a blunt nail, a kitchen match head, or similar instrument. Then the larvae in *every other cell* in the remaining rows is destroyed the same way (or two cells destroyed and one left) to leave larvae with space between them. This is then suspended flatways over a queenless hive. A simple spacer is an empty frame under the frame with the cells and a super over that. This will require angling the frames somewhat and laying a piece of cloth on top. The bees perceive these to be queen cells, because of the orientation, and build cells off of them. They should be spaced enough apart to allow cutting them out on day 14 and distributing them to either hives to be requeened (that would have been dequeened the day before) or to mating nucs.

Hopkins shim to hold the frame over the box.

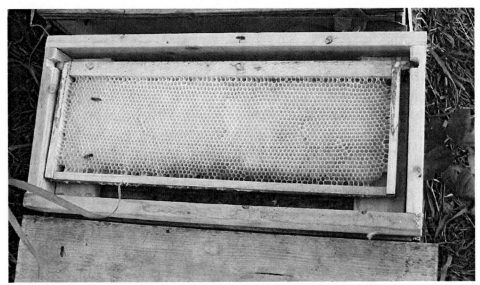

Frame of larvae in the Hopkins shim.

Cell Starter

For me the most difficult thing to get a grasp on and the most critical thing for queen rearing, other than the obvious issues of timing, was the cell starter. The most important thing about a cell starter is that it's overflowing with bees. Queenless is helpful too, but if I had to choose between queenless and overflowing with bees, I'd go for the bees. You want a very high density of bees. This can be in a small box or a large hive, it's the density that is the issue, not the total number. There are many different schemes to end up with queenless crowded bees that want to build cells, but don't ever expect a good amount of cells from a starter that is anything less than overflowing with bees.

The next most important issue with the starter is that it's well fed. If there is no flow you should feed to make sure they feed the larvae well.

Most of the rest of the complexity of the many queen rearing systems, which often seem at odds with one another, are tricks to getting consistent results under all circumstances. In other words, they are important to a queen breeder who needs a consistent supply of queens from early spring until fall regardless of flow and weather. For the amateur queen breeder, these are probably not so important as is the timing of your attempts. Rearing queens during prime swarm season just before or during the flow is quite simple. Rearing queens in a dearth or later or earlier than the prime swarm season will require more "tricks" and more work. For starters I would skip these "additions" and adopt them one at a time as you see the need.

A Cloake board (Floor Without a Floor) is a useful method. You can rearrange things so that part of the hive is queenless during the starter period and queenright as a finisher without a lot of disruption of the hive. But it's not necessary.

The simplest way I know of is to remove a queen from a strong colony the day before and cut it down to minimum space (remove all the empty frames so that you can remove some boxes and, if there are supers that are full remove those). This may even put them in a mood to swarm, but that will make a lot of queen cells. Make sure there aren't any queen cells when you start and if you use them for more than one batch be extra sure there are no extra queen cells in the hive as those will emerge and destroy your next batch of cells.

Another method is to shake a lot of bees into a swarm box aka a starter hive and give them a couple of frames of honey and a couple of frames of pollen and a frame of cells.

Beekeeping Math

Caste	Days Hatch	Cap		Emerge
Queen	3½	8 +-1	16 +-1	Laying 28 +-5
Worker	3½	9 +-1	20 +-1	Foraging 42 +-7
Drone	3½	10 +-1	24 +-1	Flying to DCA 38 +-5

Queen Rearing Calendar:

Using the day the egg was laid as 0 (no time has elapsed)

Bold items require action by the beekeeper.

Day Action Concept

-4 Put Jenter cage in hive Let the bees accept it, polish it and cover it with bee smell

0 Confine queen—So the queen will lay eggs of a known age in the Jenter box or the #5 wire cage

1 Release queen—So she doesn't lay too many eggs in each cell, she need to be released after 24 hours

3 Setup cell starter Make them queenless make sure there is a *very* high density of bees.— This is so they will want queens and so they have a lot of bees to care for them. Also make sure they have plenty of pollen and nectar. Feed the starter for better acceptance.

3 ½ Eggs hatch

4 Transfer larvae and put queen cells in cell starter. Feed the starter for better acceptance.

8 Queen cells capped

13 Setup mating nucs Make up mating nucs, or hives to be requeened—So they will be queen- less and wanting a queen cell. Feed the mating nucs for better acceptance.

14 Transfer queen cells to mating nucs.—On day 14 the cells are at their toughest and in hot weather they may emerge on day 15 so we need them in the mating nucs or the hives to be re- queened if you prefer, so the first queen out doesn't kill the rest.

15-17 Queens emerge (In hot weather, 15 is more likely. In cold weather, 17 is more likely. Typical- ly, 16 is most likely.)

17-21 Queens harden

21-24 Orientation flights

21-28 Mating flights

25-35 Queen starts laying

28 If you intend to requeen your hives, look for laying queens in the mating nucs, if found dequeen hive to be requeened

29 Transfer laying queen to queenless hive to be requeened.

Mating Nucs

Two By Four Mating Nucs Two By Four Mating Nuc

Two By Four mating nucs. Four nucs with two frames each in one ten frame sized box. Note the blue cloth sticking out. These are canvas inner covers so I can open one nuc at a time without them boiling over into the next nuc. Also note the Ready Date Nuc Calendars on the end.

A note on mating nucs

In my opinion it makes the most sense to use standard frames for your mating nucs. Here are a few beekeepers who agree with that:

> *"Some queen-breeders use a very small hive with much smaller frames than their common ones for keeping their queens in till mated, but for several reasons I consider it best to have but the one frame in both the queen-rearing and the ordinary hives. In the first place, a nucleus colony can be formed in a few minutes from any hive by simply transferring two or three frames and the adhering bees from it to the nucleus hive. Then again, a nucleus colony can be built up at any time or united with another where the frames are all alike, with very little trouble. And lastly, we have only the one sized frames to make. I have always used a nucleus hive such as I have described, and would not care to use any other."—*

Isaac Hopkins, The Australasian Bee Manual

"for the honey-producer there seems no great advantage in baby nuclei. He generally needs to make some increase, and it is more convenient for him to use 2 or 3-frame nuclei for queen-rearing, and then build them up into full colonies...I use a full hive for each nucleus, merely putting 3 or 4 frames in one side of the hive, with a dummy beside them. To be sure, it takes more bees than to have three nuclei in one hive, but it is a good bit more convenient to build up into a full colony a nucleus that has the whole hive to itself."—C.C. Miller, Fifty Years Among the Bees

"The small Baby Nucleus hive had a run for a while but is now generally considered a mere passing fad. It is so small that the bees are put into an unnatural condition, and they therefore perform in an unnatural manner...I strongly advise a nucleus hive that will take the regular brood-frame that is used in your hives. The one that I use is a twin hive, each compartment large enough to hold two jumbo frames and a division-board."—Smith, Queen Rearing Simplified

"I was convinced that the best nucleus that I could possibly have, was one or two frames in an ordinary hive. In this way all work done by the nucleus was readily available for the use of any colony, after I was through with the nucleus...take a frame of brood and one of honey, together with all of the adhering bees, being careful not to get the old Queen, and put the frames into a hive where you wish the nucleus to stand...drawing up the division-board so as to adjust the hive to the size of the colony."— G. M. Doolittle, Scientific Queen-Rearing

Queen marking colors:

Years Ending in:
- 1 or 6 – White
- 2 or 7 – Yellow
- 3 or 8 – Red
- 4 or 9 – Green
- 5 or 0 - Blue

Queen Catching and marking

Until you get the hang of it, there is always the risk of hurting the queen. But learning to do it is a worthwhile undertaking. I would buy a hair clip queen catcher and a marking tube and paint pens. Practice on a few drones with a color from a couple of years ago, or better yet the color for next year, so you don't confuse the drones with the queen. Use the current color for the queens.

My preferred method is to buy a "hair clip" queen catcher, a queen muff (Brushy Mt.) and a marking tube and a marking pen. Catch the queen gently with the hair clip. It is spaced so as not to easily harm the queen, but still be careful. If you put this and the marking tube and the paint pen (after it is shaken and started) in the queen muff then the queen can't fly off while you do this. Take the marking tube and slid out the plunger. If you move away from the hive you can lose some of the bees that are in and on the clip. Don't shake it while holding the clip portion or you may shake the queen out. If you take it in a bathroom with a window and turn off the lights you can be more assured she won't fly off. Or buy a queen muff from Brushy Mountain. Use a brush or a feather and brush off the workers as they come out and then try to guide the queen into the tube. She tends to go up and she tends to go for the light, so open the clip so she will run into the tube. If she doesn't and she runs onto your hand or glove, don't panic, just quickly drop the clip and gently but quickly put the tube over her. Cover the tube with your hand to block the light so she runs to the top of the tube. Put the plunger in. Be quick but don't hurry too much. Gently pin the queen to the top of the marking tube and touch a small dot of paint (start the paint pen on a piece of wood or paper first so there is paint in the tip already) on the middle of the back of her thorax right between her wings. If it doesn't look big enough just leave it. You need to keep her pinned for several more seconds while you blow on the paint to dry it. Don't let her go too soon or the paint will get smeared into the joint between her body sections and it may cripple or kill her. After the paint is dry (20 seconds or so) back the plunger up to halfway so the queen can move. Pull the plunger and aim the open end to the top bars and the queen will usually run right back down into the hive.

Jay Smith

Some quotes from Jay Smith (famous queen breeder and beekeeper who probably raised as many queens as anyone who ever lived)

Queen longevity:

From "Better Queens" page 18:

> *"In Indiana we had a queen we named Alice which lived to the ripe old age of eight years and two months and did excellent work in her seventh year. There can be no doubt about the authenticity of this statement. We sold her to John Chapel of Oakland City, Indiana, and she was the only queen in his yard with wings clipped. This, however is a rare exception. At the time I was experimenting with artificial combs with wooden cells in which the queen laid."—Jay Smith*

I would point out that Jay says: "This, however is a rare exception."

I think three years has always been pretty typical of the useful life of a queen.

Emergency queens:

"It has been stated by a number of beekeepers who should know better (including myself) that the bees are in such a hurry to rear a queen that they choose larvae too old for best results. later observation has shown the fallacy of this statement and has convinced me that bees do the very best that can be done under existing circumstances. The inferior queens caused by using the emergency method is because the bees cannot tear down the tough cells in the old combs lined with cocoons. The result is that the bees fill the worker cells with bee milk floating the larvae out the opening of the cells, then they build a little queen cell pointing downward. The larvae cannot eat the bee milk back in the bottom of the cells with the result that they are not well fed. However, if the colony is strong in bees, are well fed and have new combs, they can rear the best of queens. And please note— they will never make such a blunder as choosing larvae too old."—Jay Smith

"If it were true, as formerly believed, that queenless bees are in such haste to rear a queen that they will select a larva too old for the purpose, then it would hardly do to

wait even nine days. A queen is matured in fifteen days from the time the egg is laid, and is fed throughout her larval lifetime on the same food that is given to a worker-larva during the first three days of its larval existence. So a worker-larva more than three days old, or more than six days from the laying of the egg would be too old for a good queen. If, now, the bees should select a larva more than three days old, the queen would emerge in less than nine days. I think no one has ever known this to occur. Bees do not prefer too old larvae. As a matter of fact bees do not use such poor judgment as to select larvae too old when larvae sufficiently young are present, as I have proven by direct experiment and many observations."—Fifty Years Among the Bees, C.C. Miller

Queen Banks

A beekeeper can keep a number of queens in one hive if you get bees that are in the mood to accept a queen (queenless overnight or a mixture of bees shaken from several hives) and the queens are in cages so they can't kill each other. I've done these with a 3/4" shim on top of a nuc or a frame with plastic bars that hold the JZBZ cages. I put a frame of brood in periodically to keep them from developing laying workers or running out of young bees to feed the queens.

FWOF

(Floor With Out a Floor aka Cloake Board). Used to allow converting a top box on a queen rearing hive to change from a queenless cell starter to a queenright cell builder or finisher. This one is made with a 3/4" by 3/4" piece of wood with a 3/8" x 3/8" groove in it. Hang it out 3/4"or more in front and put a piece across the front under the sides to make a landing board. Cut a piece of 3/16" or 1/4" luan to slide in for a removable bottom. Coat edges with Vaseline to keep the bees from gluing it in. From left to right: The frame on a hive with the floor out. Inserting the floor. The FWOF with the floor in.

A Few Good Queens

Simple Queen Rearing for a Hobbyist

I get this question a lot, so let's simplify this as much as possible while maximizing the quality of the queens as much as possible.

Labor and food

The quality of a queen is directly related to how well she is fed which is related to the labor force available to feed the larvae (density of the bees) and available food.

Quality of Emergency Queens

First let's talk about emergency queens and quality. There has been much speculation over the years on this matter and after reading the opinions of many very experienced queen breeders on this subject I'm convinced that the prevailing theory that bees start with too old of a larvae is not true. I think to get good quality queens from emergency cells one simply needs to insure they can tear down the cell walls and that they have resources of food and labor to properly care for the queen. This means a good density of bees (for labor), frames of pollen and honey (for resources), and nectar or syrup coming in (to convince them they have resources to spare).

So if one adds either new drawn wax comb or wax foundation without wires or even empty frames to the brood nest during a time of year they are anxious to raise queens (from about a month after the first blooms until the end of the main flow), they quickly draw this comb and lay it full of eggs. So four to five days after adding it, there should be frames of larvae on newly drawn wax with no cocoons to interfere with them tearing down the cell walls to build queen cells. If one were to do this in a strong hive and at this point re-move the queen on a frame of brood and a frame of honey and put it aside, the bees will start a lot of queen cells.

Equipment

Second let's talk about equipment. One can set up mating nucs in standard boxes with dummy boards (or division boards) but only if you have the extra boxes or division boards. The advantage is that you can expand this as the hive grows if you don't use the queen. You can also build either two frame boxes or divide larger boxes into two frame boxes (commonly sold as queen castles). These need to be the same depth as your brood frames.

Method:

Make sure they are well fed

Feed them for a few days before you start unless there is a strong flow on.

Make them Queenless

So if we make a hive queenless (do what you like about having new comb or not) nine days after making them queenless these will be mostly mature and capped and be three days from emerging.

Make up Mating Nucs

At this point unless you intend to use the cells to requeen your hives, we need to set up mating nucs. The "queen castles" or four way boxes that take your standard brood frames and make up four, two frame mating nucs in one box are very good for this, but dummy boards and regular boxes can work also. In my operation these are all medium depth two frame nucs. The queen we removed earlier goes well in one of these also. We now want a frame of brood and a frame of honey in each of the mating nucs.

Transfer Queen Cells

The next day (ten days after making them queenless) we will cut out (with a sharp knife) the queen cells from the new wax combs we put in. If we used unwired foundation (or none) they should be easy to cut out without running into obstacles (as we would with wire and with plastic foundation) and can put each of the cells in a mating nuc. You can just press an indentation with your thumb and gently place the cell in the indentation. If you want you can also just put each frame that has cells on it in a mating nuc and sacrifice the extra cells (as the first queen out will destroy them). This is helpful if you have plastic foundation or you just don't want to mess with cutting out cells.

Check for Eggs

Two weeks later we should see some eggs in the mating nucs. If not, then by three weeks we should. Let her lay up the nuc well before moving her to a hive or caging her and banking her for later.

Next round just make the mating nuc queenless again the day before adding cells.

Now that these nucs are well populated by the brood the queen has laid, we can make more queens by simply making a strong mating nuc queenless and they will raise more queens. Again, it's the density of bees and the supply of food that are the issues. We can also, if they are wax combs, cut cells out and make use of multiple cells in other mating nucs as well. In this case either set up those nucs the day before or remove the queen the day before.

And that is all there is to raising a few queens.

About the Author

Michael Bush has had an eclectic set of careers from printing and graphic arts, to construction to computer programming and a few more in between. Currently he is working in computers. He has been keeping bees since the mid 70's, usually from two to seven hives up until the year 2000. Varroa forced more experimentation which required more hives and the number has grown steadily over the years from then. By 2008 it was about 200 hives. He is active on many of the Beekeeping forums with last count at about 45,000 posts between all of them. He has a web site on beekeeping at www.bushfarms.com/bees.htm

Practical
Queen
Rearing

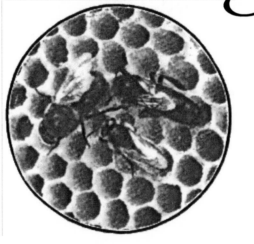

Queen, Drone, and Worker Photographed from Life. Slightly Enlarged.

By
Frank C. Pellett

Practical Queen Rearing

by Frank C. Pellett

Originally published 1918

Reprinted 2011 by
X-Star Publishing Company
Founded 1961

Transcribers Preface

This book is a good overview of most of the popular queen rearing methods. This edition has all the original plates plus two additional plates on the Hopkins method from the Australasian Bee Manual, and a three page section about the author with two more plates of the author. It also has headings added for the items that were indexed and in place of the index a more complete Table of Contents that includes those headings. It has a table of figures. It is newly typeset, not an OCR and not a digital scan. It is a very easy to read font in 12 pt.

To my good friend
M. G. DADANT

Preface

The writer has had the privilege of visiting many of the most extensive queen breeders of America, both north and south, and has tried to present, in the following pages, all the best methods of practice in use in these various apiaries. The book is small, as it has been thought wise to make the descriptions brief and to the point, rather than to elaborate them fully. Beekeepers are usually busy men, and want facts presented as simply and directly as possible in a book of this kind.

The works of Alley, Doolittle, and Sladen have been freely consulted, as well as various texts and bulletins on beekeeping. An effort has been made to make the book worthy of its title, "Practical Queen-Rearing," and methods not of practical value have largely been eliminated.

The illustrations for the most part have appeared in the American Bee Journal, many of them in connection with the author's contributions. A few have been borrowed from other works, as indicated in the text.

Beekeeping has shown a remarkable propensity toward expansion during recent months, the tendency being more and more toward specialization. The demand for good queens has taxed even the most extensive yards to the limit. It is with the hope that the methods here given will prove useful, and that the man of experience, as well as the novice, may find something of value in its pages, that this book is offered to the public.

FRANK C. PELLETT.

Atlantic, Iowa, December 27, 1917.

Table of Contents

Illustrations

Chapter I

The family of bees is an extensive one, embracing hundreds of species. On a warm day in spring, one can often see many different kinds of solitary wild bees among the blossoms of the fruit trees. Aside from their usefulness in the pollination of plants, these are of little economic importance. A little higher in the scale we find the bumble bees living together in small families of, at most, a few dozen individuals. In the tropics the stingless bees are still farther advanced in their social organization, and store small quantities of honey which is often taken from them for table use by the inhabitants of the warm countries. However, the amount of honey stored is small compared with the product of a colony of honeybees. While an extended study of the habits of the various species of wild bees would open a fascinating branch of natural history, the genus Apis is the only one that is of practical importance to the honey producer.

Apis dorsata

Much interest has been manifested in the giant bee of India and Ceylon, Apis dorsata, and at one time an attempt was made to introduce it into this country. This bee builds a gigantic comb in the open, usually suspended from a branch of a forest tree. Dorsata has a reputation of being very fierce, which Benton denies, saying they are no more so than other bees. Its habit is such that it is very improbable that it could be induced to occupy a hive, because of its single large comb, as our honeybees must do, to be properly managed.

Apis florea

In the east there is another species. Apis florea, a gentle little bee, much smaller than the honeybee. It builds a delicate little comb usually built around a twig. The quality of the honey is very good and the combs white, but the amount of honey stored in these diminutive combs is so small that they can never be of much practical importance, even though it were possible to induce them to remain in hives, which is very doubtful.

Apis Indica

There is a species in Ceylon and other eastern countries which has been domesticated with some success, Apis Indica. It is small and excitable, and generally inferior to the European races. It is known as the common East Indian honeybee. The natives hive them in small round earthenware pots, later driving them out with smoke to get the honey. Attempts to keep them in frame hives of proper dimensions have met with some success, but the quantity of honey secured is reported as very discouraging. This species is regarded as a variety of mellifica by some, rather than a distinct species. In any case it has little claim of interest to the practical beekeeper who has the better kinds.

Varieties of Mellifica.

All the honeybees known by different names, such as Italians, Blacks, Carniolans, etc., are now regarded as varieties of one species. Apis mellifica. The differences are such as naturally result from being bred for long periods of time in particular environments. Each variety has adapted itself to the particular conditions under which it lived until it is, very probably, better adapted to that particular condition, by natural selection, than any other race or variety would be. Since none of the honeybees are native to America, it can only be determined by trial which of the varieties is best suited to our conditions. The Blacks or German bees were first introduced into this country, and were very generally acclimated in all parts of the United States, before any other race was introduced. As in many localities others have since been introduced, a multitude of crosses, commonly spoken of as hybrids, have resulted. In localities where no particular attention is paid to the breeding of bees a new variety which might well be called the American bee is being developed, as a result of these crosses and the natural adaptation to a new environment. The term "hybrid" is usually used to designate any bee which is not pure, of one race or another. It is quite probable that time will demonstrate that the race which is best suited to the conditions of California is not the best for New York or Minnesota. Up to the present time, the Italians are the only ones which have been given an extended trial in all parts of the country, except the blacks, which were the first to be introduced. There is still room for extensive experiments in comparative tests of the races under the various conditions of different sections of America.

Black or German Bees.

Note: "According to the Quotations from the American Bee Journal, common bees were imported into Florida, by the Spaniards, previous to 1763, for they were first noticed in West Florida in that year. They appeared in Kentucky in 1780, in New York in 1793, and west of the Mississippi in 1797."—Dadant, Langstroth on the Honey Bee.

Black bees are very generally supposed to have been first introduced into America from Germany but very probably they came first from Spain. The native black bees of Great Britain, France, Germany and Spain are said to vary but little. The ground color of the whole body is black with the bands of whitish hairs on the abdomen very narrow and inconspicuous. F.W.L. Sladen, who was at one time extensively engaged in queen rearing in England, says that:

"In the cool and windy summer climate of the British Isles it is unsurpassed by any other pure race for industry in honey gathering, working early and late."

The blacks are easy to shake off their combs, and cap the comb honey very white, making an attractive product. Since extracted honey is coming more and more into favor, the matter of white capping is of constantly diminishing importance. One of the worst objections to the blacks is their excitable nature. When the hive is opened they run about nervously, and often boil out over the top in a most disconcerting manner. The queens are difficult to find, because of the fact that instead of remaining quietly on the comb attending to business, they run with the workers and often hide. They do not gather as much surplus on the average as Italians, under American conditions, are more inclined to be cross, and are more susceptible to brood diseases. It is a difficult matter to save an apiary of black bees, once they become infected with European foulbrood. In comparison with Italians, the latter have proven so much better that there is a very general tendency to replace the blacks with Italians and in many limited neighborhoods where beekeeping is scientifically followed, the blacks have disappeared.

The Cyprian Bee.

The Cyprian bees are in many respects similar to Italians. The pure Cyprians are said to be yellow on the sides and under parts of the abdomen, as well as having the three yellow bands as do the Italians, but the tip is very black. They are somewhat smaller than the Italians, and somewhat more slender and wasplike in appearance. According to Alley:

> "The posterior rings of the bodies of the workers are broader than those of the Italian, and, when examined, it will be noticed that the upper portion is partially black, terminating on the sides in a perfect half moon, generally two. It will also be observed that there is no intermingling of color. With pure Cyprian bees this is an invariable and uniform marking."

They also have a golden shield between the wings.

The queens are extremely prolific, but the workers are very cross and not easily subdued by smoke. After extended trial in America, they have found few friends because of this characteristic. The American beekeeper demands gentle bees. Aside from the revengeful disposition, they have many good qualities. They are said to be long lived, to build less drone comb than other races, to fly farther for stores and to be extremely hardy, wintering well. They continue breeding late in fall, and are not inclined to dwindle in spring. They build many queen cells in preparation for swarming, sometimes as many as a hundred. They defend their stores readily against robbers, and are strong and swift on the wing.

These bees are native to the Island of Cyprus, and were first introduced into this country from Europe. The first direct importation was probably that by D. A. Jones of Ontario, in 1880. It is not probable that pure stock can now be found in this country. It is thought that some strains of the golden Italians have been mixed with Cyprians in developing the bright yellow color.

The Holy-Land Bees, or Syrians.

The Holy-land bees are very similar to the Cyprians in appearance, having the golden shield on the thorax, but they show whiter fuzz rings than either Cyprians or Italians. They were introduced into this country by D. A. Jones at the same time as the Cyprians. These bees are native to Palestine, and are said to be common in the vicinity of Bethlehem, Jerusalem and other Bible cities. While they attracted much attention for a short time following their introduction, they were shortly abandoned and are no longer offered for sale in America, as far as the writer can ascertain. They are said to swarm excessively and to winter poorly, as well as to propolize badly.

Albinos

The Albinos, formerly popular, are probably of Holy-land origin, mixed with Italian, according to Root. The Albino resembles the Italian in appearance except that the fuzz rings on the abdomen are bright grey or white. Root reports them as decidedly inferior as honey gatherers.

The Italian Bee.

The Italian bee is by far the most popular race in America. It has been tried under all kinds of conditions in all parts of the country with satisfactory results. It is resistant to wax moth and European foulbrood, a good honey gatherer and gentler than the black race which preceded it.

This race was first introduced into this country from Italy.

The story of the first Importations is told by Mr. Richard Colvin of Baltimore, in the Report of the Secretary of the U. S. Department of Agriculture for 1863, as follows:

"The first attempt to import the Italian honeybee into the United States, it is believed, was made about the year 1855 by Messrs. Samuel Wagner and Edward Jessup, of York, Pennsylvania; but in consequence of inadequate provision for their safety on so long a voyage, they perished before their arrival.

"In the winter of 1858-59 another attempt was made by Mr. Wagner, Rev. L. L. Langstroth and myself. The order was placed in the hands of the surgeon of the steamer (to whose charge the bees were to have been committed on the return voyage), with instructions to transmit it to Mr. Dzierzon on reaching Liverpool; but in consequence of his determining to leave the ship to engage in other service on his arrival at Bremen, it was not done and the effort failed. Subsequently arrangements were made by which, in the latter part of that year, we received seven living queens. At the same time, and on board the same steamer, Mr. P. J. Mahan, of Philadelphia, brought one or more queens, which were supposed to be of doubtful purity. Only two or three young queens were reared by us during that fall and winter, and in the following spring we found all our imported stock had perished.

"In conjunction with Mr. Wagner I determined to make another trial, and another order was immediately dispatched. The queens, however, did not arrive until the following June. Meantime, about the month of May, Mr. S. B. Parsons, of Flushing, Long Island, received an importation of them from the northern part of Italy, some of the progeny of which he placed in the hands of the Rev. L.L. Langstroth, Mr. W.W. Carey, Mr. M.. Quinby, and other skilful apiarians, who with Mr. C.W. Rose, a subsequent importer, and perhaps some others, have bred and disseminated them pretty widely through our country."

There was much interest in the new race, and, for a long time, queens commanded from ten to twenty dollars each in some cases. The late Charles Dadant was one of the early breeders, who imported stock from Italy direct.

The Italian has been bred in America on such an extensive scale that various strains have been developed. The so-called three banded or leather colored Italians are probably more nearly typical than the goldens or five banded Italians. The Italian bee from northern Italy has three yellow bands, with pronounced bands of whitish or grey hair on each of the segments except the first and the last. It is a mild tempered bee, usually being gentle and quiet under manipulation. Unlike the blacks these bees cling closely to their combs, and the queen will often continue her egg laying when the comb on which she is working is removed from the hive and held up to the light.

It is a prolific race, and stands extremes of temperature very well. It winters well and is not adversely affected by the heat of the dry summers of the central west. The beekeeper who does not care to experiment will do well to stick to the Italians, at least until other races have been given more extended tests than have so far been given. While there are a few warm advocates of Caucasians and Carniolans, by far the greater number of practical beekeepers contend that the Italians are the best race. It is only fair to state, however, that no other race has been given the same opportunity to demonstrate its good points, and it is altogether probable that some other race may yet prove best adapted for certain climatic conditions.

Goldens

The Goldens, are the result of special breeding by selecting the queens whose progeny show the brightest color. It is thought that some strains of goldens are somewhat mixed with the Cyprians, from which ancestry came the bright color. Some breeders have paid so much attention to selecting the brightest colored individuals, regardless of other traits, that some strains are unduly cross, are poor honey gatherers and are not considered hardy. On the other hand there are strains which have been selected with due care to retain other desirable traits along with the bright color, which are gentle and productive.

Carniolans.

The Carniolans resemble the blacks but are larger, the abdomens are of a more bluish cast and the abdominal rings are more distinct. They have the reputation of being excessive swarmers, although the queens are extremely prolific. They are a gentle race and reported to be good honey gatherers, and to stand extremely cold winters. Because of their excessive swarming tendency, they are not popular with American beekeepers, but the dark color is sufficient in itself to condemn them with many who admire the bright colored bees.

It is important that they be given a fair trial in northern sections, with a hive adapted to discourage swarming, by giving plenty of room for the extremely prolific queens. The Dadant hive or Langstroth frames of jumbo depth are best suited for this purpose of any hive in the market. Since they winter well and the colonies are inclined to be populous, it would seem that they should be especially adapted to extracted honey production in colder latitudes, if the swarming tendency can be overcome.

This race is native to the province of Carniola, Austria, and was first brought to this country in the eighties. It is said that there is much variation in the markings of the bees in the province from which they came. They deposit very little propolis, and are quiet on the combs during manipulation, two desirable traits.

Caucasians.

The Caucasians greatly resemble the blacks in appearance, but they are very different in disposition. They are said to be the gentlest race of bees known. The most serious objection to them is the fact that they deposit propolis freely, being the opposite of Carniolans in this respect. They swarm freely and build quantities of burr and brace combs, which is a source of annoyance to the beekeeper. They have many desirable traits, wintering well, capping their honey white and not being inclined to drift into the hives of other colonies than their own. Since they resemble the blacks so closely, it is next to impossible to tell whether or not they are pure, which is a serious drawback to the careful breeder. A few who have tried them extensively are warm in their praises of the Caucasians and contend that they are superior to the Italians. While this may be doubted, they are worthy of a more general trial than they have so far received. A few breeders now offer queens for sale.

Banat Bees.

The Banats come from Hungary and greatly resemble the Carniolans. Some contend that they are not distinct. They are very gentle, dark in color and very prolific. They build up rapidly in spring and are said to be less inclined to swarm than the Carniolans.

Mr. T.W. Livingstone of Leslie, Georgia, had Banats, exclusively, in his apiaries and regarded them highly. He reported them as very gentle, building up early in spring and rearing brood all season.

Tunisian or Punic Bees.

This is a black race coming from the north coast of Africa. Although given a trial in America they did not meet with favor and none are now present in this country so far as known. They are bad propolizers, extremely cross, and do not winter well. They seem to have been lately given a trial in Scotland. Mr. John Anderson of the North Scotland College of Agriculture, writing in the Irish Bee Journal, October, 1917, says of them that they have some very desirable characteristics, and some that are inconvenient. He mentions the case of a beekeeper who depends solely on honey production for a livelihood (which is unusual in Great Britain), who increased forty colonies to four hundred and harvested two-and-one-half tons of honey in one season without feeding any sugar. Mr. Anderson regards the Punic bee as worthy of more attention than it has received.

Egyptians.

Bees have been kept in a primitive way for centuries in Egypt. The Egyptian bees resemble Italians in color, with an additional coat of white hairs. They are said to breed purely and not be inclined to mate with other races. They are somewhat smaller than the European races, and build somewhat smaller cells in their combs. They are reported to be cross and not easily subdued by smoke. Since they do not form a winter cluster, they are not fitted to withstand severe weather. They are said to rear large numbers of drones, and to develop fertile workers in abundance. They are not likely to prove of any value in America. In fact, they were introduced soon after the Civil War, but either perished from cold or were abandoned in favor of more promising races.

Other Races.

There are numerous other races in Asia and Africa which are as yet but little known in this country. It is hardly probable that new races superior to those already introduced will be found. The native Grecian bee is said to resemble the hybrids so common in this country, but has probably not been tried here as yet.

Chapter II

Life Story of the Bee.

In a normal colony of bees, during the summer season, will be found one queen, several thousand workers and a few dozen drones. If the bees are left to themselves and receive no attention from their owner, the number of drones is greatly increased, and often reaches the point where they consume what might otherwise be stored as surplus honey. Since there are but few readers of a book of this nature who are not already familiar with the life of the honeybee, it would seem, at first thought, that little space need be occupied in consideration of this subject. However, the volume cannot be complete without some attention to the life history of the insects, especially with attention to those points with special bearing on the subject of queen rearing.

Since the life of the colony centers in the queen, she becomes of special importance, and she receives attention from the workers worthy of her special place. Should she be removed from the hive, great excitement will shortly prevail with manifestation of serious distress on the part of the inmates. Unless she be promptly returned, the bees will prepare to replace her by starting numerous queen cells, utilizing the newly hatched larvae for the purpose.

Life of the Queen.

As stated elsewhere, all fertilized eggs laid by the queen produce female offspring. Whether these shall develop as queens or workers is determined by the environment in which the development takes place. In any case the egg hatches in about three days. Where eggs are placed in queen cells it is very doubtful whether they receive any different treatment before hatching than do the eggs in ordinary worker cells. It is after the hatching of the egg that the embryo queen receives special attention, which results in the perfect development of her sexual organs. The larger cell in which she finds herself, together with a plenteous supply of the rich food known as royal jelly, makes of her a very different creature than of her sister in the worker cell.

Fig. 1. Queen cells built under the swarming impulse.

The queen lacks the wax secreting organs as well as the pollen baskets of the worker. Neither has she the same highly developed eyes as the worker. Her period of development is much shorter, while her body is larger and quite different in appearance. Approximately sixteen days are necessary for the complete development of the queen bee from the time of the laying of the egg. Of this, three days are necessary for the egg to hatch, six days are spent in the larval stage, and seven days in completing the final transformation, during which she is sealed up in the cell. Twelve days are necessary for the last stage of development of the worker, thus requiring twenty-one days for the entire development.

Apparently the queen larvae are fed for the first thirty-six hours in very similar manner to the workers. After that time they are fed far more of the royal jelly than they can possibly consume, being left to float in the rich white substance. While the worker is fed on pollen and honey during the latter part of her period of development, the queen larvae is fed the royal jelly during the entire period of larval growth.

The Drone.

The drones are male bees and, apparently, serve no other purpose than the perpetuation of the species. Since under normal conditions a queen bee mates but once in her lifetime, but few drones are needed to serve the purpose for which they are designed by nature. In a state of nature, where colonies are isolated it may be needful that a large number of drones be reared to insure that the young queen will meet one when she goes forth to her mating flight. Where dozens of hives are kept together in a single apiary, as is the case in practice of commercial beekeeping, the beekeeper may keep the number down to the minimum, without danger that a sufficient number will not be present. Hundreds of apiaries are unprofitable because their owners fail to take the necessary care to insure the reduction of the number of drones, which consume the surplus of the colony Instead of adding to the store.

Except in the case of the queen breeder who wishes to propagate large numbers of males from choice colonies for breeding purposes, the presence of an over-abundance of drones is a serious handicap to the success of the beekeeper. The use of full sheets of foundation in the brood frames is the best Insurance against the raising of drones.

The cells in which drones are reared are similar in appearance to worker cells, except that they are larger in size. They are utilized for the storage of honey the same as are the worker cells. When the brood is developing the high arched cappings, like rifle bullets, will instantly distinguish them from the smooth capping of worker brood. Twenty-five days is necessary for the development of the drone from the time the egg is laid until it reaches maturity. Mating of honeybees takes place on the wing, and the act is fatal to the drone. He dies almost instantly, and his sexual organs are torn from his body and borne away attached to the body of the queen. After all the seminal fluid has been absorbed by the queen, the parts are removed, apparently by the workers which can sometimes be seen pulling at them after the return of the queen.

Queen Rearing in Nature.

Under normal conditions the bees build queen cells on two occasions, to supersede the old queen or in preparation for swarming. Where the old queen shows signs of falling, the bees will often build only one or two cells. When the young queen emerges, she will often be mated and begin laying without manifesting any antagonism toward the old queen. It thus happens that the

Chapter III

Improvement of Stock by Breeding.

It is highly important that every person engaged in commercial queen rearing, should make a careful study of the laws of breeding, and make a conscientious effort to improve his stock. Marvelous results have come from careful breeding of live stock and poultry, and even more striking results have attended the efforts of the painstaking plant breeders. Since bees are subject to the same laws of heredity, there is no reason why they cannot be likewise improved if the same care is given to the selection and mating of queens, that is given to other animals.

The fact that there is great difficulty in controlling the male parentage, makes the problem of breeding bees a more serious one than breeders of animals have to face. On the other hand, the possibility of several succeeding generations in a single season makes it possible to secure results in a much shorter period of time.

The beekeeper, who is intent on bettering his stock, finds it much simpler to replace his poor stock with a better grade than does the farmer who has a herd of scrub cattle or sheep. Simply replacing the queens in his colonies shortly has the effect of changing the entire stock in the apiary, since the workers are short lived. If he is not inclined to buy enough queens to replace the poorer ones in all his hives, he can very shortly rear enough on his own account to do so, if he will give the matter a little attention. If he buys even one good queen, he can shortly improve the entire stock of an apiary of one hundred or more colonies. To do this he should rear as many young queens as there are colonies in his apiary, and use them to replace the old and inferior queens. If he does this early in the season, he need give little thought to the mating of his young queens.

Fig. 2. A large average production is only secured by careful attention to the selection of stock.

If the mother from which he rears his stock is pure, all the young queens will be pure. To be sure, most of them will be mated with inferior drones, but it is a well known fact that it is only the female offspring that are affected by the mating of a queen. If her mother is purely mated, all her drones will be pure, regardless of her own mating. Within a few weeks there will be thousands of pure drones, the offspring of the young queens that have been introduced. The beekeeper should then rear a second lot of queens from a pure mother to replace all the mismated ones which were introduced early in the season. By this time, most of the drones present will be pure, and the second lot of queens will mostly be purely mated. It is thus a simple matter to replace the entire stock of a neighborhood with pure bees from the offspring of a single pure queen.

Desirable Traits in Breeding Stock.

No queen should be used as a breeder unless she is prolific, since this is of the first importance in determining the amount of honey stored. However, it is not always the most prolific colonies which store the most honey. Longevity of the bees is an important consideration, and quite possibly the difference in length of the tongues of the workers may have an important influence. It often happens that in a poor season a single colony will store a good crop, when others equally strong will get but little, or even require to be fed. The author had one such colony which made a remarkable showing for three successive seasons. The difference in production was so marked that most of the young queens reared were from this queen. A measurement of the length of the tongues of her workers showed that they possessed a slightly longer tongue than others in the apiary, or even other apiaries where measurements were made in comparison. Increased length of the tongues of the workers would place much nectar within their reach, which would otherwise be denied them. It is well worth while to have careful measurements of tongues of all colonies which make unusual showing, under adverse conditions.

In general, the breeder selects queens for breeding from colonies which store the most surplus, with little enquiry as to the particular reason therefore. Since honey is the principal desideratum of the beekeeper, he is not so much concerned in the reason why a special colony stores more, as he is in finding the particular colony.

Next to production, gentleness is a most important characteristic. It is very disagreeable to have bees that meet one half way to begin the day's work, and follow one about constantly. The fear of stings is the principal objection to beekeeping on the part of many people. While stings can largely be prevented by suitable protection in the way of veils and gloves, it is far better to select gentle stocks for breeding purposes. Where only the gentle colonies are selected for breeding stock, it is possible to very largely reduce the annoyance of stinging.

It would seem to be possible to select gentle colonies which are also good producers, and, at the same time, have other desirable characteristics.

Color should be a secondary consideration, although it is desirable to have bees nicely marked. For a time, so much attention was paid to color on the part of breeders of Italians, that everything else was sacrificed in order to get yellow bees. This was carried to such an extreme that a very general prejudice has grown up against the Goldens. While it is quite true that some strains of the Goldens are not desirable, being neither hardy nor good honey gatherers, there are strains where proper attention has been given to other points, which are very satisfactory. In general, the Goldens have a bad reputation for being ugly in disposition, yet at least one strain of Goldens is very gentle. Very much depends upon the queen breeder, and the care he uses in selecting his breeding stock. Some breeders go so far as never to use a queen for a breeder, unless the colony can be handled under normal conditions without smoke. The non-swarming propensity is also to be favored. In many localities the honeyflows are short, and, if the colony swarms at the beginning of the flow, there is little chance of harvesting a good crop. Too much care cannot be used in selecting the colonies to use for breeders. Much more attention is given to selecting the queen from whose offspring the young queens are to be reared, than is given to the parentage of the drones. The confession must be made that few breeders give any special attention to this point, although it is equally as important as far as practical results are concerned.

Control of Drones.

Since the queen is mated on the wing, and there is always the possibility that the young queen will meet an inferior drone from a distance, it is highly important that a queen breeder go to a good deal of trouble to insure that all bees within a radius of five miles of his breeding yards are requeened with pure stock of the race which he is breeding. Unless he takes this precaution, there will be much dissatisfaction on the part of his customers from receiving mismated queens.

If a breeder is so fortunate as to be within reach of a suitable place to establish a mating station where no other bees are within reach, he can do much to improve the quality of his stock. Under such circumstances, he can select his drones with the same care that he selects the mother of his queens. A colony combining as many as possible of the desirable characteristics can be carried to the isolated position where the matings are to be made and left there. A few have undertaken to rear queens on islands where no other bees are present. The broad prairies of several states offer similar isolation.

*Fig. 3. Combs built on starters only or without founda-
tion contain a large percentage of drone cells and result
in unprofitable colonies.*

Unfortunately, however, few breeders are so si-
tuated that they can control the drones thus completely. After requeening all the bees within flying distance of
the apiary, the next thing is to select the best colonies
as drone breeders and supply them with an abundance
of drone comb. This insures that large numbers of
drones will fly from these colonies, and thus increase
the chances that young queens will meet desirable
mates. Care should be used to make sure that the
combs in the brood nests of other colonies than the
breeders contain as little drone comb as possible, and
thus reduce the production of drones to the lowest
possible minimum. Traps may be used also to catch
such drones as appear in undesirable colonies. Unless
the breeder is willing to go to great length to control his
breeding stock and thus give his customers the best
which it is possible for him to produce, he should by all

means confine his attention to the production of honey or some other business. There are entirely too many indifferent queen breeders for the good of the industry.

Fig. 4. Full sheets of foundation in the brood frames insure worker combs and a minimum of drone production.

Mating in Confinement a Failure.

Some practical method of absolute control of mating has long been sought. At the University of Minnesota Prof. Jager succeeded in getting one queen impregnated artificially and for a time it was hoped that enough queens could be mated in this way for use in breeding experiments. However, after numerous trials on the part of Prof. Jager, C.W. Howard, and L.V. France, at the University, no further successful instances have been reported.

The A. I. Root Company tried some rather elaborate experiments in getting queens mated in large greenhouses, but these were likewise a failure. While enthusiasts have claimed success at different times by one method or another, their claims have generally been discredited, and up to the present, there seems little prospect of artificial control of the mating. About all that now seems possible, is to select isolated situations for the mating stations, or to limit the breeding of drones as far as possible in undesirable colonies, and encourage it in the colonies from which it is desirable to breed.

Parthenogenesis.

When the discovery was first made that unimpregnated females often are capable of producing male offspring, the public was slow to accept the fact. There was much discussion of the subject for years before it was finally accepted as a settled fact, rather common among insects. It is now well known among beekeepers that queens which fail to mate will sometimes lay a considerable number of eggs which will hatch, but all will be drones. In the same manner fertile workers produce drones which are usually smaller in size and inferior in appearance, but some very careful observers are of the opinion that they are quite capable of mating in the normal manner.

Since the mating of a queen has no direct effect on her male offspring, her workers may be hybrids, and her drones pure. It is hardly within the scope of this little book to go into detail concerning the proof of such well established facts as those above stated. These may be found in detail in several of the old text books. Those who are interested in pursuing the subject further are referred to Dadant's revision of Langstroth on the Honeybee, where a full account of the various experiments along this line are given.

The thing that we are concerned with just now is the practical effect that the facts may have upon the problems of the queen breeder, and these we have set out as briefly as possible in the foregoing pages.

Chapter IV

Equipment for Queen Rearing.

The kind and amount of equipment necessary for queen rearing will depend to a great extent upon conditions. The beekeeper who wishes to rear but a few queens for use in requeening his own apiaries, can get along very well with limited equipment. The commercial queen breeder, who expects to send out several thousand queens each year, will do well to provide a liberal amount of equipment, for, otherwise, he will be hampered and unable to get the best results. An effort is made here to describe the various systems of management, and the reader can select what most appeals to him. In general, the simpler the system, the more efficient and the larger the amount of work which can be accomplished in a given time. Several different methods are described for doing the same thing, yet it is manifestly unwise for any individual to provide himself with all the equipment described, or to undertake the various systems outlined, unless it be for the purpose of experiment rather than for practical results. Usually it is best to use modifications of equipment used for commercial honey production so that in the event of a change back to regular beekeeping the equipment can mostly be used, or sold to other beekeepers in case of giving up the work. Second hand queen-rearing equipment is difficult to sell, since there are comparatively few men engaged in commercial queen rearing.

Grafting House.

On visiting the queen breeders of the south, I was much impressed with a grafting house in common use in the queen rearing apiaries of Alabama. While it is possible to make use of the kitchen or other warm room in the house, or to do the work in the open air in warm weather, the little building shown at Figure 5 Is far more desirable. As will be seen in the picture, the building is made of matched lumber and is very tight. A seat is provided for the operator, and in front of it a bench or table running across the building and about two feet wide. This provides ample room for combs, tools, etc., and one can work in comfort and at leisure. The entire front above the table is composed of window sash, thus providing an abundance of light. Some of these grafting houses, like the one shown, are also provided with glass in the roof like a photographer's studio. It is well to provide a shutter to cover the roof in extremely hot weather, or to protect the glass during storms. A shade is also desirable for the front, to shut out too much sunlight at times. A room four by six feet is amply large for this purpose, and, by means of a small oil stove, it can be kept warm in cool weather. This is important to prevent the chilling of the larvae while grafting. Some of the more extensive queen breeders find it necessary to graft cells every day during the season, rain or shine, and during the rush days of midsummer must prepare hundreds of cells. Not the least of the advantages of this building is the protection from robbers. Where it is necessary for the operator to be at work for several hours at a time, this little building in the center of the yard is a great time and labor saver, as well as adding much to the convenience and comfort of the operator. It merits more general use. While the one shown in the picture admits more light than is necessary on bright days, the extra glass space will be much appreciated in dark and cloudy weather.

Fig. 5. Grafting house in use by southern queen breeders.

Mating-Hives.

The honey producer who rears queens only for the purpose of improving his stock or requeening his apiaries, seldom bothers much about mating-hives. When he has a lot of sealed cells ready for use, he simply kills off the old queens to be replaced and about twenty-four hours later gives each of the colonies a sealed cell. In this way he avoids the bother of introducing queens, for the young queen will emerge in the hive where she is expected to remain. From there she will take her mating flight, and, the only further concern necessary on the part of the beekeeper, is to take care to replace any queens that are lost on their nuptial flight or that fail to emerge properly.

The commercial queen breeder will require a large number of nuclei or small colonies to care for surplus queens, until they are mated and ready to be mailed to customers. There is a large variety of hives of various sizes used for this purpose. Where queen breeding is the prime object, the tendency is to use as small hives and as few bees as possible, so that the largest possible number of queens may be reared with the bees and equipment available. However, many of the most successful queen breeders find serious objections to baby nuclei and small mating boxes, and advocate nothing but standard frames for mating-hives.

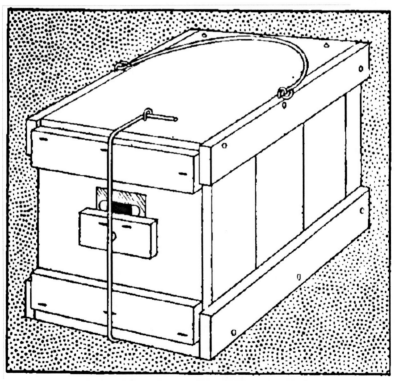

Fig. 6. Rauchfuss Mating Box.

The Rauchfuss Mating Boxes

This is perhaps the smallest mating box ever devised which has been used successfully. Beginners or
those with limited experience, are quite likely to have
much difficulty from the bees swarming out to accompany the queen on their mating flight with any small
nucleus. Even the most expert are never able to overcome this difficulty entirely. The Rauchfuss nucleus
consists of a small box with a removable front, holding
three 4 ¼ x 4 ¼ comb honey sections, figure 6. The
entrance is by means of a small round hole in the front,
which can be closed entirely when moving them, by
simply turning a small button. As devised by the inventor, one section of scaled honey is used, and sealed
brood is removed from a strong colony and cut into
squares of the right size to fill one of the remaining
sections. The presence of brood will in many cases
prevent the bees from absconding when the queen
takes her flight. When used without brood, there will
be a larger percent of loss from absconding. A cupful of
young bees taken from a strong colony is sufficient to
stock the box, when a virgin queen from a nursery cage
is run in through the entrance hole. After the box is
stocked and the young queen run in, the entrance is
stopped. When all boxes to be stocked at one time are
ready, all are carried to a point some distance from the
apiary and tied in trees, set on some convenient object,
or otherwise placed until the queens shall be mated. Of
course the entrance should be opened as soon as conditions are favorable after reaching the destination. It will
be necessary to remove the queens from these diminutive hives soon after they begin to lay. Should it be
inconvenient to do so at once, the box is provided with
a piece of queen excluding zinc which can be turned

over the entrance hole, thus preventing the queen from escaping, while permitting the bees to go afield.

The great advantage of this mating box is the small first cost, and the small number of bees necessary to stock the nucleus. They are listed at about forty cents each in lots of ten.

Fig. 7. A baby nucleus at the Minnesota University queen-rearing station.

Baby Nuclei.

The Root baby nucleus which is quite generally used is a small double hive, each side containing two frames 5 $^5/_8$ x 8 inches in size. Three of these little frames will just fill a standard Langstroth frame, and to get combs built in them it is necessary to put them in Langstroth frames, and insert them in strong colonies of bees. Some cut up combs and fit them into the little frames. Entrances to the two compartments are at opposite ends of the box. About a half a pint of bees is used to stock each compartment. This, in effect, is very similar to the Rauchfuss mating box, excepting that it is necessary to go to more trouble to get combs built especially for these nuclei. There is the same trouble from absconding, and the same danger of being robbed by strong colonies if left within reach. During a good honey flow when all conditions are favorable, it is possible to get a large number of queens mated in these little hives with a minimum of cost in bees, but during a dearth when it becomes necessary to feed to keep any kind of nucleus from going to pieces, they are likely to prove the source of much annoyance. See Figure 7.

Small Hives.

At Figure 8, we show some small hives formerly popular with queen breeders, but which have almost gone out of use. As will be seen in the picture, one is single and the other is double. The double one has entrances opening in opposite directions to avoid danger of the queen entering the wrong compartment.

These little hives hold three, and sometimes four, small frames. They are large enough to hold a nice little cluster of bees, and once established they can sustain themselves very nicely under favorable conditions. Mr. J.L. Strong, formerly extensively engaged in queen rearing in Iowa, used these mating hives for about twenty-five years with satisfaction. However, since the frame is an odd size, it is necessary either to cut up combs and fit into them, or get them built in the nucleus, so there is sometimes difficulty in getting them properly fitted out to begin with. There is really nothing to be said for them In preference to a standard hive divided into two or three compartments, and the latter can be used for any other purpose as well.

Fig. 8. Small mating hives in Strong queen yard. This type of hive was once quite generally used but is now going out of use. [From Productive Beekeeping.]

Fig. 9. Mating hives using shallow extracting frames. Achord queen yards in Alabama.

Fig. 10. Langstroth hive body adapted for four-compartment mating hive, used by J. M. Davis of Tennessee.

A few queen breeders use a shallow nucleus which is of the same length as the standard hive. In this they use shallow extracting frames. Although the frames are of the same size as those used in the apiary, the top, bottom and body must be made especially. Nuclei of this type as used by W.D. Achord, of Alabama, are shown at Figure 9. Instead of the usual hive record, short pieces of different colors are placed at the front end of the cover. The position of these pieces, which can be moved to any position at will, indicate the conditions within the hive.

Divided Standard Hives.

Fig. 11. Eight-frame hive divided into three parts; each with two standard frames, at the Jager apiary.

By far the greater number of queen breeders use the standard Langstroth hive, divided into two or more parts. J. M. Davis, of Tennessee, divides the ordinary hive into four parts. This makes use of standard hive bodies, tops and bottoms, but requires a special frame as shown in Figure 10. The two division boards that are running lengthwise are easily removed, thus leaving the hive in only two parts. In this way it is possible to unite two of the clusters at the close of the season, and leave them strong enough for wintering in that mild climate. There is an entrance at each of the four corners, each facing in a different direction. The four compartments are lettered A, B, C, and D. In opening the hive he makes it a point always to begin at A and examine each division in regular order to avoid overlooking any one of them.

At the apiary of Prof. Francis Jager where the queen breeding work of the State of Minnesota is carried on, an eight frame hive is divided into three parts, each part taking two standard frames. There is one entrance at each side, and one at an end. All that is necessary to make an eight frame hive into three nuclei, is to have two tight fitting division boards which fit into sawed slots at the ends. These must reach to the bottom to prevent the mixing of bees or the queens from passing from one compartment to another. It is necessary of course to fit the bottom board for the special purpose with entrance openings in the proper place. Our illustration (Figure 11) shows a small cover just the right size to cover one of the three compartments.

This is placed over the middle division when the regular cover is removed, to prevent the mixing of bees while the hive is open.

Both the eight and ten frame hives arranged in this manner are in general use.

Ben G. Davis, of Tennessee, the well known breeder of Goldens, is an advocate of strong nuclei which are capable of passing through a dearth or other unfavorable season without much fussing on the part of the queen breeder. With five hundred or a thousand weaklings, the queen breeder finds it a very difficult matter to carry on operations under adverse conditions.

Fig. 12. Ten-frame hive divided into two parts as used for mating hives by Ben G, Davis of Tennessee.

Mr. Davis feels that the extra cost of these stronger nuclei is cheap insurance against a poor season. Figure 12 shows his big nuclei, where a ten frame hive is divided into two parts, each with four frames. These nuclei are strong enough to store sufficient honey to winter them successfully under normal conditions, and the time saved from fussing with daily feeding and constant attention more than repays the larger investment. Then there is no trouble whatever in stocking nuclei formed in this manner. All that is necessary in order to increase the number, is to remove one or two frames of emerging brood from a strong colony, for each nucleus, give them a queen or ripe cell and let them build up slowly during the summer, as one young queen after another is mated and permitted to begin laying.

Feeders.

Some kind of feeder will be necessary to stimulate the cell-starting and cell-building colonies, at such times as no honey is coming from the field. If small nuclei are used, it will often become necessary to feed them as well. Since nearly every apiary is provided with feeders of one kind or another, it hardly seems important in a work like this to enter into a discussion of the different types of feeders in the market, and the special merits of each. The Doolittle division board feeder is very popular among queen breeders, as is also the Alexander bottom feeder. However, practically every type of feeder now in the market is in use somewhere in a queen-breeding apiary. The Penn Company, of Mississippi, use a Mason jar with small holes in the metal cover. This is inverted in a round hole in the center of the cover of the hive, Figure 13. In passing through the yard, one can see at a glance the exact amount of feed available to every colony. The feeders are easily filled and replaced without opening the hives, and, at the same time, place the feed above the cluster.

Fig. 13. Feeding with Mason jars set in the top of hives at the Penn Company yards.

Nursery Cages.

During much of the season a queen breeder with an active trade will have no use for nursery cages. Each cell will be placed in a nucleus a day or two before time for the queen to emerge, and there she will remain until removed to fill an order to requeen a colony. However, it often happens that a batch of cells will mature when no queenless nuclei are ready to receive them, and it becomes necessary to care for them otherwise for a day or two, until room can be made for them. Then some breeders make a practice of allowing the young queens to emerge in the nursery cages before placing them in the nuclei. In this case, cages will be necessary.

There is a considerable percentage of loss when queens are permitted to remain several days in the cage. Some will creep back into the cell and be unable to back out again, while others will die from other causes. Sometimes, the bees will feed them through the wire cloth, but this is not to be depended upon, and the cages must be stocked with candy to insure plenty of feed within reach. Doolittle advocates smearing a drop of honey on the small end of the cell when placing it in the nursery, in order to provide the queen with her first meal as soon as she cuts the capping of the cell. Candy is also provided to furnish food in sufficient quantity during the period that she is confined in the cage. The cages must be kept warm, of course, while the cells are incubating, and for this purpose they are usually left hanging in the hive with a strong colony. However, the bees will not keep the cells in cages sufficiently warm after the weather gets cool in late fall, nor in early spring. At such times it becomes necessary to provide a nursery heated with a lamp or other artificial heat, in which the frames of nursery cages can be hung.

Some queen breeders utilize an ordinary poultry incubator for this purpose, maintaining it at the normal hive temperature.

E. B. Ault of Texas has fitted up an outdoor cellar with artificial heat for the purpose of incubating his sealed queen cells.

Alley Nursery Cage.

The Alley cage, Figure 14, is the most popular cage, although this may be because it has been so long on the market. A nursery frame is offered by supply houses which holds twenty-four of these cages. The larger hole is just the right size to take a cell built on a cell block. The block makes an effective stopper for the hole after the emergence of the young queen. Candy for provision is placed in the smaller hole.

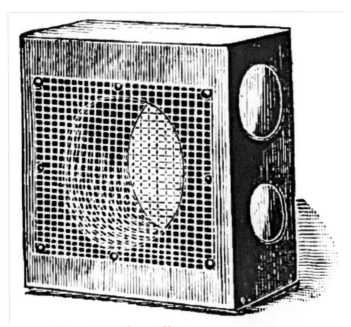

Fig. 14. The Alley nursery cage.

Rauchfuss Nursery Cage.

The Rauchfuss cage has not been long in the market, but bids fair to come into general use. Figure 15 shows the cage and Figure 16 the frame to hold about three dozen of them. This cage can be used for any purpose for which a cage is needed about the apiary. The hole at one end is large enough to take a ripe cell, while the candy at the other end can be eaten away, thus releasing the queen, and making it a desirable introducing cage.

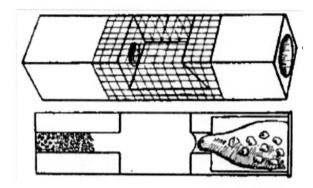

Fig. 15. The Rauchfuss combined nursery and introducing cage.

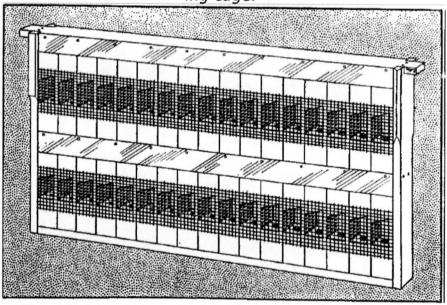

Fig. 16. Frame for holding Rauchfuss nursery cages.

Shipping Cages.

The Benton mailing cage has come into almost universal use among queen breeders. This is used as a combined mailing and introducing cage. It has been found that a small cage is desirable for sending queens in the mail, as there is less danger of injury when thrown about in the mail sacks than in a larger cage where there is more room to be bumped about. When larger cages are used, where the queen and her escort must travel long distances, as for export trade, a correspondingly larger number of bees are enclosed, thus saving each other from the shocks incident to travel through the mails.

Minor Equipment, such as cell blocks, cell protectors, etc., will be taken up in connection with the chapters relating to their use.

Chapter V

Early Methods of Queen Rearing.

Prior to the invention of the movable frame hive little progress was made in the development of bee-keeping. Commercial queen rearing as now practiced has been developed within the memory of our older beekeepers. As soon as his invention of the loose frames made the control of conditions within the hive possible, Langstroth began to experiment in the hope of being able to control natural swarming, and make necessary increase at his convenience. At that time the only known method of securing additional queens, was by means of depriving a colony of the queen. The queenless colony in its anxiety to make sure of replacing the lost mother, would usually prepare a number of cells and rear several more queens than needed. The ripe cells were taken from the hive before the emergence of the first queen, and given to nuclei or queenless colonies. As compared to present wholesale methods, this plan was crude and unsatisfactory. However, a careful beekeeper could by this means make considerable increase artificially, or provide young queens to replace undesirable ones.

Langstroth's Method.

In the first edition of his "Hive and the Honey-Bee," Langstroth describes his method of queen rearing by means of one queen in three hives. Two hives were deprived of their queens which were used to make artificial swarms or nuclei, at intervals of a week. When the first hive had been queenless for nine days, there were several sealed queen cells, which were counted, on the tenth day these were removed for use and a laying queen was taken from a third hive, C, and given to the first hive where she was permitted to lay a few days. In the meantime the second hive had been made queenless and had built cells. When these in turn were removed the queen which had been taken from the third hive, C, and placed in the first hive, was taken from the first hive and passed on to the second. The hive C, from which the queen had been taken, soon had cells ready to remove and she was replaced in her original home. Here she was permitted to stay for only a short time when she was started a second time around the circle. By keeping the queen in each hive for a period of a week at one time, sufficient eggs were laid to prevent the rapid depletion of the stock while providing a sufficient number of eggs and young larvae, to insure queen cells when she was again removed. By this simple plan he was able to get a large number of young queens and at the same time preserve the parent colonies. Whenever possible the queen cells were removed intact by taking out the frame on which they were formed and exchanging it for another from the colony, to which it was desired to give the cell. At times, however, he found it necessary to cut the cells from the combs, since several cells were often on the same comb.

For a number of years no better method was developed, and while numerous variations of the Langstroth plan were described in the beekeeping literature of the time, the only way known to secure additional queens was by means of making a colony queenless and trusting them to build cells in a natural manner. In an early edition of his "Manual of the Apiary," Cook recommended that the edges of the combs containing eggs or young larvae, be trimmed, or holes cut, somewhat after the manner known in later years as the Miller plan.

Quinby's Method.

Quinby practiced rearing queens by forming small nuclei of about a quart of bees and giving them small pieces of comb containing larvae not less than two, or more than three days old. A hole was cut in a brood comb sufficient to insert a piece of comb containing the larvae. This is described to be one inch deep and three inches long. No other brood was permitted in the hive. Concerning this plan he says:

> "I want new comb for the brood, as cells can be worked over out of that, better than from old and tough. New comb must be carefully handled. If none but old, tough comb is to be had, cut the cells down to one-fourth inch in depth. The knife must be sharp to leave it smooth and not tear it."

While practicing the method just described, he said in his book, that in many respects he preferred to rear queens in a strong colony made queenless.

The Alley Plan.

Henry Alley made a distinct advance when he developed his plan of using strips of worker comb containing eggs or just hatched larvae. Before describing his method of preparing these cells, it is best, perhaps, to outline his plan of preparing the bees to receive them so that his whole method may be clearly explained.

He recommended taking the best colony in the apiary to use as cell builders. After the queen had been found, her bees were brushed into a "swarm box," which has a wire-cloth top and bottom, to admit the air.

"The bees should be kept queenless for at least ten hours in the swarming box, else the eggs given them for cell building will be destroyed. Soon after being put into it they will miss their queen and keep up an uproar until released."

The bees in the swarm box were kept in a cool room or cellar and fed a pint of syrup. In the meantime the old hive has been removed and a queen rearing hive placed on the old stand. At night the bees are returned to the new hive on the old stand and given cell building material provided as follows:

In the center of the hive containing the breeding queen an empty comb has been placed four days previously. This will now contain eggs and hatching larvae. The bees are carefully brushed off this comb and it is taken into a warm room to be cut into strips. With a thin, sharp knife, which must be kept warm to avoid bruising the comb, the comb is cut through every alternate row of cells. After the comb has been cut up into strips, these are laid flat on a table and the cells on one side of the midrib are cut down to within a quarter of an inch of the septum as shown in Fig. 17. Every alternate egg or larva is crushed by means of a match pressed gently into the shallow cells and twirled between the thumb and finger, Fig. 18. This gives room enough for a queen cell over each remaining one, Fig. 19. A frame containing a brood comb with about one-half cut away is used as a foundation for the prepared strip. The uncut side of the strip is dipped into melted beeswax and at once pressed against the lower edge of the comb. It is necessary to use care to have this melted wax of just the right temperature so as not to destroy the eggs by overheating them, while at the same time warm enough to run readily and stick to the dry comb. The shallow cells, those which have been trimmed, open downward in the same position as a natural queen cell built under the swarming impulse.

Fig 17 Comb cut down for cell building, by Alley.

Fig. 18. Every alternate egg is crushed with a match twirled between the fingers.

Fig. 19. Queen-cells by the Alley plan. [From Productive Beekeeping.]

The care of the cell building colonies, emerging queens, etc., is the same by this method as any other and will be found in detail further on. See page 63. Aside from the strips of prepared cells, no brood will be given to the queenless bees, and they will concentrate their attention on building cells, with the result that a considerable number of fine cells will be secured.

Chapter VI

Present Day Methods.

While most queen breeders of the present day use some modification of the Doolittle cell cup method, a few still cling to the Alley plan or some modification of it. J.L. Strong, a well known queen breeder of Iowa, who has but recently retired, continued to follow the Alley plan in detail until the end of his queen breeding career. Mr. Strong was a beekeeper for half a century and engaged in commercial queen rearing for about twenty-five years. The Davis queen yards in Tennessee use a modification of this method, using drone comb instead of worker comb. This necessitates grafting, as with artificial cell cups.

The Davis Method of Using Drone Comb.

At the Davis yards in Tennessee, a modification of the Alley plan is used. Instead of cutting down worker comb in which eggs have already been laid as in the Alley plan, they cut down fresh drone comb wherever available. This necessitates grafting of larvae the same as in the cell cup method later to be described. Strips of new drone comb are cut down, as already described, and fastened to wood supports. Royal jelly is taken from queen cells the same as in the cell cup method, and a small quantity placed in each drone cell which it is desired to use. Worker larvae from the hive occupied by the breeding queen are then carefully lifted from their cells by means of a toothpick or grafting tool, and placed in these prepared cells. Every third or fourth drone cell can be used in this manner. These cells are given to strong colonies to be built, the same as by the Alley plan or cell cup plan.

Fig. 20. A batch of finished cells grafted with drone comb at the Davis apiaries.

Fig. 21. Cutting away cells built on drone comb.

Mr. J.M. Davis has tried about all the systems so far given to the public during the nearly fifty years that he has been engaged in queen breeding. After giving the Doolittle cell cup method an extended trial, he abandoned it in favor of the plan above described. By this plan, it is possible to get large batches of fine cells, although it becomes necessary to have combs drawn above excluders and without foundation, in order to get a sufficient supply of drone comb for the thousands of cells which are built in a yard, doing an extensive business. Figure 20 shows one batch of 37 finished cells by this method. Cells built by this plan are not as convenient to remove and place in nursery cages or mating nuclei as those having the wood base. These must be cut apart as in Figure 21. This also necessitates some special means of carrying them about to avoid injury to the tender occupants. For this purpose a block with 24 holes bored in it is used at the Davis apiaries. As the cells are cut from the frame they are placed in the block, in the natural position. The block is easily carried from hive to hive while placing the ripe cells. Figure 22.

The cell block enables the queen breeder to carry a batch of cells right side up without danger of injury.

Natural Built Cells by the Miller Plan.

What has, of late, been known as the Miller method of rearing queens, was probably not entirely original with him, but has been used in more or less the same form for many years. However, Dr. C.C. Miller has given the method new prominence, and brought it forcibly to public attention. In offering it, he did not even claim to be putting forth anything entirely new, but presented it as a very satisfactory method for the honey producer to provide himself with a limited number of queens with little trouble. The plan was so simple that it made an instant appeal, and has been widely published and generally used under the name of the Miller Plan. The author probably can present the matter in no other way so well as to copy Doctor Miller's original article concerning it from the American Bee Journal, August, 1912:

"Yet it is not necessary to use artificial cells. The plan I use for rearing queens for myself requires nothing of the kind. And it gives as good queens as can be reared. I do not say that it is the best plan for those who rear queens on a large scale to sell. But for the honey producer who wishes to rear his own queens I have no hesitation in recommending it. I have reared hundreds of queens by what are considered the latest and most approved plans for queen breeders; and so think that I am competent to judge, and I feel sure that this simple plan is the best for me as a honey producer. I will give it as briefly as possible.

"Into an empty brood frame, at a distance of two or three inches from each end, fasten a starter of foundation about two inches wide at the top, and coming down to a point within an inch or two of the bottom bar. Put it in the hive containing your best queen. To avoid having it filled with drone-comb, take out of the hive, either for a few days or permanently, all but two frames of brood, and put your empty frame between these two. In a week or so you will find this frame half filled with beautiful virgin comb, such as bees delight to use for queen-cells. It will contain young brood with an outer margin of eggs. Trim away with a sharp knife all the outer margin of comb which contains eggs, except, perhaps, a very few eggs next to the youngest brood. This you will see is very simple. Any beekeeper can do it the first time trying, and it is all that is necessary to take the place of preparing artificial cells.

"Now put this "queen cell stuff," if I may thus call the prepared frame, into the middle of a very strong colony from which the queen has been removed. The bees will do the rest, and you will have as good cells as you can possibly have with any kind of artificial cells. You may think that the bees will start "wild cells" on their own comb. They won't; at least they never do to amount to anything, and, of course, you needn't use those. The soft, new comb with abundant room at the edge, for cells, is so much more to their taste that it has a practical monopoly of all cells started. In about ten days the sealed cells are ready to be cut out and used wherever desired."

This plan is especially useful to the novice or to the beekeeper wishing for but a few queens at one time. It is simple, easy and never failing under any normal conditions.

Our illustration, Figure 23, shows this method with four strips of foundation used to start, instead of two as Doctor Miller suggests in his article.

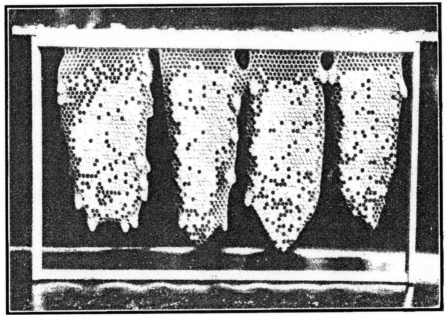

Fig. 23. Queen-cells built naturally by the Miller plan.

Big Batches of Natural Cells by the Hopkins or Case Method.

Many extensive honey producers who desire to make short work of requeening an entire apiary, and who do not care to bother with mating boxes or other extra paraphernalia, make use of the Case method, which has been somewhat modified from its original form. This method is advocated by such well known beekeepers as Oscar Dines of New York and Henry Brenner of Texas. The plan was first used in Europe.

To begin with, a strong colony is made queenless to serve as a cell building colony. Then a frame of brood is removed from the center of the brood nest of the colony containing the breeding queen from whose progeny it is desired to rear the queens. In its place is given a tender new comb not previously used for brood rearing. At the end of four days this should be well filled with eggs and just hatching larvae. If the queen does not make use of this new comb at once, it should not be removed until four days from the time when she begins to lay in its cells. At that time nearly all the cells should be filled with eggs and some newly hatched larvae.

This new comb freshly filled is ideal for cell building purposes. The best side of the comb is used for the queen cells and is prepared by destroying two rows of worker cells and leaving one, beginning at the top of the frame. This is continued clear across the comb. We will now have rows of cells running lengthwise of the comb, but if used without further preparation the queen cells will be built in bunches, that it will be impossible to separate without injury to many of them. Accordingly we begin at one end, and destroy two cells and leave one in each row, cutting them down to the midrib but being careful not to cut through and spoil the opposite side. Some practice destroying three or four rows of cells, and leaving one to give more room between the finished queen cells. (see Fig 24b.)

We now have a series of individual worker cells over the entire surface of the comb, with a half inch or more of space between them. The practice varies somewhat with different beekeepers beyond this point. However, this prepared surface is laid flatwise with cells facing down, over the brood nest of the queenless colony, first taking care to make sure that any queen cells they may have started are destroyed. In general, it is recommended that the colony be queenless about seven days before giving this comb. By this time there will be no larvae left in the hive young enough for rearing queens, and the bees will be very anxious to restore normal conditions. Some beekeepers simply take away all unsealed brood, rather than leave the bees queenless so long.

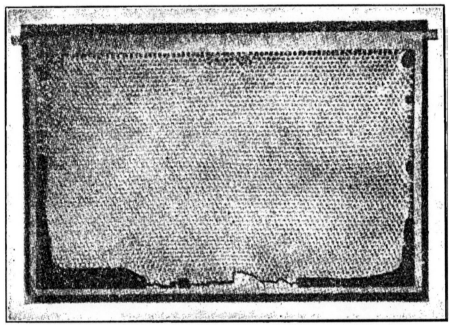

Fig. 24a. Frame for holding comb horizontally above brood-nest for getting queen-cells by the Case method.

As generally used, this method requires a special box or frame to hold the prepared comb. This is closed on one side to prevent the escape of heat upward and to hold the comb securely in place. Figure 24a. Some kind of support is necessary to hold the comb far enough above the frames to leave plenty of room for drawing large queen cells. It is also advisable to cover the comb with a cloth which can be tucked snugly around it, to hold the heat of the cluster. By using an empty comb-honey super above the cluster, there is room enough for the prepared comb and also for plenty of cloth to make all snug and warm.

Fig. 24b. Comb Prepared for Queen Cells

Fig. 24c. Comb of 80 good cells

Strong colonies only should be used for this, as for any other method of queen rearing. If all conditions are favorable, the beekeeper will secure a maximum number of cells. From seventy-five to one hundred fine cells are not unusual. (see Fig. 24c.) By killing the old queens a day or two before the ripe cells are given it is possible to requeen a whole apiary by this method with a minimum of labor. According to Miss Emma Wilson, it is possible to get very good results by this method, without mutilating the comb, although it is probable that a smaller number of queen cells will be secured. By laying the comb on its side as practiced in this connection, the cells can be removed with a very slight effort and with a minimum of danger.

The Doolittle Cell Cup Method.

Nine queen breeders in every ten, it is safe to say, use the Doolittle cell cups. While it is possible to rear queens on a commercial scale by other methods, few queen breeders care to do so. One can control conditions so nicely by the use of artificial wax cups and can determine so nearly how many cells will be finished at a given time, that this method is in all but universal use in commercial queen breeding apiaries. Most of the extensive queen breeders count on turning out queens at a uniform rate, increasing the number as the season advances to keep pace with the probable demand. It is of no advantage to a breeder to produce five hundred ripe cells at a time when he has market for only a dozen queens. He estimates as nearly as possible the demand for the season and establishes a sufficient number of mating nuclei to care for the queens as they emerge. During the height of the season a queen is only permitted to lay enough eggs to enable the breeder to satisfy himself that she is fertile and otherwise normal. Queens thus follow each other in rapid succession in the various mating boxes, throughout the season.

It was the difficulty of keeping up a dependable supply of queens to supply his increasing trade that led G. M. Doolittle of New York State to experiment with artificial cells. The successful outcome of his extended experiments has largely revolutionized the queen trade. They have already been in use for about thirty years. One can make from one hundred and fifty to two hundred of these wax cups per hour, so perhaps this plan can be followed as easily as any from the point of time required in the various operations. Dealers in bee supplies now list these artificial cells for sale at a small price, and many buy them already prepared. They can be used either with or without a wood cell base. When used without the base they are attached to wood strips by means of melted beeswax. However, the wood base is very generally used, since the cells can be changed about with much less danger of injury. A sharp pointed tack is imbedded in the base, which makes it very easy to attach them to frame supports on which they are inserted into the hives. Figure 25 shows a frame of newly prepared cells ready for the hive. It will be seen that a strip of foundation is used above the wood supporting the cell cups. This will soon be drawn by the bees and filled with honey. More often the beekeeper cuts away part of a comb already drawn for use in this way.

Mr. Doolittle used a wood rake tooth as a form on which to mold the cells. Lacking this, a round stick about the size of a lead pencil, but with carefully rounded end, may be used. Beeswax is melted in a small dish over a lamp or on a stove of moderate heat. It must not be kept too hot, otherwise it does not cool rapidly enough. A mark should be made on the stick nine-sixteenths of an inch from the end, and the stick dipped into water to prevent the wax from sticking. After giving it a quick jerk to throw off the water it is then dipped into the melted wax up to the mark. The dipping is done quickly, twirling the stick around as it is lifted out to distribute the wax evenly. As soon as the wax is sufficiently hardened, it is dipped again, this time not quite so deep. The form is thus dipped again and again, each time lacking about a thirty second of an inch of going as deep as before, until the base of the cell is sufficiently thick to make a good cell.

Fig. 25. Frame of prepared cups by the Doolittle method.

These artificial cells answer the purpose as well as those built by the bees, and if other conditions are normal the bees accept them readily. If wood blocks are used they are now ready to be attached to the blocks, or if not, direct to the wood strips. Figure 25.

For use, it becomes necessary to supply each cell cup with a small amount of royal jelly, and then with a toothpick or grafting tool carefully lift larva, not to exceed thirty-six hours old, from a worker cell and place it on the jelly in the prepared artificial cell.

Chapter VII

Preparation for Cells.

Whether one uses the Alley plan or some of its modifications, or the Doolittle cell cup method, certain stages of the process of getting the cells built may be the same. A supply of royal jelly will be necessary to begin with only where grafting, or changing the larvae from worker cells into prepared cells, is practiced. The preparation of colonies for building cells, finishing them and caring for them until ready for emergence of the young queens, is very similar by any of these methods.

There are numerous variations of the treatment of colonies in preparation for cell building, and several of these will be described in an effort to treat the whole subject in a comprehensive manner.

Getting Jelly to Start With.

If the beekeeper wishes to start cells early in the season before there has been any preparation for swarming, it is sometimes difficult to secure a supply of royal jelly readily; especially is this true when the colonies are still weak from wintering. The first thing to do is to look carefully for supersedure cells, when making the spring examination of the apiary. Failing queens may be replaced at any season, and one or two cells will be built in anticipation of the supersedure. If a cell is found, this difficulty is at once disposed of, providing it is at the proper stage. The royal jelly is found in the bottom of the queen cells and is a thick white paste, very similar in appearance to the paste ordinarily used for library purposes or mounting photographs. Sometimes, when it is quite thick, it is desirable to thin it slightly by the addition of a small quantity of saliva or a drop of warm water. Only a minute amount of jelly is placed in each of the prepared cups, so that a well supplied queen cell will provide a sufficient quantity to supply thirty to fifty of them.

If no cells containing jelly are found, the usual plan is to remove the queen from a vigorous colony and permit them to start cells. The author very much dislikes to remove queens except when absolutely necessary, and prefers some other plan. A simple way is to place a wire cloth over the top of a strong colony in place of the cover. On this set a hive body containing at least three frames of brood in the various stages, being sure that there is no queen on the frames, and that there is plenty of newly hatched larvae. All adhering bees should be left on the combs. The cover is then placed over all and the hive left closed for two days, when there will be an abundant supply of royal jelly available.

The Author's Plan.

The author, not being engaged in queen rearing commercially, can choose a favorable time for rearing such queens as are necessary to make increase or for requeening. While the method seldom fails even under unfavorable conditions with him, it is very possible that it might not be satisfactory under some conditions.

To begin with, the queen is found and placed, on the comb on which she is, in an empty hivebody. Sometimes the remainder of the space is filled with empty drawn combs, sometimes one or more frames of brood are added, as circumstances dictate. The hivebody containing the queen is then placed on the hivestand in the position where the colony had already been placed. Above the hivebody containing the queen is placed a queen excluder, to prevent the queen from going above. If the weather is warm so that there is no longer any danger of chilling brood from dividing the cluster into two parts, an empty set of extracting combs is placed over the excluder. The original hive containing most of the brood is now placed on top of this empty chamber. Twenty-four hours later the bees are given a frame of cellcups containing larvae. These cups are placed in the hive in the same manner as usual, except that they have no royal jelly. A thin syrup made with sugar and water or honey thinned with water is then poured freely over the tops of these frames. The worker bees gorge themselves freely with the syrup and, since the brood in the upper chamber is so far from the queen below, the bees are easily stimulated to start queen cells. Usually from one to three of these dry cells will be accepted, and two days later will furnish an abundant supply of royal jelly for grafting purposes. A second lot of cells is now prepared with jelly, and these are given to the bees in the upper story in the same manner. Syrup is poured over the frames as freely as before, with the result that a large portion of the cells are likely to be accepted. The author does not claim that the idea is altogether original with him, but simply outlines it as his method of practice. Feeding the bees freely at the time of giving a batch of cells is rather common practice among the queen breeders in certain localities. By this method, it is easily possible to secure a supply of royal jelly without dequeening a colony or interfering with the laying of the queen. If it is too cold to place an empty super between the brood nest and the brood in the upper story, the plan will usually work with only the excluder between. After the weather becomes warm enough, it is easily possible to continue building cells indefinitely above the same colony, by lifting the brood above as fast as sealed in the brood nest. The young bees emerging in the upper chamber continue to supply nurses as needed. It will be readily apparent that to be successful this plan requires a strong colony.

Fig. 26. A batch of cells by the cell-cup method.

*Fig. 27. Larvae not to exceed thirty-six hours old
should be used for grafting.*

Transferring the Larvae.

Some beekeepers make a practice of placing a
frame of cellcups in the hive over night in advance of
the grafting. The idea is that the bees will work them

over, smooth and polish them, thus placing them in more attractive condition for the acceptance of the prepared cells. The author has never been able to convince himself that this plan brings enough better results, in practice, to justify the extra trouble, where large numbers of queens are to be reared.

The cellcups are placed in the wood bases and fastened in place as shown in Figure 25. Commercial queen breeders usually have two or three bars of cells in each frame instead of only one. About fifteen cellcups to each bar is not unusual, so that with a liberal number accepted it is often possible to get from thirty to forty finished cells in each batch. Figure 26.

At this stage the grafting house described previously is very desirable. The queen cells from which the royal jelly is to be taken, together with the prepared cellcups and a frame of newly hatching brood from the breeding colony are now taken to the grafting house or into a warm room for the final preparation. For transferring the jelly and the larvae, there are specially prepared tools in the market. These look very much like knitting needles with one end flattened and slightly bent to one side. However, one can do very well with a quill cut down to a strip about a sixteenth of an inch in width, with the end bent in similar manner. Even a toothpick can be made to serve quite well.

An ingenious device for transferring larvae is described by John Grubb of Woodmont, Pa. He uses a small stick of wood about three-sixteenths of an inch thick and four inches long, one end of which is whittled down to a long tapering point. A long horsehair is doubled, then twisted together, and doubled again. Both ends are laid on the stick, the circular center extending beyond the end. Fine thread is wrapped around the hair and the stick, to hold all firmly. The doubled hair makes a circle about a tenth of an inch in diameter beyond the pointed end of the stick. With this horsehair spoon it is an easy matter to lift a larva from a cell and transfer it to a cellcup. It is easily and quickly made and materials necessary are usually within easy reach.

First a bit of royal jelly is placed in each cellcup, and then a larva about twenty-four to thirty-six hours old is carefully lifted from its cell and placed on the jelly. There is some difference of opinion as to the proper age of larvae, but all agree that larvae more than three days old should never be used. Nobody holds that better queens can be reared from larvae two days old than from younger larvae, although some think that as good results can be obtained. The majority seem to favor larvae from twelve to twenty-four hours old, with some strong advocates of thirty-six hours as the proper age. Figure 27. However, it may safely be said that twenty-four to thirty six hours is as old as larvae should be for this purpose. Probably up to this age as good or better results will be obtained as from the use of younger ones.

Something has already been said about the importance of selecting the breeding stock carefully. This is a vital matter if good results are to come from the breeder's work. The larvae used in grafting should be the product of the best queen available.

At Figure 27 we show the magnified larvae in the cells at about the proper age for grafting. Sladen recommends that larvae not quite as large as a lettuce seed be used. With a little experience one will soon come to tell readily the approximate age of the larvae by the appearance.

Chapter VIII

Getting Cells Started.

For building cells one must have strong colonies, Figure 28, and to insure this condition, one must have his bees in good shape in early spring. While it is often advocated that stimulative feeding be resorted to early, in order to build the colonies up to a sufficient strength, the author inclines to the belief that colonies in two stories will build up just as rapidly if there is an abundance of sealed honey in the hive, as is possible with stimulative feeding. Sometimes it seems that uncapping a portion of the honey has a stimulating effect, but feeding in small quantities, for the purpose of stimulating the bees to greater activity, rarely seems necessary until the time comes to give them the cells. At this time feeding is often needed in order to get large batches accepted and finished. When honey is coming in from natural sources, feeding is, of course, unnecessary.

The real problem is to get the bees into the right temper to accept the cells readily, and finish a large portion of them properly. This point has been touched upon rather indirectly, already, under the discussion of the various methods. A strong colony which is preparing to supersede the queen is very desirable at this time. Such a colony will accept cells readily and will supply them with royal jelly abundantly. No better cells can be had than those built in a supersedure colony. It will pay to look through the apiary very carefully in search of such a colony, rather than to resort to artificial conditions. A colony which is preparing to swarm, will do very well, also, only they must be watched carefully, to make sure that a swarm does not issue as soon as the cells are sealed. When a colony is found to have queen cells already built which contain eggs or larvae, these cells may be destroyed and a frame of prepared cells given. Little attention need be paid to the presence of the queen, for she will not disturb the new cells under such conditions.

Fig. 28. Only strong colonies should be used for build-ing or finishing cells.

If no colony is to be had which is already in the cell-building notion, it then becomes necessary to stimulate the cellbuilding instinct artificially. There are several methods of doing this.

Removing Queen and Brood.

Probably the most generally practiced method is to take a strong colony, and remove the queen and all unsealed brood. Empty combs and those which contain only honey and pollen are left in the hive. The queen should be placed in a nucleus, or given to another colony where needed. All bees should be carefully brushed from the combs containing the brood in order to leave as large a force of nurse bees as possible. The brood is then given to another colony to be cared for.

About ten or twelve hours later the bees will be in the mood to build queen cells. Being without brood, the nurse bees will be abundantly supplied with food for the larvae, and will accept a batch of prepared cells very eagerly.

When giving the cells, it is well to follow the practice of some of the most extensive breeders and feed liberally at the moment, to insure a larger portion of cells accepted. For this purpose an ordinary garden sprinkler serves very well. Thin sugar syrup is sprinkled freely over the tops of the frames as described previously. The bees gorge themselves in cleaning up the syrup and anxiously seek larvae to be fed. This method of feeding is desirable at the time of giving cells by any method.

Some breeders leave the prepared cells in the colony to which they are first given until they are sealed. However, a larger number of first class cells will usually be secured by working two colonies together, one as a cell-building colony and the second as a cell-finishing colony. The cell-finishing colony should be equally strong with the cell-starting colony, but not all the brood is taken from it. At the time that the brood is taken from the first colony, part of the brood is removed from another, and the remainder raised above an excluder, leaving the queen in the brood nest below on one frame of brood, and with empty combs in which to continue laying. This we will call the finishing colony.

Twenty-four hours after the prepared cells have been given to the queenless and broodless bees in the cell-starting hive, we will probably find most of the cells partly built, and the larvae abundantly supplied with royal jelly. If we leave them as they are, some of these cells are likely to be neglected, so that not all will come to maturity. We may now safely remove these cells and after carefully brushing off the nurse bees with a feather, give them to the cell-finishing colony, placing the frames above the excluder. By this time the bees in the second colony will have been forty-eight hours separated from the queen which still remains below the excluder. Since no eggs have been laid with the brood above for this period, the bees are in much the same condition as a colony with a failing queen and accordingly accept the cells as readily, as a rule, as a supersedure colony will do.

When the batch of started cells is taken from the starting colony, it is given a second lot of newly prepared cells. This may be repeated regularly for some time. However, the same bees cannot serve as nurses for very long and it will be necessary to supply the starting colony with frames of sealed brood ready to emerge at frequent intervals if the same colony is used as a cellstarting colony for more than ten days. Usually the number of cells accepted in each batch will soon begin to diminish, so that it will be desirable to prepare another colony for this purpose after eight or ten days.

There is a great difference in individual colonies as to the number of cells built, and it sometimes becomes necessary to experiment a bit to find the best colonies for this purpose. Some colonies will build double the number of cells that others will build. An extensive breeder will find it necessary to have several cell-building colonies at one time. Figure 29 shows a strong cell-finishing colony at the Davis apiary in Tennessee.

The Swarm Box.

Alley used much the same plan as above described, except that he first found the queen and then shook all the bees into a swarm box which is made by placing a wirecloth bottom and cover on an ordinary box of suitable size. The bees were smoked before shaking them into the box to induce them to gorge themselves with honey, and then they were confined in the box from morning until evening. The wirecloth admits plenty of air and by the time the bees are placed in a hive for cell building, they will recognize their hopelessly queenless condition, and be ready to accept the prepared cells with little delay. Alley gave eggs in strips of natural comb, instead of the prepared cells, it will be remembered, but the principle is the same. He left the bees queenless in the swarm box for at least ten hours. He also fed the bees syrup while confined in the box.

Rearing Queens in Oueenright Colonies.

The author prefers to rear queens in a queenright colony, since it is not so difficult to maintain normal conditions over a long period of time, and the bees are not so sensitive to fluctuations in weather conditions or honeyflow. It is not always possible to make a success with the first batch of cells given by this plan, but once accepted the same colony can be kept busy rearing cells for weeks, or even all summer if desired.

Fig. 29. A strong cell-finishing colony.

One plan which is followed by successful breeders is to select a strong colony for cell building. Remove the cover, and put a queen excluder in its place. Then take enough frames of brood from several different colonies to fill a second brood-chamber above the excluder, leaving one vacant space. Care must be used to make sure that no queen is on the frames placed in the second story. The vacant space is left as near the center of the colony as possible, and a few hours later a frame of prepared cells is placed there, feeding the bees with syrup from the sprinkling can at the time the cells are given. If this first batch of cells is not readily accepted try again the following day. After four days a second batch can be given, and a new batch every four days thereafter. By this plan the cells are left with the colony until ready to be given to the nuclei. It only becomes necessary to add two or three frames of sealed brood every week to provide the colony with plenty of young bees for nurses, to continue cell building indefinitely. About ten to fifteen sealed cells can be secured from a single colony every four days by this plan. If a heavy honeyflow comes on, it may become necessary to add supers between the brood nest below and the cell-building chamber above, since the old queen continues to lay in normal manner below the excluder. By this method the cell-building colony will give a crop of honey as well as queens. The addition of so much brood from other colonies will keep the cell-building colony very strong throughout the season. Of course, frames of honey must be removed from time to time as frames of brood are given, and, during a good flow, it may become necessary to remove frames of honey quite often to prevent crowding in the cell-building chamber.

Feeding.

During a dearth of nectar it often becomes necessary to resort to stimulative feeding to induce the bees to continue cell building by any of these methods. Of course, a queenless colony will build some cells under almost any conditions, but to get good cells in sufficient numbers, a fresh supply of food must continue coming to the hive daily. If there is none in the field a pint or more of thin syrup should be fed daily, preferably at night, to prevent robbing.

Chapter IX

Care of Finished Cells.

About four days after the prepared cell cups are given to the bees, the finished cells will be sealed, Figure 30. If the weather is warm they may be placed in cages and transferred to other colonies for safe keeping, until time for the young queens to emerge. However, in cool weather, there is danger that the young queens will be chilled and injured, if the cells are placed in cages so that the nurses can no longer warm them by direct contact. Most breeders leave the cells to the care of the bees until the evening of the ninth day. The cells are then caged or given directly to the nuclei, where they are to be mated. In general, it is better to place the cell at once in the nucleus.

Fig. 30. Finished cells by the Doolittle method.

Great care must always be used in handling sealed queen cells. Any slight jar is likely to dislodge the nymph from its bed of royal jelly and injure it seriously. The bees which may cluster about the cells may be driven off by smoking them or by carefully brushing them away. The longer the cells are left undisturbed, the less the danger of injury to the young queens. The bees should never be shaken from a frame containing queen cells.

It is necessary to separate cells built by the Alley plan by cutting with a sharp knife. The knife should be kept warm to get best results. Otherwise, instead of cutting freely it may simply crush the wax and injure a cell. Figure 21 shows how the cells may be cut apart.

It is important, also, to keep the cells right side up at all times. Some breeders use a cell block such as may be seen at Figure 22. This enables the breeder to carry a whole batch to the apiary to be placed, one at a time, in the nuclei, without danger of injuring them.

It often happens that a batch of cells will be ripe and the nuclei not yet ready to receive them for one reason or another. In that case, candy should be placed in the nursery cages, and the cells placed in them on the ninth or tenth day after the cells are given to the bees. It should be remembered that the queens will emerge on the fifteenth or sixteenth day after the eggs were laid. Should a virgin queen emerge before the cells are removed and cared for, she is likely to destroy at once all that remain. Thus all the beekeeper's labor is for naught.

It is necessary to exercise some care in extremely hot weather to avoid overheating the cells when carrying them about in the hot sun. Well known queen breeders admit having lost valuable cells on more than one occasion by overheating through exposure to direct sunshine on a hot day.

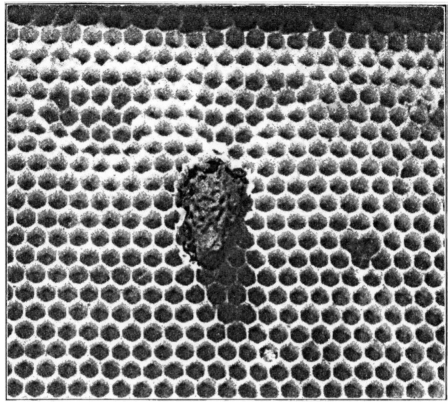

Fig. 31. Placing cell in nucleus without brood.

In placing the cells in the nuclei the cell should be gently pushed into the side of a comb just above the brood, if there is brood. However, it often happens that no brood is present in a nucleus when a cell is placed. In that event it should be set into the comb near the center of the hive. Figure 31.

Use of Cell Protectors.

If a nucleus has been queenless for twelve hours when a ripe cell is introduced, there will seldom be any need for using protectors. However, it often happens that the breeder will have ripe cells ready which he wishes to place as fast as the queens are removed. When the bees destroy queen cells they do so by opening the cell at the side, and never from the end. Taking advantage of this fact a wire protector has been made which remains open at the end, thus permitting the queen to emerge without further attention, Figure 32. By the time the queen is ready to emerge, the bees will discover the absence of the old queen and the newcomer will be welcomed.

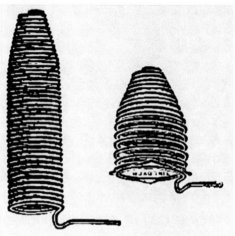

Fig. 32. Cell protectors.

Formation of Nuclei.

In the chapter on equipment for queen rearing, the various styles of mating boxes and hives have been described. If the standard hive is used, the formation of nuclei is a simple matter. As many colonies as may be needed to make the desired number of nuclei are broken up, and the combs together with adhering bees are placed in the nuclei. One frame with the old queen is left in the old hive, and it is usually well to leave a second frame of brood with her, to enable her to build up the colony again more rapidly. The rest of the space is filled with empty combs. One frame of brood and bees, together with one empty comb or one containing honey, is placed in each nucleus, Figure 33. The entrance is then stopped with grass to prevent the escape of the bees for several hours. By the time they have gnawed their way out they will become accustomed to the new condition, and most of them will remain in the new position. Unless the frame given is well supplied with brood, it is desirable to give two frames to each nucleus.

Fig. 33. A queen-mating yard composed of standard hives, each divided into two parts.

A day or two later sealed queen cells may be given safely. As the season advances, the demand for queens increases, and the breeder will find it necessary to increase the number of mating nuclei. As each queen is allowed to lay for a short time in the nucleus before caged for shipment, many of the nuclei will build up rapidly. From time to time one will be found which can spare a frame of brood and bees as already described. At the close of the season these nuclei are united to make them strong enough to winter as full colonies.

Stocking Mating Boxes or Baby Nuclei.

Much difficulty is sometimes experienced in getting the bees to stay in these small hives. The plan usually recommended is to shake the bees into a wire-cloth cage and confine them there for several hours. Four or five hours later run in a virgin queen among them. At nightfall, shake them into the mating box and leave them undisturbed for a few days. Some of the old bees may return to their former hive the next morning, but most of them are likely to remain. There is some danger that they may swarm out with the queen when she comes out for her mating flight. However, after one queen has been successfully mated and there is some brood in the little hive, there will be less trouble with the next one. These little hives must be watched to make sure that they do not at any time become short of food, otherwise they sometimes swarm out and leave the brood.

The available space is so small that the queen can be left but a very short time. The two little combs are soon filled with eggs and with no more room to lay, the queen may lead a small swarm, and thus desert the hive.

Another plan of stocking these hives is to shake a lot of bees from several hives into a box with wirecloth top and bottom similar to the Alley swarm box, and keep them confined for several hours. It is desirable that these bees be brought from a distance, if possible. When ready to stock the mating hives, wet the bees enough to prevent flying readily and dip them out with a tin dipper, turning a sufficient quantity into each compartment. A supply of virgin queens is ready at hand, and as each compartment is filled, a virgin is dropped into a dish of honey and then into the compartment with the bees. The entrance is opened at night to prevent the loss of bees before the excitement subsides. This is the plan practiced at the Root yards.

Chapter X

Combining Mating with Making of Increase.

Fig. 34. A queen-rearing apiary in Tennessee.

The usual methods of artificial increase, such as division or formation of nuclei to be built up, weaken the colony to a considerable extent. Should the season prove unfavorable after nuclei are formed, it may be necessary to feed them for a long period of time, only in the end to find it necessary to unite them again to get them strong enough for winter. Getting queens mated in an upper story is not new; yet there are some elements in the following plan which differ somewhat from methods previously given to the public.

Fig. 35. Queen-mating nuclei under the pine trees of Alabama.

The author has experimented to a limited extent in the hope of finding a plan which takes nothing from the parent colony, other than the honey necessary to rear the brood composing the new colony. There is no risk, since the old colony is not weakened by removing part of the field force, and the division is not made until the new colony is strong enough to shift for itself under almost any conditions. The following plan comes near realizing this ideal, having been uniformly successful in a limited way, even under unfavorable conditions. This is the outgrowth of a system of swarm control in the production of extracted honey, as described in Productive Beekeeping.

If the extracted honey producer can keep his colony together during the season, he should be able to get maximum results. Some increase is necessary in most any apiary, with any kind of system, to replace such colonies as are lost through failing queens, poor wintering or other causes, even though the beekeeper does not care to make extensions.

If the bees can be kept from swarming and the young queen be mated in a separate compartment, she can rear her own colony in due time, and they can be removed without reducing the product of the old queen, whose progeny will remain with the parent colony.

To begin with, when the colony becomes populous, place the queen on a frame of brood in an empty hivebody and fill out with empty combs. This is set on the regular hivestand occupied by the colony. The working force coming from the field will find their queen with an abundance of room in which to lay. This is the system of swarm control advocated by Demaree to this point. Now place a queen excluder over the hivebody containing the queen, and over this, a super of empty combs. On top of these is set the original hivebody containing the brood. A hole is bored in this upper hivebody to give the bees an extra entrance above. About twenty-four hours later a ripe queen cell is placed in the upper story with the brood. The queen will emerge in a day or two, and, in due time, will leave the hive on her mating flight, by way of the augur hole. Within a few days more she will be laying in the upper hivebody, while the activities of the bees will continue without interruption in the lower story. Within three weeks all the brood from the old queen (in the upper story) will have emerged. The brood which now appears in the upper story is a net addition to the resources of the colony, and, when the upper story is nearly filled with brood, it can be removed and placed on a new stand without checking the work of the colony.

To illustrate: A strong colony was given a queen cell as above described on May 21st. On July 14th, the upper hivebody with a young queen and seven frames of brood were removed to form a new colony. The strength of the parent colony was little affected apparently. Possible swarming had been prevented, temporarily at least, by the Demaree plan of placing the old queen in the empty hive below. There were two colonies better than any parent colony and swarm we had that season. In this way there had been no risk or loss. The new colony was not removed from its parent until both were provided for, neither was the possible crop cut short by dividing the working force of the parent colony at a critical time.

After three years of success with this method the author feels confident that it will prove successful on a large scale. Both queens can be left in the hive until the close of the honeyflow if desired, but there is little to be gained by leaving the queen above after her chamber is filled with brood. If both are left in the hive until late in the fall, one of the queens is likely to disappear.

If desired, the process can be repeated as soon as the upper story has been removed, as by this time the old queen will have filled the lower story with brood again. By beginning early, it should be possible to make two and possibly three new colonies, without reducing the honey crop from the parent colony to a serious extent.

This same plan might be used for the purpose of mating additional queens while making some increase, by the breeder who wishes to accomplish both ends at the same time. The method is particularly valuable to the honey producer who wishes to make some increase or rear queens for use in his own apiary, without reducing the honey crop. If increase is not especially desired, the same plan can be worked for the purpose of superseding queens. When the young queen has become nicely established in the upper chamber, the old queen can be removed from below and the position of the bodies reversed. It would be well to permit both queens to continue laying until the height of the honeyflow, in order to get as large a field force as possible for storing the crop.

Chapter XI

Shipping Queens.

The Benton cage, Figure 36, is almost universally used in this country for shipping purposes. So generally is it used, that it is as staple as any other item of bee-keeping equipment, and can be purchased through any dealer in supplies. While cages ready stocked with candy are offered for sale, most queen breeders prefer to make their own candy and thus save some cost, as well as making it fresh as needed.

Fig. 36. The Benton mailing cage

Making the Candy.

Candy for queen cages is made of honey and sugar. Under the postal regulations it is necessary to boil the honey for at least thirty minutes, unless the apiary from which it and the queens are taken has been inspected by some duly qualified officer, who is authorized to issue a certificate of health.

Care must be used, also, to make sure that the sugar used contains no starch. Powdered sugar is used for candy making, and some powdered sugar contains starch, which is detrimental to the bees confined for long in the cages.

Heat the honey and stir in as much of the powdered sugar as can be mixed in by stirring. When no more can be added by stirring, spread the powdered sugar on a mixing board and remove the dough to the board and mix it like a batch of biscuits. Some experience is necessary to determine when it is just the right consistency, neither too hard nor too soft.

According to Root, boiled honey as required by the postal regulations, does not give satisfactory results where queens are confined for long journeys. Since the idea of the regulation is to prevent the spread of disease through the honey; he recommends the use of invert sugar as a substitute for the honey.

Another kind of candy made without the use of honey, is used by some breeders. This is made by using 12 pounds of granulated sugar, 1½ pounds candy-makers' glucose, 1¼ quarts of water and $1/3$ teaspoonful of cream of tartar. The cream of tartar and glucose are added to the water and heated together in a kettle. The sugar is added after the mixture comes to a boil, stirring continually while putting in the sugar. After the sugar has all been dissolved, stop stirring and let it heat to 238 degrees. Then remove from the fire and let cool to 120 degrees, and stir again until it looks like paste, when it is ready for use.

Caging the Queens.

With a few trials, one will shortly get the knack of catching a queen off the comb by her wings. Holding the cage open end downward in one hand, it is easy to so place her head in the opening that she will catch her front feet on the wood, and readily climb up into the cage. When she goes in, the thumb should be placed over the opening until a worker is caught, and ready to follow in similar manner. The novice at queen rearing often makes the mistake of placing too few bees in the cage with a queen. It is well to place as many workers in the cage as there is room for, without crowding, especially if the journey to be taken is a long one. As a rule the queen will be the last to die, if the bees are in normal condition when placed in the cage. It often happens that queens received from a distant place are still alive, with all their attendant workers dead in the cage. Of course the queen would not much longer survive after the workers were all dead. If the candy is properly made and sufficient in quantity, a queen will often live for several weeks in a cage, with sufficient attendants.

After queens are caged they should be placed in the mails as quickly as possible to avoid confining them longer than is necessary. Although they live for a considerable time in the cages, one can hardly believe that the confinement is conducive to the health of the queen, and the shorter the time necessary to get her to her destination, the better.

What the Buyer has a Right to Expect.

When a man sends his money for a queen in response to an advertisement, he has a right to expect that no inferior queens be sent, even though he buy

untested stock. Some breeders have the reputation of
sending out mismated queens that have been laying for
a period long enough to show the fact, as untested
queens. While few breeders guarantee that untested
queens will be purely mated, to knowingly send out
mismated stock, to fill orders for untested queens, is
certainly dishonest. It is needless to say that no reputa-
ble breeder would do so. The breeder who expects to
establish a paying business has no asset so valuable as
the confidence of his customers, and this is only secured by sending out good stock and standing ready to
be more than reasonable in making good any losses.

The buyer has reason to expect that he will receive pure queens, carefully reared; that the breeder
shall maintain his mating nuclei in localities as free as
possible from impure stock, and entirely free from
disease.

Grading.

There is a great difference in the practice of different breeders in the way queens are graded. Some
advertise only three grades, untested, tested and select
tested queens. Others make five or more grades, adding select-untested queens and breeders. There should
be some effort made to establish a standard by which a
buyer can tell in advance what he is likely to get from
an order for any one of these grades.

In general, an untested queen is one which has
been mated and has been permitted to lay for a few
days, but not long enough for the emergence of the
workers. Breeders who make it a rule to send out all
queens which are reared, regardless of quality, are not
likely to build up a permanent business.

No poor queens should be sent out in any case,
except by special understanding, and then not for

breeding purposes. There is a small demand for queens for scientific purposes, which can be supplied with mismated or otherwise inferior stock without injury to anyone. Such queens should never be sent to a bee-keeper for introduction into normal colonies for honey production.

A tested queen is generally one which has been permitted to lay until her workers begin to emerge, and thus by their markings demonstrate the pure mating of their mother. She should properly demonstrate other qualities also. Select tested and extra select-tested or choice select-tested queens are, of course, queens which for one reason or another are more promising than the average of tested queens. Too many grades offers a good chance for the breeder to get an extra price from a buyer, without giving an equivalent in value. It is very true that queens showing unusual traits are worth more than the general run of queens, but it is difficult to grade the different degrees of behavior into a half dozen different classes and always give a uniform value.

Virgin queens, are of course, unmated queens. While there may be occasionally a good reason for the purchase of virgins, as a rule the practice is not to be encouraged. The difficulty of introducing a virgin after she is several days old and consequent danger of loss, is one good reason why they should not be shipped. The danger that they will become too old for mating before a favorable opportunity is offered, is another. The breeder who confines his business principally to the sale of untested queens, and who gives good value for the price asked, is the one who has the fewest complaints.

Much dissatisfaction arises from the sale of breeding queens at high prices. The buyer who pays five or ten dollars for a breeding queen, will too often expect too much of her, and, consequently, be disappointed.

Then it often happens that a queen old enough to demonstrate her value as a breeder, will be superseded shortly after her introduction into a strange colony.

Queens that have been laying heavily suffer seriously from the confinement in a small cage and the journey through the mails. Often they will never do as well for the buyer as they have done previously, and he is inclined to feel that he has not been treated fairly. As a rule, the same money invested in young untested queens, will bring far better results to the buyer, as well as being better for the seller. If a half dozen young queens are purchased from a breeder with good stock, at least one of them is quite likely to prove excellent. The best queen that the author ever has known he secured as an untested queen at a nominal price. There is probably no extensive queen rearing yard which would part with as good a queen for fifty dollars after she had demonstrated her value. In fact, she would not be for sale at any price, for she would be too valuable as a breeder. Yet the chances are that, after she had demonstrated her ability by outdoing everything else in the apiary for three successive seasons, she would be superseded within a few weeks after being sent through the mails.

Buyers should bear in mind that old queens which have laid heavily for one or more seasons, cannot be expected to repeat their former performances after a journey by mail. Such queens can only be shipped safely on combs in a nucleus, where they can continue laying lightly for some time. Someone has compared the sudden checking of the work of a laying queen, with the shipment of a cow, which is a heavy milker, without drawing her milk for several days. Neither can be expected to be as good again.

Chapter XII

The Introduction of Queens.

In order to be successful in the introduction of queens, it is necessary to overcome the antagonism of the colony toward a stranger. It must be borne in mind that, normally, a strange bee will be recognized as an enemy or a robber and at once driven out or killed. In order that the queen be welcomed as a member of the community, it is necessary that she be permitted to acquire the colony odor, and that she become somewhat familiar with her new surroundings so that she will not manifest, by her own excitement, the fact that she is a stranger. There are many indications of the colony odor and, in the absence of proof to the contrary, it is safe to assume that the bees depend upon this common odor as a means of identification of the members of the community.

There are many different methods of introduction of queens, which are followed with greater or lesser degrees of success. All these methods may be divided into two classes: those which depend upon the confinement of the queen until she acquires the common characteristics of the hive, as the cage methods; and those which create such an abnormal condition and so much confusion in the hive, that the undue excitement of one or more individuals will not be noticed, as the smoke or other direct methods.

Under the first plan, the bees are at first antagonistic to the new queen, which is recognized as a stranger, but are unable to reach her because of the barrier furnished by the screen covering the cage. After a time the bees recognize the fact that no other queen is present in the hive, the antagonism disappears, and she is accepted as the natural mother of the community.

Under any method in the second class, the colony is thrown into a state of excitement and uproar, to such an extent that the agitation and fear manifested by the new queen, upon finding herself in a strange hive, will not be apparent to the bees, since they are also in a state of confusion. By the time the excitement subsides, the foreign odor of the new queen will no longer be apparent, and she will settle down to the business of egg laying as though she had always been present in the hive. By this method it is the usual way to remove the old queen either shortly before or just at the time the new queen is introduced.

Details of Cage Methods.

All the variations of the cage method are compa-ratively simple. The old queen is first removed from the hive and the new one is introduced in a cage. Figure 37.

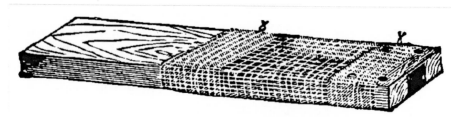

Fig. 37. The Miller introducing cage.

Probably the safest method of all is the placed alone in a cage that covers a small patch of emerging brood. The emerging bees are, of course, friendly enough, and within two or three days she will be laying in her small enclosure and surrounded by a small group of attendants who found her present when they emerged. The cage is then carefully removed, and the comb replaced in the hive with as little disturbance as possible. Such a cage is made with a piece of ordinary wirecloth about four inches square, sometimes smaller. Each of the four corners is cut away for about three quarters of an inch. The four sides are then bent down, forming a wire box open at the bottom. The queen is placed under this and the wire pressed into the comb. It is well to have a few cells of sealed honey inside the cage, although the bees are likely to feed the queen through the meshes of the cage. When this plan is used in a hive where no brood is present, some newly emerged workers should be placed in the cage with the queen. The attitude of the bees toward the queen will determine when it is safe to release her. If on opening the hive, the cage is found to be covered with a tight cluster of bees, she would be balled immediately if released. When the bees are found to be paying but little attention to her presence, it is usually safe to remove the cage.

The Benton mailing cages are stocked with candy before the queens are confined. Usually there will be quite a little of this candy still left, at the time the queens are to be introduced. If so, all that is necessary is to remove the old queen, remove the paper across the exit hole which is filled with candy, and leave it to the bees to remove the candy, and release the queen. It is likely to require from one to three days to remove the candy, and by that time, there is little danger to the new queen. If but little candy remains, the paper should be left over the hole for a day or two before removing. When the paper is removed, if the candy is almost gone, a little broken comb honey may be pushed into the cavity. Bees are likely to be friendly to the queen which has been caged in the hive for two days, and the bees which remove the honey are likely to gorge themselves to the point of quietude.

Fig. 38. A Mississippi queen-rearing apiary.

Some beekeepers by going to a little extra trouble, insure success by this method. When new queens are ordered they cage the old queens in the hive until the newcomers arrive. The old queens are then destroyed, and the new ones placed in the same cages and replaced in .the same hives. The cages have already acquired the hive odors, and the bees have become accustomed to the presence of their queens in the cages. By the time the candy has been removed, there is a very small element of danger.

Direct Introduction.

The easiest time for direct introduction of queens is during a heavy honeyflow. At such a time the bees will be in a constant state of activity because of the wealth of honey coming in, and queens can be introduced with a minimum of danger. At such times, the author has gone to the hives to be requeened, caught the old queens and run in the new ones, with little effort to disarrange the affairs of the community, yet the plan, worked with entire success with colony after colony. Many of the direct methods which are so successful during a honeyflow, must be followed very carefully under other conditions, or failure will result.

There are several of the direct methods, familiarly known as smoke method, flour method, water, and honey methods, etc. The same principle underlies them all. In every case the object is to develop such an abnormal condition within the hive, that the change of queens can be made without the fact being discovered by the bees.

The smoke method has recently been exploited as something new. Some of the details of the practice are all that is new, for Alley described a similar way of introducing queens by means of tobacco smoke as long ago as 1885. He directed as follows:

"When tobacco smoke is used to introduce them, throw some grass against the entrance to keep the smoke in and the bees from coming out. Blow in a liberal amount of smoke, and then let the queen run in at the top through the hole used for the cone-feeder."

The method as advocated by A. C. Miller does not anticipate the use of tobacco, but the ordinary smoke always available to the beekeeper with a lighted smoker. He describes his plan as follows:

"A colony to receive a queen has the entrance reduced to about a square inch with whatever is convenient, as grass, weeds, rags or wood, and then about three puffs of thick white smoke—because such smoke is safe—is blown in and the entrance closed. It should be explained that there is a seven-eighths inch space below the frames, so that the smoke which is blown in at the entrance, readily spreads and penetrates to all parts of the hive. In from fifteen to twenty seconds the colony will be roaring. The small space at the entrance is now opened; the queen is run in, followed by a gentle puff of smoke, and the entrance again closed and left closed for about ten minutes, when it is reopened, and the bees allowed to ventilate and quiet down. The full entrance is not given for an hour or more, or even until the next day."

Fig. 39. A queen-rearing apiary in Georgia.

Neither of the smoke methods above given, nor, for that matter, most of the direct methods, are entirely reliable under adverse conditions. The great advantage in the use of such a method is the saving in time. Some queen breeders of the author's acquaintance have used the smoke method extensively for this reason, and with good success. Introducing a queen which is taken from a hive or nucleus and given at once to another, is a much simpler matter than the introduction of a queen which has been caged for a week and probably travelled several hundred miles in a mailbag, where she had opportunity to acquire all kinds of foreign odors. The experienced man will soon learn when he can with safety depend upon a short cut, and when there is danger in doing so.

Honey and Flour Methods.

These methods are similar except that in one case honey is used and in the other case flour is the medium. The honey method is used with good success in introducing virgins to bees in packages, after they have been confined for a few hours. The queen is simply dropped into a cup of honey and entirely submerged in it, and then dropped in among the bees, which at once proceed to clean her up. For introducing queens into full colonies, this plan does not always succeed.

Where the queen is covered with flour, she may be accepted or not, depending much upon other conditions. Where the honey method is used, the queen is much more likely to be accepted if the honey in which she is dipped is taken from the hive to which she is to be given, at the time of her introduction.

Water Method.

This method requires a little more trouble, but is generally successful according to reports, and also according to the author's experience. The bees are shaken from the combs, and sprinkled with water until they are soaking wet. The new queen is wet likewise and dropped on the pile of wet bees in the bottom of the hive. The combs are then replaced and the hive covered.

Neither of these methods is attractive, since it hardly seems like proper treatment to give a valuable queen.

Introduction of Virgins.

A newly emerged queen while she is still downy, say within half an hour of the time of her emergence, can be run into any queenless colony or nucleus with safety. The bees are apparently conscious that any bee of such a tender age is incapable of harm, and she is accepted as a child of the community. For such, it is not necessary to provide any artificial stimulus of any kind; smoke, flour, or water are all unnecessary.

Virgins that are four or five days old are more difficult to introduce, than are fertile queens. Alley recommended dipping the virgin in honey, thinned with a little water as above described, and then dropping her into the queenless hive. He wrote that virgins could only be introduced successfully to colonies that had been queenless for at least three days. It has often been advised to leave colonies queenless for this period before introducing fertile queens, but the author prefers to give a fertile queen immediately on removing the old queen. With virgins there is a larger element of danger.

Chapter XIII

Spreading Disease from the Queen Yard.

It is an unfortunate fact that much of the responsibility for the present wide-spread prevalence of foulbrood must be laid at the door of the careless queen breeder. Foulbrood has been introduced into many localities by the purchase of queens from diseased apiaries. The queen breeder cannot use too much care in keeping his apiary and his locality free from disease. In any event, queens should not be mailed to purchasers from an apiary where disease is present. In our present state of knowledge of European foulbrood, it is uncertain in just what manner the disease is spread, but it is very probable that a queen bee, taken from a diseased colony, might be the means of introducing it into a healthy colony, even though no honey or bees accompany her.

It is reasonably certain that there is little danger of the spread of American foulbrood, except in the honey from diseased colonies. The postal regulation which requires that honey used to make candy, to stock queen cages to be sent through the mails, be boiled for thirty minutes, is supposed to meet all requirements. While this may be true, as far as American foulbrood is concerned, it is not sufficient protection for the purchaser, from European foulbrood or paralysis.

The late O.O. Poppleton related something of his experience with paralysis, to the writer. For a time he had serious losses among his bees from this disease. He was finally able to trace the trouble to the introduction of queens from the yards of a well known breeder. By requeening all his yards with a different strain of bees, he was able to eliminate the disease. Later he introduced the same disorder to his apiaries again with queens from another source. On investigating the matter, he was surprised to learn that the man from whom he bought the new lot of queens, had previously purchased a breeding queen from the breeder from whom he had first contracted the disease. It accordingly became necessary to requeen his apiaries with new stock, a second time, to get rid of paralysis.

Fig. 40. A Minnesota queen yard.

Diseases of adult bees are, as yet, but little understood; but it is quite probable that there are several different diseases, all of which are known under the general name of paralysis. It is very evident that this trouble, whatever its nature, is widely disseminated by the sale of queens and bees in packages. The trouble has long been prevalent in the south, especially in

Florida, but, of late, it is becoming common in many northern localities. It has attracted special notice in Wisconsin and Washington. In dry and warm seasons it is not serious, but in cold and damp summers becomes a serious problem.

Cases have been called to the writer's attention, where all the bees introduced from a certain locality have died with this disorder, while the stock which had previously been present in the apiary remained in healthy condition.

If the business of queen breeding is to remain permanent and profitable, it is highly desirable that concerted action be taken, looking to the control of shipment of queens or bees from diseased apiaries. The buyer should be assured that he will not endanger his future prospects by buying queens anywhere that they are offered. About the only solution that seems readily apparent is federal supervision of queen yards. An increasing number of expert beekeepers are being employed in the extension service of the United States department, and these could be used to inspect all queen-breeding apiaries at least once each year, in connection with their other work.

Several of the states have provision for the inspection of queen-breeding apiaries, and withhold certificates of health from apiaries where disease is found. . However, not all the states have inspection and those that have do not have uniform regulations. The shipment of both bees and queens is becoming so general that uniform interstate regulations are very desirable.

In the opinion of the author, the future of the business depends very much on the attitude which the queen breeders, as a class, assume toward this question.

About the Author

Frank Chapman Pellett

From Professor Paddock's address at the dedication of the Frank Chapman Pellett Memorial:

"Frank C. Pellett was born on July 12, 1879 on a farm three miles northeast of Atlantic, Iowa. He was the son of Ambrose and Ellen Chapman Pellett. He attended rural school but later he was forced to leave school because of a condition of health which, always after, was to hamper his activities. Yet books were his ever-constant companions. His was a life of study and meditation. He formed his own ideas and plans and followed them throughout his life.

"In 1902 he moved to the small Ozark town of Salem, Missouri. He was much interested in the extensive stands of native pine timber and the wild life found therein. Here he operated a fruit farm and read law in the office of Hon. A. D. Gustin. He was admitted to the Missouri bar in 1905 and practiced law in Salem for two years. He could not develop sufficient interest to continue his effort in legal work but decided to dedicate his life to those things which he loved so much-nature, wildlife, birds, plants and bees. His home farm near Atlantic was set aside over forty years ago as a wildlife preserve to foster native plants and native wildlife. This farm, in later years, developed into the Honey Plant Test Garden which he developed in conjunction with the American Bee Journal.

"In 1912 he was appointed State Apiary Inspector of Iowa which office he held for five years. In 1915 he became associated with the American Bee Journal as Field Editor and later Associate Editor until his death.

"Mr. Pellett was made an honorary vice-president of the Apis Club of England and an honorary member of the Bee Kingdom League of Egypt. He was a long time honorary member of the Rotary Club of Atlantic and the Kiwanis Club of Hamilton, Illinois, and the Tri Beta of Carthage, Illinois. He was Literature and American Men of Science. In 1947 he received the National Skelly Award for Superior Achievement in Agriculture. The National Association of State Garden Clubs Award was presented with a special medal by the Iowa State Horticultural Society. He was a fellow of the Iowa State Horticultural Society, the Iowa Academy of Science, and the Royal Horticultural Society of England. He was also a fellow of the American Association of Advanced Science, and the American Association of Economic Entomologists.

"Mr. Pellett was the author of thirteen books on beekeeping, honey plants, birds and flowers of the wild, horticulture, and other nature subjects. He was co-author of several other books, including his chapters in the 1946 and 1949 editions of The Hive and the Honeybee. His greatest contribution and service to the beekeeping is his book American Honey Plants.

"He was an early contributor to the program of the Iowa Academy of Science. The first contribution appeared in the printed proceedings of the Academy for the meeting held at Grinnell, Iowa, on April 29 and 30, 1910. This was entitled "Mammals of Iowa, A Preliminary Annotated Catalogue of Recent Mammals of Iowa". Separates followed on "Nest Boxes for Woodpecker", "Food Habits of the Skunk", "Life History and Habits of Polistes Metricus, Say", "The Harvest Mouse of Iowa', and "Butterflies of Chance Occurrence in Cass County". These titles indicate the wide variety of interest of Frank C. Pellett and his ability to make himself fairly familiar with each of these fields which now employ specialists. It has well been said that Frank Pellett was of the "old school" of naturalists."

Frank Chapman Pellett

Scientific Queen-Rearing

As Practically Applied;
Being A Method by which the Best of
Queen-Bees are Reared in
Perfect Accord with Nature's Ways

G. M. Doolittle

Scientific Queen-Rearing

by G. M. Doolittle (1846-1918)

Published in 1889 by Thomas G. Newman & Sons

THIS BOOK
IS
AFFECTIONATELY DEDICATED
TO
Mr. Elisha Gallup,
MY TEACHER IN BEE-KEEPING,
FROM WHOM I LEARNED MY FIRST
LESSONS IN QUEEN-REARING, AND WHO
TRUTHFULLY CLAIMED, THAT AROUND
THE QUEEN CENTERED ALL THERE WAS
IN APICULTURE.

Preface

For many years I have been urged to write a book on beekeeping, and almost scolded because I did not do so. My excuse for not doing so has been, that there were many exhaustive treatises on this subject already before the public, written by Messrs. Langstroth, Quinby, King, Cook, Root and others-hence there was no reason for thrusting more books upon the world, which had nothing for their subject-matter but the general outlines of bee-keeping.

To be sure, there are two little books in pamphlet form bearing my name, viz.: *The Hive I Use*, and *Rearing Queens*, which have been published, but these are only a compilation from articles which I have written for the different bee-papers.

As all bee-keepers of to-day are aware, I have given all of my best thoughts, on the subject which the desired book would cover, free to the world through our many bee-papers, so, had I complied with the request made, the matter in the book would have been mainly that which I had written before; and owing to this self-same cause, the reader will perhaps find some fault with the present work.

Finally, the urgent requests of my friends for a book became so numerous, that I decided to hold back from the public, a part of my experiments and research, along the line of Queen-Rearing, (as in this branch of our pur-suit I have taken more interest, and gave to it more thought and study, than to all else pertaining to apicul-ture), and when that research and the experiments were completed, give all which I had dug out regarding Queen-Rearing, to the public, in book form. The culmination of this decision is now before you, and the reader can decide whether I have made a mistake, or not, in the undertaking.

Although I have given to the book the title of *Scientific Queen-Rearing*, there is much in it that is not scientific, as the reader will soon discover, and some lofty minds may pass it by in disdain on this account.

It is not a manual, giving in terse, sharp periods, the greatest amount of accurate information in the briefest space. My style, I fear, is often like my bee-yard, which in looks is irregular, while it attempts something useful. I never could be pinned down to systematic work. I always did like to work at the bees near a gooseberry-bush, full of ripe, luscious fruit, or under a harvest apple-tree, where an occasional rest could be enjoyed, eating the apples which lay so temptingly about. Do we not all need an occasional relaxation from the severer duties of life? If so, the rearing of Queens for our own apiary, gives us a change from the all-important struggle for honey, whereby we can get money.

In brief, it is my sincere conviction, that something to relieve the monotony of every-day life is good for humanity, and it is my wish to diffuse this belief as widely as possible.

I frankly admit that the following pages are very much the same in character, as if I had taken the reader by the arm, from time to time, and strolled about the Apiary and Shop in the time of Queen-Rearing, and chatted in a familiar way on the topics suggested as we passed along.

At the outset, I shall undoubtedly be met by those inevitable "Yankee questions" - Does Queen-Rearing pay? Would it not pay me better to stick to honey-production, and buy the few queens which I need, as often as is required?

I might answer, does it pay to kiss your wife? to look at anything beautiful? to like a golden Italian Queen? to eat apples or gooseberries? or anything else agreeable to our nature? is the gain in health, strength, and happiness, which this form of recreation secures, to be judged by the dollar-and-cent stand-point of the world?

Can the pleasure which comes to one while looking at a beautiful Queen and her bees, which have been brought up to a high stand-point by their owner, be bought? Is the flavor of the honey that you have produced, or the keen enjoyment that you have had in producing it, to be had in the market?

In nothing more than in Queen-Rearing, can we see the handiwork of Him who designed that we should be climbing up to the Celestial City, rather than groveling here with a "muck-rake" in our hands (as in "Pilgrim's Progress"), trying to rake in the pennies, to the neglect of that which is higher and more noble. There is something in working for better Queens which is elevating, and will lead one out of self, if we will only study it along the many lines of improvement which it suggests. I do not believe that all of life should be spent in looking after the "almighty dollar;" nor do I think that our first parents bustled out every morning, with the expression seen on so many beekeepers' faces, which seem to say, "Time is Money" The question, it seems to me, in regard to our pursuit in life, should not be altogether, "How much money is there in it?" but, "Shall we enjoy a little bit of Paradise this side of Jordan. However, being aware, of the general indifference to Paradise on either side of Jordan, I will state that I have made Queen-Rearing pay in dollars and cents, having secured on an average about $500 per year therefrom, for the past five years; and that all may do as well, I proceed at once to describe the ground over which I have traveled, and tell how it is done.

Before doing so, however, I wish to say that all along the way I have picked up a little here and there, so that most of the credit for that which is valuable in this book (if there is any value in it), belongs to some one else besides Doolittle. It has been picked up in such little bits, that I hardly know to whom I should give credit, so I will simply say, that the most of the suggestions which I have received, have come through the bee-periodicals, and quite largely from the reports which they have given of different thoughts dropped at many bee-conventions.

G. M. DOOLITTLE.
Borodino, N. Y.

Table of Contents

Table of Figures

CHAPTER I. INTRODUCTORY REMARKS.

First Colony my Father Bought

When I was about seven years old, my father procured some bees by taking them of a neighbor on shares. I remember, almost as if it were yesterday, how animated I was, as he and the neighbor of whom he took the bees, came near the house with the hive suspended on a pole between them, by means of a sheet tied at the four corners. The hive was deposited on a bench a few rods from the back-door of the house, one cold morning in early April, where it was thought that it would be a good place for them to take up their abode for the future.

My curiosity about these bees hardly knew any bounds, and although that day was cold and dreary, I was often out by the hive to see if I could not catch a glimpse of some of the inmates. The first warm day on which they took a general flight, my delight was great, to see them "cut up their antics" about the hive, as I termed it; and when the first pollen appeared, or when they began to go into the hive with "yellow legs," (as father always would speak of the gathering of pollen, even in the later days of his life), I was near the hive, an interested spectator.

As the days passed by, I became all anxiety about their swarming, and many were the questions which I plied father with, in regard to how this was done. In answer to some of these questions, he told me that the Queen led out the swarm, undoubtedly getting this impression from seeing the young Queens of after-swarms out on the alighting-board with the first bees, when a second or third swarm issued. On the mention of the Queen, I wanted to know all about her, but it was very little that father could tell me, except that he often saw her with the swarm. As only box-hives were in use in this locality, it was no wonder that he knew so little regarding this all-important personage of the hive, as now viewed from the present stand-point of our profession.

My anxiety for swarming-time to come was so great that it seemed that it never would arrive, and when it did come, the impression which it made upon my mind was so lasting, that, as I write, I can almost see those bees whirling and cutting circles in the air, seemingly thrice as large and active as is a swarm, in my later years.

After they had clustered, were cut down, and brought to the empty hive, my anxiety to see the Queen became paramount to all other interests which this exciting time presented; and when, as the last half of the swarm was going in, she was seen, although only a brown German Queen, I thought her very majestic in appearance, and the sight well worth all the hunting we had done to find her.

Time passed on, and in a few years the apiary had grown so that swarming was quite frequent, and had somewhat lost its novelty; yet there has been no time in my life but what it has had very much of interest to me.

My First Swarm of Bees

During one swarming season, a third swarm issued, and in alighting it separated into three parts, so that none of the little clusters had more than a quart of bees, while one had scarcely more than a tea-cupful. Father was about to put all of the three clusters into one hive, but I finally persuaded him to let me put the little one into a small box that I had, and see what I could do with them. In getting them into the box, I saw three Queens go in, which excited my curiosity very much. I remember of planning how such swarms, which had many Queens, might be multiplied to great numbers; but to say that any idea of Queen-Rearing entered my head at this time, would savor of imagination.

The little colony built three pieces of comb a little larger than the hand, but soon after cold weather came, the bees died, as father had said they would, when he let me try the experiment. In a year or two more, that dreaded disease- foul brood-appeared in the apiary, and as father knew nothing about how to control it, all the bees were soon gone.

Years went by, with little or no interest on my part regarding bees, except as a runaway swarm passed over my head while at work in the field, or as I and some of the neighboring boys robbed bumble-bees' nests; till at 17 years of age, in time of sugar-making, a bee-tree was found, by hearing the roaring of bees on their cleansing flight, as I was going to visit a neighbor's sugar-bush, not far away. The next warm day I went out looking for bees, and, before noon, I found another bee-tree. These trees were left until fall, when they were cut, but in falling they so scattered the bees and comb, that with the little know-ledge I then had, I thought that I could not save them.

American Bee Journal; King's Bee-Keepers' Text-Book; Quinby

Twenty years ago I cut one of my feet so badly that I was confined to the house nearly all winter, and as reading was my chief amusement, it so chanced that I picked up *King's Bee-Keepers' Text-Book*, which I had purchased the year I found the bee-trees, because the advertisement about it said that it told "how to hunt bees." As soon as I began to read this book, I contracted what is known as the "bee-fever," which took so strong a hold of me that I was not satisfied till I had borrowed and read Langstroth's book, and purchased Quinby's work, besides subscribing for the *American Bee Journal*.

In the spring I purchased 2 colonies of bees, from which originated my present apiary. This was in the spring of 1869, and as that was a very poor season, I secured only one swarm from the 2 colonies, and very little experience along any line of the pursuit, except that of buying sugar and feeding up these 3 colonies for winter.

The next June I went to see a man who kept some Italian bees (the first I had ever seen), who lived about four miles from me. When I arrived, I found him at work at Queen-Rearing, so I was all interest at once. He showed me all that he knew of Queen-Rearing, during my frequent calls on him that summer, and the next spring I went into partnership with him in the Queen-business, he rearing the Queens, and I doing the selling-doing this by taking the Queens around to the bee-keepers who lived within 10 or 15 miles of us, and introduced them into the apiaries of those who would buy. I remained in partnership with him during the next year, and, as a whole, I made it profitable, for I not only got some cash out of it, but at the end of that time I had a full knowledge of the old plans of Queen-Rearing. During this time I had partially Italianized my own apiary, so the next year I started out on "my own hook" in the Queen-business, although not doing much at it in the way of selling to outside parties, till some years later.

My First Loss of Queens

After losing nearly, or quite, one-half of my Queens, one spring, owing to their poorness in quality, I began to study up plans for the rearing of better ones, which study I have kept up till the present time. Into this branch of our pursuit I have put all the thought and energy at my command, as well as to apply the accumulated thoughts of others, as, expressed in our bee-papers, till I think that

I can truly say there is much in the following chapters never before given to the public.

Alsike Clover.

CHAPTER II. IMPORTANCE OF GOOD QUEENS.

Upon no other one thing does the honey part of the apiary depend so much, as it does upon the Queen. Give me a good Queen-one which can be brought up to the highest production of eggs, just when we want them-and I will show you a honey crop, if the flowers do not fail to secrete nectar; but with a poor Queen-one that you must coax for eggs, to little or no purpose, at the right time-the flowers often bloom in vain, even when the honey-secretion is the greatest,

I have had in my apiary, at different times, Queens that with all the coaxing which I could bring to bear on them during the forepart of the season, would not lay any more eggs previous to the honey harvest than were needed to keep the spring strength of the colony good, so that when the yield of honey was at its height, there would not be one-fourth the number of bees to gather it, that there should be. When the yield of honey came, then these Queens would begin breeding, so as to get plenty of bees in the hive just as the harvest closed, only to eat up the little honey that the few laborers there were in the harvest had gathered. The more Queens of this sort a bee-keeper has, the worse he is off. This is a peculiarity of the Syrian races of bees, but many poorly-reared Queens act in the same way, no matter to what race they belong.

Then, again, I have had queens which would not be coaxed to till more than three or four Gallup frames with brood at any season of the year; so that at no time were there laborers enough in the hive to make a respectable showing, no matter how much honey there was in the field. Others would appear very prolific for a short time, but just when I wanted them the most, and when I supposed that all was going well, an examination would show that they had died of old age, even when they might not be more than six to twelve months old. This would cause a break in the production of bees, at a time when every day of such production would count many pounds of honey in the honey harvest.

From the above it will be seen, that in no one thing in bee-keeping does quality count for as much as it does with the Queen or mother-bee. Of course, if we are to only count our colonies, then a poor Queen is better than none; and there are other times when she is such, as in holding a colony together till we can get a better one; but I repeat, that an apiary with all poor Queens, is worse than no bees at all. When we come to fully realize the great achievements which can be obtained with a really good Queen- one that will give us from 3,000 to 4,000 workers every day for a month previous to the honey harvest, we, as apiarists of America, will put forth more energy along this line of our pursuit than we ever yet have done.

Look at that colony you had one spring, which gave you 100, 200, 400, 600, or even 1,000 pounds of honey (one or two reports of nearly a thousand pounds of honey from a single colony have been given in the past, while the reports of those giving from 400 to 600 are numerous), and see why it did so well, while the average of your whole apiary did not come up to one-half that amount. Why did that colony do so well? Simply because it had a large working-force of the right age, and at the right time, to take advantage of the honey-flow when it came. And how came it to have such a force at the right time? Because the Queen was a good one, doing her part just when she should, and not at some other time. Why did the others fail of doing the same thing? Either because they did not have good Queens, or because the owner failed to have the Queens do their duty, when they should have been doing it.

All Queens Should be as Good as the Best

"But," says one, "Can I get all colonies to do as well each year, as my best colonies do?" I will answer that by asking, what is there to hinder? If all are in the same condition as the successful one, would they not do equally well? Most assuredly they would. So then we see that the trouble lies in not having the colonies all equal with the one which did so well. The reason that all are not in the same condition, devolves primarily upon the Queen; and secondly upon the strength in which the colonies come through the winter.

Poor Queens the Cause of Winter Losses

Of late, I have inclined to the opinion that on the Queen rests, to quite an extent at least, much of the cause of our wintering troubles. One thing is certain, if we cannot have all colonies exactly alike, we can approximate very nearly to it-much nearer than many imagine, if we work for that object, along the line of bringing the Queens to as nearly perfection as possible, and cease the breeding of cheap Queens-that class which "do not cost the apiarist anything."

If there is anything in which I take some little pride, it is that since I began to breed my Queens for good quality, and for that only, this variation of yield of honey from different colonies has grown less and less, till, at the present time, the average yield of honey from each colony in the apiary is very nearly alike, while fifteen years ago some colonies would give 75 per cent more honey than would others.

What a few of our best Queen-breeders can do, all can do, if. they will only put the same interest into their work along this line; and one of the objects of this book is to tell those who desire, how they can, by careful attention to the rules laid down herein, become breeders of the best of Queens, as such are of great importance to the amateur as well as the specialist.

Basswood or Linden.

CHAPTER III. NATURE'S WAY OF REARING QUEENS.

The Creator of all things looked over His work after He had finished it, so we are told, and pronounced it "GOOD;" hence we could reasonable expect that at that time all things created by Him were of the highest type of perfection. He now told all animated nature to "multiply and replenish the earth." For this reason, we find a disposition in our bees to swarm, and although during the last century men have tried with great persistency to breed this disposition out, or make a hive which would accomplish the same thing, yet so far that disposition stands defiant toward all of these unnatural schemes, and just as some individual is ready to cry "Eureka," out comes a swarm, and all of our plans lay prostrate in the dust.

Many have been the reasons given, to account for bees swarming, such as the hatred of an old Queen toward the rival inmate of a queen-cell, which the bees had succeeded in getting, in spite of her frowns and anger; the hive being too small to hold the accumulated thousands of workers, insufficient ventilation, etc.; yet in my opinion none of these things ever caused a swarm to issue, in and of themselves, for everything in nature is held obedient to the command of Him who controls the Universe. And I rejoice that this is so; for I firmly believe that better results can be obtained where bees swarm, than would be the case if we could breed out the swarming trait.

Queens Reared Under the Swarming Impulse

A new swarm goes to work, with an energy never possessed by the bees at any other time (unless it is by the parent colony) immediately after its young Queen gets to laying. This swarming trait also produces Queens of the highest type of perfection, not being equaled by any except those reared under one other of Nature's conditions, which will be spoken of at length in the next chapter. Many have been the claims made, that Queens reared by different methods, are *just as good* as those reared under the swarming impulse; but I have yet to hear it claimed that Queens so reared are any better than are those reared where the swarm issued under the conditions which Nature designed that they should.

Swarms Issuing Before Queen-Cells are Built

I am met here, by the claim that many colonies of Italian bees swarm, without any preparations for swarming being made, by way of providing queen-cells before the swarm issues, as is usually the case; and that Queens reared under such circumstances, where there are but few bees in the hive to feed, nurse, and keep the royal occupants of the cells warm, certainly cannot be as good as those reared under the superior planning of the skilled apiarist.

Artificial Queen-Rearing

I freely admit that Queens reared by some of the plans of "artificial" Queen-Rearing, may excel such Queens, but I claim that the first-named conditions are not such as Nature originally designed that they should be. I do not believe that an isolated colony (as all colonies are isolated, except by the intervention of man) ever cast a swarm under such conditions. It is the compacting of colonies together in large apiaries, which bring about such results, thereby causing what is termed "the swarming-fever"—where swarms issue under the most unfavorable circumstances imaginable, sometimes even swarming without a Queen, thus leaving the parent colony brood-less, and without means from which to provide them-selves with a Queen. After careful watching in my own apiary for years, and closely questioning other parties, I have yet to find where the first swarm of the season, from any apiary, has ever issued previous to the sealing of the first Queen-cell.

But, says one, "You are always crying Nature! Na-ture!! Don't you know that man's intelligence, by opposing Nature's laws at the right time, can get ahead of her ways, and thus secure better results?" No; I did not know any such thing; nor do I believe it. It is only as the intelli-gence of man moves along harmoniously with the laws of Nature, that any improvement can be expected. Is not this true?

Suppose I cut one of my fingers quite badly; and when it stops bleeding I wonder what I am to do with it, to have it get well as quickly as possible. While I am thus wondering, along comes a man of superior (?) intelli-gence, and he says: "I see you have cut your finger. I am glad I happened along, just at this time, for I have a *salve* which will heal up that wound at once, and by your using it, your finger will be as well as ever in a few days. This salve of mine has the greatest healing-properties of any salve known." Reader, do you believe that the salve has any healing quality, or that my finger will ever be as perfect as before? I do not. All that any salve can claim as doing, is to assist Nature to make the best of a bad job, for it is Nature that does the healing, not the salve. Would I not have been better off, had I not cut the finger?

Again, some day during the month of June, I chance to run against the body of a choice apple-tree, with the hub of my wagon-wheel; and in doing so I knock a patch of bark off the tree, as large as my hand. Along comes a man familiar with grafting, and applies some grafting-wax, saying: "This will heal over the place, and make it as good as ever." Do you believe it? Will not the tree always show a scar? In the knocking off of that bark, the apple-tree received a shock, or something which was contrary to its nature, and as soon as the first effect was over, every power that was in the tree was brought to bear on this place to remedy the damage, and it was only wherein the wax kept off the warring elements which would work against the repairing of the damage, that the wax did any good. Just so with anything that goes against Nature's laws. The first thing to be done, is to get rid of the antagonizing force, and as soon as that is done, Nature tries to get back to the spot where she was before, as nearly as possible, and as quickly as she can.

Let a man take a drink of whisky, and in a little while you will see him cutting up all manner of antics (that otherwise he would not have thought of doing), when you call him "drunk." What is the matter? He has taken something into his system that is not moving harmoniously with Nature, and Nature is trying to "kick" out this antagonizing force. If so much antagonizing force (whisky) has been taken, that Nature has to kick vigorously to expel it, the man is kicked over, for the time being; but after this force has been expelled, Nature begins the work of healing, and the man "rights" up again, but never gets back to where he was before.

Rearing Queens Unnaturally

Now apply this to the bees: Along comes the antagonizing force—the apiarist—who is going to rear Queens intelligently (unnaturally), and kills the mother of the colony. What is the result? The whole colony acts for the first few hours very much as did the man after having drank the whisky. What is the trouble? Why, Nature is "kicking," that is all. After becoming reconciled to fate, the bees, through Nature's law, go to work to repair the damage done, and, as in all of the other cases, she does this as quickly as possible, even where only eggs are given from which to rear a Queen. Under these conditions, antagonizing forces come in, and I do not believe that the "wound" can be made as good as ever, any more than they could in the three illustrations which I have used. Queens partially deficient in some points will be the result.

Some claim that this last is a natural condition for a colony of bees to be placed in, but I take exception to any such claim; for there are very few colonies that ever find themselves in such a condition, without the interference of man. During all of the ages, up to within about a century past, how different the method whereby Queens were produced, which have stood the test of thousands of years.

Let us look at Nature's plan for a moment or two, so as to see how it does compare with the above work of many of our apiarists. We find that Queen-Rearing and swarming are only done during a period when both honey and pollen are being gathered from the fields. When this condition of things prevails, the bees are getting strong in numbers, and soon embryo queen-cells are started, in which the Queen lays the eggs which are to produce the royal occupants.

Queens Depositing Eggs in Queen-Cells

Some claim that it is not fully settled that the Queen deposits the egg in the queen-cells at the time of natural swarming; but my assistant once saw her do it, and so have others, while the position of the eggs in the cells prove it, even had no one ever witnessed her in the act. Others claim that the Queen lays the eggs in worker-cells, along the margin of the comb, when the bees build queen-cells over them. Any one familiar with the inside of a bee-hive, should know that such reasoning is fallacious, for the embryo queen-cells are often formed many days before eggs are found in them, as thousands of bee-keepers can testify.

Feeding the Larvae

These eggs remain in this form for about three days, when they hatch into little larvae which the bees now begin to feed. Some claim that royal jelly is placed around these eggs before they hatch, but if this is true, it is something that I have never seen, although I have watched this matter very closely for years. Neither do I find that the little larvae are fed much more plentifully, during the first 36 hours of their existence, than are larvae which are in worker-cells; but when of about this age, the bees begin to feed them so liberally on royal food, that they actually float in it during the rest of their growth; this supply often being so great that there is left a lump as large as a pea, of partially dried food, after the Queens emerge from their cells, while all of their operations are conducted leisurely, for the bees are in no haste for a Queen, as their mother is still with them in the hive. There is no hurrying up to replace a loss, thereby using old larvae, or scantily feeding the same, as is done when Nature is antagonized; but all is done by a system reaching perfection. If a cold, bad time comes on now, they do not hesitate to tear down the cells, and wait for a favorable time to come again for them to "multiply and replenish the earth."

All of this shows us that the bees are only obeying the laws which govern the economy of the hive, instead of a force outside of that economy, which compels them to make good a loss that man has brought about. It seems to me that we can always consider it safe to go according to the teachings learned by a close observation of our "pets," and unsafe to go contrary to the rules and laws which govern them. At least this is the belief which I have always had, and along this line has been my study, while trying to find out the best plans whereby Queens of the highest type could be reared. I have never succeeded in rearing Queens which pleased me every time, till I commenced to work in harmony with Nature's plans. When I learned so to work, I found that ray Queens were improving all the while, and to-day I am well satisfied that I have made a great improvement in my stock, beyond where it was ten years ago.

White Clover.

Abnormal Queen Rearing

Besides what is known as the "swarming-plan," the bees have another way of rearing the best of Queens, which, together with the former, are the only plans by which Queens are reared, except where the bees are forced to do so by some abnormality of the colony. My experience goes to prove that where such abnormality exists, Queens which are then reared do not come up to that high standard that they do where reared as Nature designed they should be. However, there are very few Queens reared, except when the colony is in a normal condition, only as the colony is interfered with by man; so that we find the usual plans adopted by nearly all Queen-Rearers of the past, going in the direction of these few exceptions, rather than along the line which Nature designed.

Forcing Process of Rearing Queens

While rearing Queens by the "forcing process" (at times when they could not be reared by natural swarming), I came across a colony in early spring, which had, as far as I could see, a good Queen, yet on the combs there were two very nice queen-cells under way, with little larvae floating in an abundance of royal jelly. As queen-cells which were formed in my Queen-Rearing colonies, when worked by the "forcing process," were not supplied in this fashion with royal jelly, I decided to keep watch of this colony, and see if I could not learn something.

In due course of time these cells were sealed, when, to all outside appearances, they were just as perfect as I had ever seen in natural swarming; while the cells which I was compelling the bees to build by taking their Queens away from them, did not so appear. One of the cells I transferred to a nucleus, just before it was ready to hatch, while the other was left where it was, to see what would become of the matter. The Queens hatching from both cells, proved to be every bit as good as any Queens I ever reared in the height of the honey harvest, by natural swarming, even although it was by dint of coaxing that I could get Queens reared at all by the "forcing" plans, as adopted fifteen years ago; while none of the "forced" Queens would compare with these two in beauty, vigor or length of life.

Soon after the young Queen which was left in the old hive commenced laying, the old mother began to decline, and, in the course of a week or two, was gone; yet had I not opened this hive for a month, at this time, I would never have known that a change had taken place as regards the Queen, from the appearance of the brood which was in the hive.

Human Interference with Nature

Right here let me say, that from all of my experience with bees, I am led to conclude that 999 Queens out of every 1,000 reared, where man does not interfere with the bees, are reared by one of these two plans; yet there are those persons among our number, who claim that they are along the line of Nature; or rearing Queens by a still better plan than these two, where they take away the Queen from a colony at any time they think best, and compel the bees to rear others, often when it would be the last thing the colony would wish to do. Gentlemen, your position is not a consistent one, nor is it one that you would adopt along any other line except Queen-Rearing; and I hope that this Book will open your eyes, so that in the future you will try to be in accord with the wants of the bees, and thus be rearing Queens of superior quality, instead of those which cannot be other than inferior.

Building Queen-Cells when the Queen is in the Hive

To return: After I had this experience with the colony that had "two Queens in a hive," (which was a surprise to so many fifteen years ago, when it was thought that no colony ever tolerated but one laying queen at a time), I began to watch for a like circumstance to occur, which happened about a year from that time. In the latter case, as soon as I found the cells, they were sealed over, and not knowing just when they would hatch, I at once cut them out and gave them to nuclei. In a few days I looked in the hive again, when I found more cells started, which were again cut off and given to nuclei, just before it was time for them to hatch. In this way I kept the bees from their desired object for some two months, or until I saw that the old Queen was not going to live much longer, when I left one of the cells, which they had under headway, to mature. By this plan I got about sixty as fine Queens as I ever reared, and laid the foundation for my present plan of securing Queens, which is about to be given in this book.

As time passed on, I was always on the lookout for such cases of building Queen-cells, with the old Queen present in the hive, where there was no desire to swarm, and in this way I have secured hundreds of splendid Queens with which to stock my own apiary, and to send to those who wished Queens of the best grade. If there is any difference between Queens reared by this last of Nature's ways, and those reared by natural swarming, that difference is in favor of Queens reared to replace the old mother, when she shall get past being of use to the colony; so that I have no hesitation in pronouncing

Queens thus reared, of the highest grade which it is possible for the intelligence of man, combined with the natural
instinct of the bees, to produce.

Two Queens in a Hive

Having decided that Queens thus reared were superior to any other, the thing next to be done, was to get
some plan that the bees would accept, whereby Queens
could be so reared just when and where the apiarist
desired. To accomplish this, I have studied hard and
worked faithfully, putting into it all of my best thought for
some six years past, till I have perfected a plan whereby I
rear Queens by Nature's best method, in the same hive
with a laying Queen, and that, too, just when and where I
wish to have them reared, having Queens in a single
colony in all stages of development, from the just-hatched
larvae to virgin Queens and those just commencing to lay.
All about how to do it will be told in this book, but before
doing so, I wish to take the readers over some of the
ground which I have traveled, so that they can see some
of the steps taken; for in thus seeing, perhaps some new
thoughts may be suggested to them, which will lead in
other directions from what is here given, which, when
followed out by some other person than myself, may be of
great help to the bee-fraternity.

CHAPTER V. OLD METHODS OF REARING QUEENS.

My first experiments at Queen-Rearing were tried in 1870. During the month of July, a second swarm issued having two Queens, and as I saw them on the alighting-board of the hive, the thought came to me that here was a chance to save a nice Queen, which, when she got to laying, could be used to replace an old one that I had in the apiary. According to this thought, I detached the alighting-board, as soon as I saw one of the Queens go in, and took the board with all the adhering bees, to another hive, in which I put an empty comb, hiving the little lot of bees and Queen in it.

At night I read up on Queen-Rearing (as far as I could with the books which I then had), from which I found that the way to rear Queens was, to place little colonies in small or nucleus hives, they having frames from four to six inches square. As I wished to do things as they should be done, I went to work the next morning, and constructed a little hive that held three frames about five by six inches. Into these frames I fitted comb, then I went to the hive containing the little swarm, and shook them out of it into this small hive. In due course of time, the Queen commenced to lay, and was used as I had designed.

I now began to look up what I was next to do with the little colony, and found all that was necessary was, to leave them alone, when they would start two or three queen-cells, which would be well taken care of, and make just as good Queens, where but a small nucleus was used, as would a larger number of cells in a full colony. Queen-Rearing now looked very simple and easy to me, so I left the nucleus to mainly care for itself for the next five days.

From time to time, as I looked into the little hive (for I was so anxious about the matter, that I could not keep away from it) I expected to find queen-cells started, but every time I opened the little box, not a cell could be found.

The fifth day after I took the Queen away, a bee-keeper came along, who was considered quite a Queen-breeder in those days, and to him I told the story of my trial at Queen-rearing. He asked to see the little colony, and when I showed it to him, he quietly cut a hole in one of the combs where the smallest larvae were to be found, saying, that "now they would start some cells," which they did. He also said that, "while the most of the Queens then reared were reared in just such little nucleus-hives, yet he believed that it was better to rear them in full colonies, as he thought Queens thus reared were better fed, and that the warmth of a full colony was conducive to a better development of the royal occupants of cells built, hence we secured more prolific and longer-lived Queens."

In due time one of the cells hatched in the nucleus, and the Queen was so small, and so poor, that I decided if I must have such inferior Queens as that, I would let the bees do their own Queen-Rearing, as I had done in the past.

When the next season came, I found myself again longing to "dabble" in Queen-Rearing, so at it I went, although I never again tried the nuclei boxes in doing so; for when I came to look into the matter more thoroughly, I was convinced that the best nucleus that I could possibly have, was one or two frames in an ordinary hive. In this way all work done by the nucleus was readily available for the use of any colony, after I was through with the nucleus.

In trying this the next time, I simply took the Queen away from the colony I wished to breed from, at a time when there was plenty of honey and pollen in the fields, for by this time some were opposing the plan of rearing Queens in nucleus boxes, and also claiming that the only proper time for rearing Queens was when plenty of honey and pollen were to be had by the bees, as it was natural for the bees to rear Queens only at such times. I succeeded in getting a fine lot of cells from which some extra-nice Queens were obtained-as I then considered them.

This caused the Queen-Rearing "fever" to run high, which, together with my procuring some Italians, caused me to work at it many times during the summer, although I determined not to spoil my prospects of a crop of honey, by using too many colonies in the business. Although using Italian bees for Queen-Rearing, (as it was then claimed that black nurses would contaminate the young Queens) yet, during this summer, I succeeded in getting as high as 157 queen-cells built on one comb, while the usual number built by one colony would be only from three to twenty, on all of the combs in a hive. If I could have had the Syrian bees at that time, the number of cells might not have been so much of a wonderment to me. I thought this a great achievement, and something well worth being proud of, so I told my neighbors about it, and gave it to some of the bee-papers also.

Loss of Queens

All went on "swimmingly" till the spring of 1873, when, without any cause, as far as I could see, one-half of all the Queens that I had in the apiary died, leaving the apiary in poor condition for the honey season, which caused me to meditate a little on what could be the reason of such a wholesale death of my beautiful Queens. A careful looking into the matter revealed that of all the Queens that had died, two-thirds were those which had been reared the previous season, while not one had died from those that had been reared by natural swarming.

What seems strange to me now, in looking back over the past, is, that all of these Queens died so suddenly and the bees made no effort at superseding them. They all had brood in abundance for the time of year, and the first I knew that all was not right, was when I would find them dead at the entrances of the hives. After this I began to try other plans of Queen-Rearing, none of which pleased me any better than the one I had been using.

About this time there came a general dissatisfaction with most of the plans, then termed "artificial Queen-Rearing," and the reason given for Queens so reared not living any longer, or doing any better, was that such Queens were not reared from the egg for a Queen, but were fed worker-food for a time, and queen-food afterward, thus producing a bee that was part worker and part Queen; hence it could not be as good as a bee that was a perfect Queen in all her parts.

Then came the following process, which I often see given at the present time, as one by which prolific and long-lived Queens can be obtained:

Take a frame of new comb, and put it in the colony having the breeding Queen, leave it there till you see the first little larva hatched, when it is to be taken out, the bees shaken off, and then placed in an empty hive that is to be put on a stand of a populous colony, after moving the colony away. This is to be done in the middle of the day, when plenty of bees are flying. After trying this method of procedure a few times, I came to the conclusion that it was one of the very poorest ever given to the public, for the Queens so reared were very nearly, if not quite, as inferior as were those reared in the little nucleus boxes.

And how could it well be otherwise? for by such a plan only field-bees were obtained as nurses, while Nature designed young bees to do this work. While in early spring old bees do nurse brood, by their being brought up gradually to it, yet in this case, bees that had gone out in search of honey, with no idea of ever again being called upon to nurse brood, and with a good mother in the hive when they left, were suddenly confronted with starving larvae from which they must rear a Queen at once, while chyme or royal jelly was the most remote thing which their stomachs contained. This is one of the many plans which go almost in direct opposition to Nature's laws, and one that I claim should never be used, if we wish to have our bees improving, instead of retrograding.

I might give many other ways by which good Queens are said to be reared, which are as inconsistent with the best quality in Queens, as darkness is when compared with daylight; but I forbear. I have only gone over this ground of the past, to show how Queens used to be reared, and how some bee-keepers still rear them, so that those who read the methods, soon to be given, may compare them with those formerly used, and see how we have been advancing along this line of our pursuit. I wish to say to any reader of this book, who is still practicing any of the old, poor plans: Don't do it any longer; for you must know, it seems to me, that only inferior stock can result from the longer continuation of such practice.

CHAPTER VI. LATER METHODS OF REARING QUEENS.

After testing all of the then known plans, as given in the previous chapter, and becoming disgusted with them, I turned my attention toward natural swarming, as a means by which to rear Queens in the future. Looking toward the end of getting as many Queens from this source as possible, I began stimulating my best Queen-Rearing colonies early in the spring, by some of the many methods given to accomplish this work, so as to get them to swarm early, and then by hiving the new swarms from these colonies, on frames of brood, kept them swarming till late in the season, so that, as a rule, I could get, in this way, all of the Queens that I wished to use in my own apiary.

If, at any time, I was likely to fail of this, I would take a piece of comb containing little larvae, from my best Queen, and after shaving off the cells down to one-eighth of an inch of the septum of the cell, with a thin, sharp knife, so that I could see the larvae plainly, I would go to a hive having an inferior Queen, that was preparing to swarm, and after removing the larvae from the queen-cells that. they had under way, I would by means of a goose-quill tooth-pick, having its point broad and curved, [Fig. 1], lift the little larvae from the piece of comb I brought, and put them down in the royal jelly which the larvae from the inferior Queen was enjoying only a few moments before. Some take a frame (brood, bees and all) from the hive having their best Queen, and, when ready, lift the larvae from the bottom of full-depth cells, but it bothers me to see to do this. Where it can be done, it saves cutting and otherwise injuring the combs, while the bees protect the larvae from being chilled, should the day be cool. The cells thus operated upon were marked, by pushing 1 ½ inch wire-nails through the comb near them, so that if the bees constructed other cells, I would not be deceived.

Fig. 1.-Tooth-Pick for Transferring Larvae.

In this work I often found partly-built queen-cells with nothing in them, or perhaps some would contain eggs, which, when I found them, I would take out, substituting the larvae in their places. As a rule, I would be successful with these, as well as with those that were put into the cells that contained royal jelly, but now-and-then a case would occur when only those placed in royal jelly would be used.

Right here I wish to say, that only the best of tested Queens should be used as Queen-mothers-Queens known to possess all the desirable requisites that make a good Queen; and, as we must often cut the combs, to get the little larvae for transferring, it is better to have the poorest combs in the hives with these Queens, so as not to spoil the good combs in the apiary; or, if preferred, we can keep these best Queens in a very small colony, so that the bees will fill the holes made in the combs when taking out brood, by building in worker-comb, as such small colonies always will do, if fed sufficient for this purpose. By thus working I obtained good Queens, although it required much work, and probably I should never have worked out other plans, had it not been that at about this time I began to have calls for Queens, from abroad.

This placed me in a position where I must have some other process of Queen-Rearing, or refuse to take orders for Queens. As I wished to please all who desired some of my Queens, I began experimenting, and soon brought out the following method, which I will give at length, as I have to still rear some Queens by it early in the spring, and late in the fall, when there are not enough bees in the hive, or when they are too inactive to use the new plan to be given in the next chapter. However, I use wax-cups (having royal jelly placed in them), as described later on, instead of embryo queen-cells, as will be spoken of here.

Transferring Worker-Larvae to Queen-Cells

In changing larvae from worker-cells to queen-cells, as given above, the thought occurred to me, that if the bees would take the larva when put into a perfectly dry queen-cell, on the combs of a colony preparing to swarm, they ought to do the same when placed in a like condition in a queenless colony.

Previous to this, I had often changed larvae in queen-cells started in a queenless colony, taking out those that the bees were nursing, and substituting others from my best Queen, where the bees had plenty of royal jelly in the cells, and secured good Queens by this plan, which is now used by very many of our best Queen-breeders. Good Queens are reared in this way, but the point about it that I do not like is, that the number of cells which will be started is very uncertain, while they are scattered about in different parts of the hive; and worse than all, the combs have to be badly mutilated in cutting out the cells, or else much time spent at the queen-nursery, watching for the Queens to hatch; for if this is not done, many of them will be destroyed.

But, how should I get the embryo queen-cells, in which to put the little larvae? was the first thought which confronted me. I remember that away back in some of the bee-papers, some one had proposed making queen-cells to order, on a stick, for a penny a piece, and why could I not so make them? It would do no harm to try, I thought; therefore I made a stick, so that it would just fit inside of a queen-cell, from which a Queen had hatched, and by warming a piece of wax in my hand, I could mould it around the stick, so as to make a very presentable queen-cup. While doing this, some one happened along, who wished to see some of my Queens, so I went out in the apiary to show them. In doing this, I noticed some queen-cups [see Fig. 2] which had been just started by the bees, and it was not long before I saw where the embryo queen-cells could be procured in plenty, if I saved all I came across in my manipulations with the bees. When I returned to the shop, I had about a dozen of these cups, that I had clipped off the combs, while showing my friend the Queens, which, with the 5 or 6 artificial cells that I had made, gave me plenty for a trial.

Fig. 2.-Embryo Queen-Cup or Queen-Cell.

To fasten these to the combs, I melted some wax in a little dish, over a lamp, when, by Dipping the base of the queen-cups in the wax, and immediately placing the cup on the comb, it was a fixture. So as not to spoil a good comb, I took an old one, such an one as had been damaged by mice, or one that had many drone-cells in it; and to have the cells built in the centre of the comb, as I wished them, a piece was cut out as large as a man's hand, at the desired place. I now turned the comb bottom side up, and fastened as many queen-cups as I wished queen-cells built, along the now under side of the hole that I had cut, and, after having transferred a little larva into each cup, the comb was returned to its former position. [See Fig. 3.]

After taking the queen and all of the brood away from a populous colony, I substituted this prepared frame for the queen and brood. Upon looking the next day, to see what the result was, I found that the bees had destroyed all the larvae but one, and that was in one of the cups that I had taken out of a colony.

Before I forget it, I will here say, that in all of my efforts at this time, to get the bees to use any of the cells that were made from beeswax, I made an entire failure; for, out of hundreds tried, not a larva could I get accepted, even when I gave a colony none other, save cups thus made. However, later on I learned how to make the bees use them, as will soon be given. Not being willing to keep a colony queen-less for one queen-cell, I gave back their brood and Queen; then I sat down to study out the reason why I had made a failure.

Fig.3-Comb with Queen-Cups.

The result of this study convinced me that no colony would immediately go to rearing Queens after the old Queen had been taken away from them. At the expiration of three days from the time the Queen is taken away from a colony, the bees usually have numerous queen-cells under way, but rarely before; while, in the above case, I had expected the bees to start them at once.

I now went to another populous colony and took its Queen away, together with one comb, when a division-board feeder was placed where the comb was taken out. At night I fed the colony a little warm syrup (as they were not getting much honey at the, time), and continued this nightly-feeding for eight days.

Three days after taking the Queen out, I went to the hive and took all of the brood away, but left the other combs having honey and food, arranging them close up to the feeder, leaving a place between the two central combs, for the prepared frame to be inserted. The hive was now closed, when the bees were shaken off the combs of brood, and the brood given to a colony which could care for it.

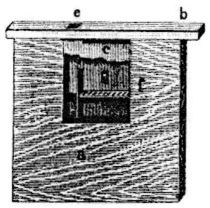

Fig. 4.-The Division-Board Feeder.

Royal Jelly

On these combs were numerous queen-cells, which showed that the bees were secreting or producing an abundance of royal jelly. As I wished this jelly to accumulate in-the stomachs of the nurse-bees, I took the brood away from them this time, before I put the little larvae into the queen-cups. In this way a colony will be prepared to rear as good Queens as can possibly be reared, when no Queen is present in the hive while the cells are being built, and is ahead of any other way that I ever tried, where the Queen is to be taken away.

It will be seen that an hour before they were feeding thousands of worker-larvae besides the queen-larvae, when, all at once, they are obliged to hold the accumulating chyme, and feel a great anxiety for a Queen, as will be shown by their running all over the hive, flying in the air, and otherwise telling of their distressed condition, when you come with the prepared frame to put it in the hive. By now supplying them with from 12 to 15 little larvae, all cradled in queen-cells, upon which they may bestow all the provisions and caresses that they were bestowing before on a whole colony, it could hardly result otherwise than in producing as good Queens as could be produced by any plan not exactly in accord with Nature's ways.

On placing the frame in the hive, on this my second trial, I had great confidence of success, while the next day on opening the hive I was assured of it, by seeing all of the queen-cells accepted, except those that I had made of beeswax. These accepted cells were completed in due time, and from them I obtained Queens which were as good mothers as any I had ever had up to this time, outside of Queens reared by natural swarming. I now used this plan for many years, and if properly done, it never fails of giving fairly good Queens. At all times when honey is not coming in abundantly, feeding is resorted to, and when the mercury is lower than 85° in the shade, all operations with the larvae are done in a room of that degree of temperature, or a little higher.

Age of Larvae to Use for Queens

But I think that I hear some one ask, "How old a larva do you use? and, how about the occupant of a cell being fed royal food, from the time it is hatched from the egg?" I have conducted many experiments to see how old a larva may be, before being placed in the royal cell, and yet have it produce a good Queen. Some who advocate that Queens should be reared from the egg, claim that, in natural swarming, royal jelly is deposited around the egg before it hatches, so that the little larva literally swims in jelly from that time till after the cell is sealed up; and also that where an egg or little larva is selected, from which to rear a Queen in a queenless colony, adjoining cells are torn down, so as to make room for a large amount of royal jelly at the start.

I have carefully watched, time and time again, to find out if an egg laid in a queen-cell was treated any differently for the first four days (after it was deposited in such cell by a Queen), than an egg laid in a worker-cell, and as yet I have failed to find any difference; so if any bee-keepers have seen what is described above, they have seen something that I have never been able to discover.

I also find, that where a colony is made queenless, the little larva is floated out with royal jelly, till near the end of the cell, when a queen-cell is built out and downward over the comb, rather than that the bees tear away cells, as described; especially is this the case with old combs.

At this time of hatching, the nurse-bees begin to feed the little larvae; but, so far as I am able to judge, the larva in a worker-cell is surrounded by three times the food it can use, for the first 36 hours of its existence. Somewhere from this, to the time the larva are three days old, the bees begin to stint them as to food, so that the organs are not developed as they would be if fed abundantly during the rest of their larval period.

I also claim that the food fed to all larvae, up to the time they are 36 hours old, is exactly the same, whether the larvae are designated for drones, Queens, or workers; and that the difference comes by the queen-larva being fed large quantities of this food, all of its larval life, while the others are fed sparingly later on, or else a different kind of food given after they are 36 hours old. Some experiments which I have conducted point in this direction, but, as yet I have not completed them fully enough to warrant the giving of them here.

If the above is correct (and I firmly believe that it is) it will be seen that the larva in a worker-cell has all of its wants supplied for the first day-and-a-half, and is developing towards a Queen just as fast, prior to this, in a worker-cell, as it possibly could in a queen-cell, surrounded by ten times the food that it can consume.

Hundreds of experiments in using larvae from three hours old, up to those of 36 hours, prove that Queens from the former are in no way superior to those from the latter, while the bees always choose the latter, where the power of choice is left to them. As all of my plans of rearing Queens require the changing of small larvae, I have dwelt thus largely upon this very important point, so that the reader might know just where I stand in this matter. Years of success in producing the best of Queens, together with the result of many experiments, conducted by some of our best Queen-breeders, go to prove that I am correct in the above conclusion.

Time of Hatching

A little practice will enable any one to know about how old the larvae are, by glancing at them in the bottom of the cells. Bear in mind that a larva but thirty-six hours old is a small affair, as the rapid growth is made at the latter end of its life; and if you think that there is any chance of a mistake on your part, in not knowing larva of that age, or younger, you should put a frame in a hive and watch for eggs, then watch for the eggs to hatch into larvae, when, by looking in-the cells from twenty-four to thirty-six hours afterward, you will know to a certainty, just how such as you should use will look. If you have been as correct as to the age of larvae used, as you should be, all of the Queens will hatch from the prepared cells, in from eleven-and-a-half to twelve days from the time the frame was given to the queenless colony. An expert can judge so closely that he can figure the time of hatching to within three or four hours.

In taking care of these cells, I generally do it on the afternoon of the tenth day, if it is pleasant; for if deferred till the eleventh day it necessitates taking care of them on that day, no matter what the weather may be.

As soon as the cells are taken from the hive, I go to the colony which had the brood given them (when it was taken from this colony in preparing it for Queen-Rearing), and take three frames well filled with brood, on one of which is the Queen, and place them back in the now queenless and broodless hive, being particular to see, in putting the frames of brood in the hive, that the Queen is on the center comb, so that the bees which go with her will surround and protect her, till all the bees become thoroughly mixed.

If it is early spring, I shake the bees in front of their own hive, from off the combs which do not have the Queen on, so as to keep this colony as strong as possible; for in three days from this time, this colony is to go through the same course that the other did, in rearing more Queens.

At the end of three days, all of the brood that was left, is to be carried to the hive which now has the Queen, so it will be seen that no colony will lose over thirteen days of time, by this process of Queen-Rearing, before it is back in nearly as good condition as ever. If we wish to save still more time to this colony, some colony can be kept purposely to care for the cells by keeping it queenless and giving brood occasionally to keep up the population. Into this colony the frames of queen-cells can be placed as soon as sealed, thus keeping none of the

Queen-Rearing colonies Queenless for more than eight days.

Some tell us that a queenless colony will rear four or five lots of queen-cells before the young bees get too old to rear perfect Queens; but I say, do not rear but one lot of cells from any colony, at one time, if you wish to have good Queens.

The reason is obvious, why a second lot of cells will not be as good, if you will take pains to read over what was said about getting the colony in condition to rear good Queens. It would be nearly as bad as causing all old or field bees to rear Queens, for the nurses have now been six days without anything that should cause them to prepare chyme, hence they have none in their stomachs to feed the little larvae, so they must go to work to produce more before they can do this work. By the time they could give them chyme, the larvae would receive a check from which they would never recover, even if they could be fed as much and as good food afterward, which is unreasonable to suppose.

The "Alley" Plan

Since I adopted the above method, what is known as the "Alley plan of Queen-Rearing" has been given to the public; but after a thorough trial, I fail to find any point wherein it is superior to the one given above, while in some few points I consider it inferior.

However, good Queens can be reared by the Alley process-very much better ones than those reared by any of the old plans that were used by most of the Queen-breeders before he gave his to the world. For this reason, Mr. Alley should have a prominent place assigned him, among the ranks of those who have done much to advance the cause of apiculture during the Nineteenth Century.

CHAPTER VII. THE NEW WAY OF REARING QUEENS.

While rearing Queens, as given in the last chapter, I became anxious for some plan by which I could get Queens reared by natural swarming, so that the cells would be all on one comb, and in a shape to care for as easily as were those which were built from the queen-cups that I gave to queenless colonies. For years I had practiced taking the larvae out of queen-cells, which the bees had under way, and substituting larvae from my best Queen, by the transposition process; but in all of these cases I had to take up with the cells where the bees had built them, besides, in many instances, after going over all of the combs in a hive I would find only three or four cells, so I had to do a great deal of work without receiving much benefit from it; while in cutting off these cells, I was obliged to mutilate many of my very best combs. This did not please me, so I set about seeing what could be done by way of having cells built where I desired them.

To this end, I prepared cells the same as I had done in giving them to queenless colonies, after which I placed the frame in a hive where the colony was preparing to swarm. I then waited two days, when I opened the hive, hoping that the bees had taken the larvae which I had given; but in this I was disappointed, for every one of them had been removed, while, much to my surprise, I found that every cell but two, contained an egg deposited therein by the Queen. In this I had gained a point, even if it was not just what I had been looking for.

I now watched these cells, to see what would become of them, and found that they were treated the same as others are-the colony swarming on the sealing of the first cell. As these cells were brought to perfection, I was not long in comprehending that in this I had a plan that would give me the cells all on one comb, the same being reared by natural swarming, and by it I could secure at least twice the number that I had ever been able to obtain before. By getting the colonies, having the best Queens, to swarm early, and keeping them at it as late in the season as possible, I could rear fully four times as many splendid Queens by this process as before, besides having the cells in such shape that every one of them could be saved with very little trouble.

In this way I kept on until I found that I could not find sufficient embryo queen-cells to keep up with my now-increasing calls for Queens, hence I must manage in some way to increase the supply of these, or else go back to the old way of Queen-Rearing for a part of my supply. The latter I very much disliked to do, which led me to go over the ground of making cells again, as I had formerly done.

Making Queen-Cups

While thinking of this matter, it came to me--why not dip the cells, the same as my mother used to dip candles? This thought so waked me up, that I wondered at myself for not thinking of it before, and immediately I had some wax in a small dish, over a lamp, to melt. While this was melting, I hunted up the old stick that I used in forming the cells at my first trial, which was nothing more than a tooth out of a common hand hay-rake. This tooth was now fitted to a queen-cup, as perfectly as I could do it with knife and sand-paper, while a mark was made around the tooth where the open end of the cell should come, so that I could know just how deep I wanted it to go in the wax, to give me the desired depth of cell.

By this time the wax had melted. I then got a dish of cold water, and after dipping the end of the stick in the water (up to where it was marked, or a little deeper), and giving it a quick jerk, to throw off the water not needed, it was quickly lowered into the wax up to the mark, and as quickly lifted out, twirling it around and around in my fingers, so as to cause the wax to be equally distributed over the wood. I now had a film of wax over the stick, so frail that it could not be handled, but in it I saw the commencement of a queen-cell, which would, I was sure, be a boon to my fellow bee-keepers, for the wax much resembled the very outer edge of a queen-cup built on new comb.

I then dipped it again, not allowing it to go as deep within one-sixteenth of an inch as before, and in twirling the stick after taking it out, the end having the wax on was held lower than the other, so that the lower end, or the base, would be the thickest, as the wax would flow toward the lowest point. As soon as the wax on the cell was cool enough to set, it was again dipped, not allowing it to go as deep in the wax as it did the previous time, by about a thirty-second of an inch, when it was cooled as before. In this way I dipped it from six to eight times, when I had a queen-cup that pleased me, as the outer edge was thinner than the bees made theirs, while the base was so thick that it would stand much more rough usage than would cells built by the bees. I now held it in the water, twirling it so that it would cool quickly, and, when cold, it was very easily taken off of the stick or form, by twisting it a little. It could then be fastened to a comb, by dipping in melted wax, the same as I did with one of the cups.

I had now solved the mystery of queen-cell making, and to make them quickly, I made more sticks, so that as soon as the wax had set on each, they could be laid on the table to cool, by placing them on a block [Fig. 5] having little notches in it.

Fig. 5.-Some of the Paraphernalia Used by a Queen-Breeder.

[EXPLANATION OF THE ABOVE ENGRAVING:- Beginning at the right hand side, we have 1st, a mailing-cage used in shipping Queens; 2nd, the three forming-sticks, laid on the notched-block while the wax is cooling; 3rd, the dish for cold water; 4th, the lamp having the dish of wax on top; 5th, a wire-cloth cage used' in introducing Queens; 6th, (near the front edge of the table) the ear-spoon used in scooping up the royal jelly; 7th, the stick used to place royal jelly in the queen-cup; 8th, a queen-cell protector, showing a hatched cell in the same, with the stopper in place; 9th, (at the back side of the table) a stick having wax queen-cups attached, showing their position on the stick.]

In this way I could be dipping right along, while the wax on several sticks was cooling. I finally found that three were as many as I needed, for if the thin film first formed became too cold before it was dipped again, it did not work so well in taking the cells off.

Later on I dipped the stick deeper in the wax than at first, as I found that the bees would not reject so many of the cells when this was done. I find by measuring, that I now dip the sticks in the wax nine-sixteenths of an inch the first time (measuring from the extreme point), and dipping less and less each time, as before stated, so as to get the base of the cell very thick, which I consider a great advantage. A convenient way to get the right depth, is to raise one side of the lamp a little [Fig. 5], so that the wax will be deeper in one end of the dish, than at the other. Dip in the deep end first, having the wax deep enough in this end so that it will come to the right point on the stick when the end strikes the bottom, and keep going toward the shallow end, as you proceed.

By holding the stick, when lifted from the wax, at different angles while twirling it, the cell can be made heavy at any desired point. To keep the wax at the right depth, add a little occasionally, putting it in that part of the dish immediately over the lamp, so that it will melt quickly. To secure the best results, keep the wax just above the melting point, for, if too hot, it requires many more dippings to get the same thickness of cell, besides bothering in other ways.

The question now before me was, would the bees accept of these cups the same as they did the natural cups that I clipped off the combs? I feared not; and in this I was right, as the first trial proved. This was a disappointment to me, although I had thought it might be so.

Putting Royal Jelly into Queen-Cups

While studying over the matter, it came to me one night as I lay awake-why not put some royal jelly into these cups, the same as there was in the cells that I had always been successful with, when transferring larvae in the swarming season? This seemed so reasonable that I could hardly wait for the middle of the day to come, when I could try it. At 10 o'clock the next day, I had a dozen cells prepared, each having some royal jelly in it; then larvae were placed in the royal jelly, the same as I always did, when using the transposition process in swarming-hives. In this way the larvae had an abundance of queen-food, even though the workers did not feed them in from two or three hours to half a day, which was quite a step in advance of setting them in embryo queen-cups, with only the food that I could take up with the tooth-pick, as previously stated.

To get the royal jelly in the cups, I dipped a little of it out of a queen-cell that I took from a colony building cells, taking one that was nearly ready to seal, as such had the most jelly in it. After the large queen-larva was thrown out, the whole jelly was stirred up in the cell, so as to get all of one thickness, for that in the bottom of the cell will be found much thicker than that about the larva.

When thus mixed, the jelly was taken up with the ear-spoon on a pair of tweezers [Fig. 5], which was then transferred to the hollowed-out end of a little stick, by drawing the bowl of the ear-spoon over the end of the stick, until about one-eighth inch in diameter of the jelly was standing on the end of it. The end of the stick having the royal jelly on it, was then lowered into the bottom of the wax-cup, when it was twirled a little so as to make the jelly stay on the bottom of the cell, the same as it does when the bees place it there.

The amount of jelly used for each cup, was about the size of a "BB" shot, when on the end of the stick, before lowering into the cup, or one-eighth inch in diameter, as stated before. To get the first jelly of the season, a colony must be made queenless; but after this, it is secured by taking one or two of the prepared cells at any time before they are sealed up.

Brushing Bees off the Combs

Having the frame in readiness, it was given to a colony that was preparing to swarm, and left for two days. When I opened the hive this time, and drew out the prepared frame, you can imagine my pleasure, at seeing 12 as nice queen-cells under headway as I ever saw, all looking like so many queen-cells built out of new comb- they were so light colored. In three days more, these 12 cells were capped, and, in due time, 12 as splendid Queens as I ever saw, hatched from them. There was now no need of searching combs for embryo queen-cells, for I had something very much better, and something which would stand more rough usage than the other cups ever would endure.

Queen-Cells Built on a Stick

My next idea was to have all of my queen-cells built on a stick, or piece of frame-stuff, the same as I had read about; so when I again made some, instead of taking the cup off the form, I only loosened it enough so that it would slip off the stick easily, when it was again dipped in the wax and immediately placed on a mark on the piece of frame-stuff, which mark I had designated as a place for a cell [Fig. 6]. In an instant the cup had adhered to the frame-stuff, when the forming-stick was withdrawn. This cell was placed near one edge of the stick, which was one inch wide, one-fourth of an inch thick, and long enough to crowd between the side-bars of one of my frames. The cell was also placed near the centre of this stick, as to its length, but close to one side of it, as to width.

The next cell or cup was placed one and one-half inches to the right of the first, while the third was placed the same distance to the left, and so on until six were on the stick, I then put the next on the opposite side from the other six, and half way between them, so that when I had six more cups on, or 12 in all, the cells alternated with each other, which gave more room to each cell when occupying a given space, than would have been given, had I placed each along the centre of the stick [Fig. 5].

To get the designated places for cells, set the dividers the distance apart that you desire the cells, and after having put one of the points where you desire to have the first cell, "walk" them along until you make the number of point-marks that you want cells, and, in dipping, set a wax-cup over each point-mark.

Having the cells thus fixed, helped in several ways, as it gave the bees a better chance to cluster among them in building, while it gave me a better chance to manipulate the cells in transferring the royal jelly and larvae to them, taking them off, etc., and especially in getting them off the stick; for, when fixed in this way, all I had to do was to push gently on the outside of the base of the cell with my thumb, then off it came, without any danger of injuring it in any way.

Affixing the Wax-Cups to the Stick

After dipping cells for some time, the forming-sticks will get so coated with wax that the water will stand in drops, instead of flowing freely over them, thus causing the cups to stick so as to spoil them in taking off. When this is the case, the stick is dipped in the water, and immediately placed between the second and third fingers of the left hand, close up to the hand, and twirled around once or twice, which causes the water to spread out over the stick; when it is dipped in the wax, and will work again as well as ever. When working continuously I am able to make from 150 to 200 cups in an hour, so it will be seen that but little time is required to make all that will be needed in any Queen-Rearing establishment.

Fig. 6.-Affixing the Wax-Cups to the Stick on which they are to be built.

I now had easy sailing all along the line, for all I had to do was to prepare the frame with these wax-cups, put in the royal jelly, which was now very easily done by laying the stick of cells bottom-side up on a table [Fig. 5] or chair before me, while the royal jelly was being dished into each; and before removing from the table, transfer the larvae taken from the hive, having my best breeding Queen, into each cell. When all was ready, this stick of prepared cells was crowded between the side-bars to a frame of comb, which had been previously cut so that the cells would come in the center of the comb, while the ends of the stick slipped through a slot cut in the comb for them.

Fig. 7.-Frame Showing how the Stick of Queen-Cups is Fitted into It, and How the Cells look after being Built from the Cups. The Illustration is less than one-fourth Size, as the Frame Used is for the Gallup Hive, which is 11 ¼ x 11 ¼ Inches.

I keep several old combs, say 10 or 12, for this purpose, which are used as often as cells are needed. This is a great convenience, and saves destroying or mutilating a valuable comb every time we wish cells built.

By doing the whole in a warm room, I was independent of the weather, for in carrying the frame to the hive, I wrapped it in a warm flannel cloth, when it was at all cool outside. In this way I was sure to get a large proportion of the cells completed, whether used in a colony preparing to swarm, or in a colony which I had fixed for Queen-Rearing.

Marking Experimental Hives

Before carrying to the hive, a slight shaving was taken from the top-bar of the frame, and the date placed on it, so I could know just when these cells would hatch. If the figures on it read 7-20, I knew that the cells were prepared July 20, and should be cared for on the 30th of that month. I now wrote in a small book kept for this purpose, 7-20 placed in hive 40 (if that happened to be the hive they were placed in), so that by looking at this book at any time, I knew where each frame of cells was, and when put there. If cells were started with larvae from any but my best Queen, the name of the Queen was also placed on the frame and in the book, so that I knew just what I was about at all times. By looking in this book, I knew just when and where I should go to get these cells, so that none were destroyed by the Queens hatching before I expected.

Keeping the Larvae Warm

In getting larvae for Queen-Rearing, when very cool, I placed the piece of brood under my clothing near my body, as soon as cut from the comb, and kept it there while carrying it to the room, so that there was no danger of chilling the larvae; or, if handier, as was sometimes the case, I placed it in a box having a heated iron in it, as will be explained farther along. A small room especially adapted for the handling of cells, brood, etc., which can be kept warm in cool weather, is almost a necessity to the man who makes a business of Queen-Rearing. When only a few Queens are reared, the family kitchen can be used, as there is almost always a fire there-providing the "better half" is willing.

After working successfully along this line, with both colonies and those made queenless for Queen-Rearing purposes, I chanced one day to find a colony which was about to supersede its Queen, and gave evidence of being one of those colonies which might have two Queens in a hive. I was not long in deciding to try my process of cell-building in this colony, so I at once destroyed the queen-cells which the bees had started, having royal jelly in them, using the royal jelly to put in the wax-cups I had prepared for them. I soon had the frame of prepared cells in the colony, when I waited rather impatiently for the next two days, till I could see what the developments would be.

At the end of two days, I went to the hive, and upon lifting out the frame I found that eleven out of the twelve cups prepared, had been accepted, and were now on the way to completion, as perfect queen-cells. Two days before the cells were ready to hatch, I had them photographed, and I have given the reader a fairly good picture of them, [Fig. 7], which also represents these cells as built in an upper story over a queen-excluding honey-board, as about to be described.

To say that I was delighted with this success, would hardly express it; I was almost happy in thinking that I could now get queen-cells of the best type, and by the quantity, as long as I could keep that old Queen alive at the head of the colony, for it was an exceedingly strong one.

Upon leaving this colony, I went to another strong one, and removed its Queen, so that I might have a place where I could put these cells as soon as they were sealed, for safe keeping, till they were old enough to give to nuclei; for my object was to keep this colony rearing Queens constantly, as long as the mother lived. As soon as the cells were sealed, they were removed and placed in the queenless colony, and another prepared frame given to them.

This last frame was accepted the same as the other, and, if my memory serves me rightly, this colony completed eleven sets of cells, before the old Queen gave out entirely. At any rate, they reared so many that I saw the plan was a complete success, for all of these Queens were of the highest possible type, as to color, size and fertility; while the amount of royal jelly put in each cell was simply enormous, so much so that one cell taken from the frame would have jelly enough to start the twelve cups on a prepared frame at any time; while large quantities were left in the bottom of each cell, after the Queens had hatched out.

Securing Large Crop of Comb Honey

Before going farther, I must digress a little. In January, 1883, I met Mr. D. A. Jones, of Canada, at the "North Eastern" bee-convention, which was held that year at Syracuse, N. Y.; and in a private talk, regarding how to best secure the greatest yield of comb honey, he told me that he had found that he could get the most honey by having it built in the brood-chamber of the hive, fixing things so as to have the sections surrounded with brood, as it were.

To briefly describe the plan, as I now remember, it was as follows: When the honey season arrived, the brood-chamber was divided into three parts, the central one having five combs in it, which were to be enclosed with perforated-zinc, or, what we now term, queen-excluding division-boards. These five frames were to contain, as far as possible, only hatching brood, so that the Queen might have room to lay, as the brood hatched out. On either side of this small brood-chamber, sections were to be placed, by hanging in wide frames filled with them to the amount which it was thought that the colony required; while beyond these wide frames of sections, the remainder of the frames of brood, taken out when reducing the brood-chamber, were to be placed, having an equal number on each side.

Will it Pay to Bear Two Queens?

In ten days, the hive was to be opened, and the Queen hunted, and, when found, the frame she was on was to be put outside, when the remaining four frames were put over to take the place of those outside of the sections, while those outside were to be placed in the brood-chamber, and the frame having the Queen on it returned. By treating the colony in this way every ten days, the Queen furnished as much brood as she otherwise would, while the sections were kept in the middle of the hive (so to speak) all of the while, which caused the bees to work assiduously to fill up this vacant space in the brood-nest, thus giving more honey than could be obtained in any other way.

This looked so reasonable to me, that I accepted it at once; but with the usual caution which I have always thought best to use, where trying something new, I made only two hives to be worked on this plan the following season. Without going into the details farther, suffice it to say that on account of pollen in the sections, and some other difficulties, the plan did not succeed with me as I had expected, so it was given up.

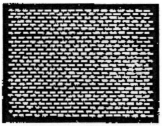

Fig. 8.-Queen-Excluding Division-Board.

Bees Starting Queen-Cells

There was one thing that I learned, however, which started me on the road to a new discovery along the line of Queen-Rearing; which was, that in every case, where unsealed brood was placed outside of the sections, the bees would start from one to three queen-cells, and unless I cut them off, the Queens hatching from them would supersede the old one, or else a swarm would be the result, when the combs having the cells on, were placed back with the Queen again. This I did not like, as it was too much bother to look over the combs carefully, every ten days, in addition to the other work, when a change of combs was to be made.

One of the colonies so tried, had one of my best Queens in it, and when I came to cutting off the cells, I was not slow to see that they might be made to form no small part in my Queen-Rearing business. However, the cutting of nice combs to get these cells stood in the way of my desiring to get all of my Queens in that way, and besides, all the Queens so reared did not please me, for the colony was often so spread out with sections between the brood, that the necessary heat to get good Queens was not always present.

I now began using the queen-excluding metal [Fig. 8], between the upper and lower stories of the few hives that I worked for extracted honey, and in one or two cases, brood from the lower story was placed in the upper one, over the queen-excluding honey-board. Again I had queen-cells built as in the former case, which were cared for as well as any I had ever seen, although, as a rule, but two or three would be built on one lot of brood.

In thinking the matter over one night, while I was awake, doing some planning for the future, it came to me that these cells were built under precisely the same conditions, that the cells were when the bees were thinking of superseding their Queens, at which time I was enabled to get the best of queen-cells built. To be sure, the Queen below was a good one, but as she could not get above, the brood that the bees had there did not increase any, so they concluded that they must have a better Queen in this part of the hive; hence they went to work to produce one.

One thing that I had always noticed was, that where the bees had their own way in the matter, where cells were built to supersede a Queen while she was still in the hive, they never started more than three or four cells, while one or two were more often built than otherwise. That the bees only built about the same number in these cases of brood above a queen-excluding honey-board; and, also, that I have never known a swarm to issue, simply from having queen-cells in such an upper story, when none were below, shows that they consider the conditions the same as in case of supersedure.

Having become satisfied that I was right on this point, the next step was to see if the plan which had proved so successful with the colony about to supersede its Queen, would work above the queen-excluding honey-board; and if it would, I would be a step farther in advance than I had ever been before; for in it I saw something of great value to the bee-keeping fraternity in the future.

A frame of queen-cups were now prepared as before, and to make sure of success, if such a thing were possible, I raised two frames of brood (mostly in the larval form) above, so as to get as large a force of nurse-bees about the prepared cells as possible, to properly feed the queen-larvae. The prepared frame was placed between the two having brood in them. In two days I examined this frame, and found that my conclusions were right, for every cup had grown to a half-built queen-cell, while the little larvae were floating in a quantity of royal jelly, that more than half filled the cell. These were finished in due time, and from them hatched Queens which were every whit as good as any I had ever seen.

I now had things brought to where I was master of the situation, so that I could rear the best of Queens, just when and just where I wanted them, and that, too, with a laying Queen in the hive at all times, so there would be no loss in honey-production, to any apiarist, while rearing Queens; and the beauty of it all was, that these cells were all on a stick, so that they could be made use of without injuring any of my good combs, or in any way endangering any of the occupants of the cells.

Not knowing to what extent this plan could be carried, and yet secure good Queens, I went slowly at first, not giving any one colony a second prepared frame, till after the first had been removed and more brood placed above. As I leave the cells where they are built, till they are nearly ready to hatch, or for ten days, this, took five colonies to give me a lot of queen-cells every other day, as I desired them, during the height of the season.

The next season, wishing to see how much there was in the plan, I put in a prepared frame as soon as the first cells were sealed, and then another as soon as these were sealed, and so on indefinitely. As far as I could see, the last lot of cells were as good as the first, although, as a rule, I did not get quite as many accepted. It was a rare thing that the bees finished less than nine out of twelve prepared cells during the first season, while the bees would frequently build and properly care for the whole twelve. In crowding them so fast, they would sometimes give me only five or six, yet, as a rule, they would average about eight, so that I really gained nothing by thus crowding things.

For this reason, I now kept along the line of work followed the first season, till the past summer, when, to see what might be accomplished, preparatory to writing this book, I gave a prepared frame to a colony every two days, and, while they did not complete as many cells on each frame as formerly, when I gave them less often, yet some cells would be built on every frame.

In this way, I had in one colony having a laying Queen below, queen-cells in all stages of progress, from those just ready to hatch, down to larvae that the bees had just commenced to feed, by adding to the royal jelly which I had placed in the cups; and, besides this, I had Queens kept in nurseries in different parts of the upper story. I also had in this same upper story, Queens just hatched, and some just commencing to lay, by having a part of the upper hive formed into nuclei, by using perforated-metal division-boards, as will be explained farther on. It will be seen that there is scarcely any limit to what can be accomplished by this method of Queen-Rearing, and queen-fertilizing.

However, as a rule, I think that a little better Queens can be reared by the way I worked the plan the first season, for the cells are better supplied with queen-food, where unsealed brood is placed in the upper story every ten days-enough better, in my opinion, to pay for the extra work.

Again, I would not put over twelve wax-cups on a stick, for if more are used, the young Queens are not fed quite as well. In my experiments I have used as high as twenty-four cups, and had every one accepted and finished; but unless the colony was an extremely populous one, I did not get quite as good Queens as when only twelve were used.

To show what may be accomplished by this method, I will say, that in the honey harvest I have prepared sticks having from four to eight cups on them, the sticks being made of the right length to crowd into any section of the hive that I was using, so as to keep it from falling down, when the section was placed between two others, in which the bees were at work over a queen-excluding honey-board; when upon going to the hive at the end of ten days, I would find as nice queen-cells, nearly ready to hatch, as any one needs to see.

I have also had Queens fertilized and kept till they were laying, one in each section of the hive, yet this plan of producing Queens in sections, is not to be recommended, as it spoils the section from being first-class for honey, ever afterward.

Caution to Queen-Breeders

By way of caution, I wish to say, that if a Queen is by any means allowed to hatch in the same apartment where the cells are, these cells will at once be destroyed. If the bees with such a Queen, are shaken off the combs, so as to get her out of the way of more cells being built, and the bees are allowed to enter the hive below, with this Queen (as would be natural for any apiarist to do), this young Queen will destroy the one below, no matter how prolific or how valuable a Queen you may have there. This queer procedure, I bring to bear on all Queens that I wish to supersede, as will be explained farther on.

Another thing: In the fall (or in this locality after August 15th), when the bees begin to be inactive, or cease brood-rearing to any great extent, the warmth generated in the upper story will not be great enough to produce good Queens, and as the season draws toward a close, no queen-cells will be completed, unless we feed sufficiently to arouse the whole colony into activity, and keep them thus all the time while the cells are being built. At such times I have had cells that usually mature in eleven-and-one-half days, to be from sixteen to twenty days before hatching; while the Queens would be almost black, from the same brood which produced very yellow Queens during June, July, and the first half of August. In all parts of the country where fall flowers abound, I presume just as good Queens will be reared in September by this plan, as at any time; but as we have no fall flowers here, I cannot be positive on this point.

At any time when the bees work the cells very slowly, I know no better plan than that which was given in the last chapter, only I use the wax cells, prepared with royal jelly, as given before. Then in early spring, before the colonies are strong enough to go into the upper stories, we must use that way also, for there are not enough bees in any hive to use the above method, which I believe to be the best of all. However, much can be done by way of getting a colony strong, by giving sealed brood, so this last process can be used much earlier in the season, than otherwise would be the case.

In this chapter, I think that I have given something of great value to the fraternity; and if it shall lead to the universal rearing of a better class of Queens, than have formerly been reared, I shall be well paid for all my efforts in this direction.

CHAPTER VIII. GETTING THE BEES OFF THE CELLS.

It would hardly seem that a chapter should be devoted to this subject, but from what I know of others doing along this line, and what I used to do myself, I am satisfied that many Queens are materially injured before they are out of the cells, and many more are killed outright. We frequently see in print, the instruction given, when handling frames after the bees have swarmed and before the queen-cells have hatched, that to get the bees off the combs, the "frames should be shaken," or "the bees shaken off in the usual way." Many write, asking why their queen-cells do not hatch, or why so many Queens hatch with crippled wings, or having a dent in one side of the abdomen. When answering such an enquiry, and asking for more particulars, these almost always reveal that they have shaken the frame having the cells on, to get the bees off.

One man came a long distance for some brood from my best Queen, from which to rear some Queens to cross with his stock; and after securing some 50 or 60 splendid cells from this brood, by the transposition process, he came after more brood, telling me that out of the whole lot, only three Queens hatched, and only one of those being perfect. Upon asking him how he got the bees off the frames having the queen-cells on them, he said he shook them off the same as he always did. It does not seem that any one would be so thoughtless, yet there are hundreds who do not stop to think of these little things.

Queen-Cells Must Not be Shaken

I wish to emphasize the words, *never shake the bees off a frame having queen-cells on it, nor in any way suddenly jar it!* for queen-cells are much more liable to injury while on the frame, than they are when taken from where they were built. The reason for this lies in the fact that when the comb is moved, there is a heavier body than in a single cell; hence the heavy body takes more force to move it than does the lighter one, which is apt to give it an accelerated speed, and when suddenly brought to a stop, it causes a concussion much greater than the operator dreams of; while this concussion sets the Queen pupa to tumbling around in its cell, to a very damaging extent, and one nearly, if not quite as great, as would come to it had the tree in its forest home, as provided by Nature, been suddenly blown over by some thunder-gust. Under such circumstances, no one would expect queen-cells to hatch. Care should always be used in handling cells, but especially before they are removed from the combs or frames. My way of doing this is as follows:

Having lifted the frame with the cells on it, from the hive, it is carefully put down near the entrance, letting it rest on the bottom-bar, so that it will occupy the same position that it did in the hive. To thus fix it without killing the bees, which may be on the under side of the bottom-bar, I lower it so that it will touch the short grass which grows about the sides of the hive, and then, by drawing the frame endwise toward the entrance, the bees are all brushed off on the grass, so that none are left where they would be killed. This is the way I always do, when putting the frame on the ground for any purpose.

Having the frame near the entrance, and leaning up against the hive, I smoke the bees thoroughly, so as to cause them to fill themselves with honey, and while they are doing this, I arrange the interior of the hive, when it is closed. I then smoke the bees some more, and if they seem inclined to leave the comb and run into the hive, I keep smoking them until all have run in; but if they are loath to leave the comb, which they usually are, I take hold of the frame and raise it a foot or so above the entrance, when, with one of the soft bee-brushes, now used by most beekeepers, or with the feather end of a quill from a turkey's wing, the bees are brushed off, brushing about the cells very carefully.

Fig. 9.—"Yucca" Bee-Brush.

If the bees have filled themselves with honey, they are rolled off the comb with the brush very easily, and seldom offer to sting; but if you undertake to brush them, when the frame is first lifted from the hive, you will find that they will stick to the comb and cells "like tigers," while their fury will be scarcely less than would be shown by the tiger herself, if pushed around while being robbed of her cubs.

Where frames of queen-cells are obliged to be handled on a cool day, or in a rain-storm, it is sometimes best to take them, bees and all, into the warmed room where the work is done with cells. When there, remove the cells as quickly as possible, allowing the bees to remain on the comb, which they will generally do if we do not smoke them very much while removing the frame from the hive. Upon reaching the room, they will begin to fill themselves, and if spry, we can remove the cells while they are filling, and carry them back to the hive without losing any.

Once having the cells in the warm room, we need be in no hurry, for if the room is of the right temperature (from 85° to 95°), the inmates of the cells are advancing toward maturity just as fast as they would be in the hive.

Cleome in Bloom.

CHAPTER IX. WHAT TO DO WITH THE QUEEN-CELLS.

Having the queen-cells all built, and being nearly ready to hatch, with the bees all off from them, the next thing we want to know, is what to do with them.

There are three ways generally employed in using them, the one used most being to give the cells to nuclei or queen-less colonies; next, putting the frame having the cells on, in a lamp-nursery, leaving them there, and taking out the Queens as fast as they hatch; and, lastly, putting each cell in a separate cage having food in it; while the cages are so arranged, that from 12 to 24 of them will just go inside of a frame, filling it solid the same as a comb would; when the frame of cages is placed in the center of a full colony of bees, where it is left till the Queens hatch, when they are to be disposed of as is thought best.

The last is called a queen-nursery, and has the advantage over the lamp-nursery, in the fact that no watching is required to keep the Queens from killing one another, should several hatch during the night, or when the apiarist was at other duties. As I wish to say something about each method, I will speak of them in the order named above.

As a chapter will soon be given on forming nuclei, all that I shall say in this, will be in regard to how it is best to get the cells into these nuclei, or into a full colony, if you wish to put them there. Bear in mind, that if you wish to be sure that the cells are not torn down, you must wait about giving them to any colony, whether weak or strong, from twenty-four to forty-eight hours after their laying Queen has been taken away; or, in other words, the colony must be queenless for that many hours, before it is time to take the cells from the colony where they were built.

Some claim that an unprotected queen-cell can be given to a colony, at the time a laying Queen is taken from the same; but all of my experience goes to prove that where this is done, nine out of every ten cells, thus given, will be destroyed. I am not alone in this, for many prominent apiarists write me that they can do no better. However, I will say, that it may be that I do not have my colonies small enough, for a colony so small that there is not over 200 bees in it, will accept of a Queen or a queen-cell much more readily than will a colony of 20,000 workers.

In giving queen-cells to colonies, if the weather is warm enough so that there is no danger of chilling the cells, or say when the mercury is at 75° or above, I take the frame of cells in my hand, and go to the nuclei with it (unless there is danger of robbing, from scarcity of honey in the fields), and put the frame of cells down next to the hive containing the nucleus or colony, the same as it was placed at the entrance of the hive in getting the bees off, so as to be sure that the cells are not injured in any way.

I then open the hive, and, having selected the spot on the comb where I wish to put the cell, I place my thumb on the base of one of the queen-cells and gently bear down, when it cleaves off the bar of wood to which it was fastened, as easily as need be. Now, instead of cutting the combs and fitting the cells into them, the way we always used to do, all I have to do is to place the cell against the side of the comb where I wish it to stay, and gently push against the base of the cell, when it sinks into the comb far enough to make it a fixture. After this comb is lowered into the hive, and the, next comb brought up to it, this next comb touches the cell on the opposite side, so that it cannot fall out, even if the bees do not stick it fast, which they are almost sure to do, even if the other comb does not touch it.

Here is another advantage that these cells have over those entirely made by the bees, in that the wax is so thick at the base of the cell, that it is impossible to dent them with any reasonable handling; also this mode of putting on the comb keeps the nice worker-combs from being injured, as was the case with the old process, given in nearly all of our bee-literature.

Basket for Carrying Queen-Cells, Robber-Bees

If robber bees are so that they will follow around, trying to get at the frame of honey with the queen-cells, the same is taken to the shop, or other room where I work, when the cells are taken off the bar of wood, in the same way as described before, and placed in a little basket that I have for the purpose, as fast as they are removed.

Why I use a basket in preference to anything else, is because in warm weather this lets the air around the cells, it being of open-work, so that when the sun strikes on them there is not nearly the danger of over-heating that there otherwise would be. In a tin dish, when left in the sun, it takes only a moment or two to spoil the in-mates of the cells by over-heating, while on hot days I have to be careful, even with the basket. Upon going to the apiary, the cells are now taken from the basket and used as before.

If the weather is very warm, or the colony to which the cell is given, is a strong one, I do not take out a frame when inserting the cell, but simply spread the top-bars apart a little, and crowd the cell lightly between them so that it will stay. This works equally well, if we are sure that the colony is strong enough to cluster around it, so as to keep it warm; but in all cool weather the other is the safest plan; as in the first case I always put the base of the cell just above the brood, so that the point comes down on the brood, which insures the bees caring for it, as surely as they care for the brood.

If the weather is cooler than 75° in the shade, I heat one of the largest weights to my scales (any iron will do) so warm that I can just hold it in my hands, when it is dropped into a wooden box of sufficient size to let it slip into it, and over the weight is placed several thicknesses of felt, cut to fit the box. The cells are now placed on the felt, when a cover for the cells, made by sewing two or three pieces of the same felt together, and having a little handle attached, is placed over them. The cells are now protected from the cold, so that I can go right on with my work, regardless of the weather.

When I go to a hive, all I have to do, is to lift the felt cover by the handle, take out the cell and put back the cover. If you have more cells than you think the iron will keep up the necessary heat for, all you have to do is to leave the rest in the warm room until you can return and heat the iron again.

Oil-Stove for Work-Shop, Protecting Queen-Cells from the Cold

An oil-stove is very handy to keep this room warm, or to heat the iron or anything else with, and has the advantage in allowing you to control the temperature of the room perfectly, by turning the wicks up or down. However, any stove, or any room having one window in it, will answer all purposes, for any one who has not things fitted for this special work.

What I have to say about the lamp-nursery will be brief; for, to tell it just as it is, I have very little use for it, and none whatever in the way it is generally used. I do not like to be obliged to watch as closely as is necessary to keep the Queens from killing one another while hatching; besides, the gain when used in this way, is very little if any, where the cells can be left in the colony building them, until they are nearly ready to hatch, as is the case with the plan of cell-building which has been given. By the old ways, where the colony was kept nearly, or quite idle, from the time the cells were sealed till they were ready to hatch, the time saved to the colony by taking them to the lamp-nursery as soon as sealed, undoubtedly paid; but where the work of the colony is going on just the same, whether there are cells in the hive or not, the case is different.

The only time when I use it, is early in the spring or late in the fall, when I have to get the cells built in queenless colonies, and then I only use it in connection with the queen-nursery, which does away with all watching, or sitting up at night to see to the Queens.

Virgin Queen Traffic

Were it not for the traffic in Virgin Queens, which is now growing, and bids fair to assume gigantic proportions, and the few times when we have more queen-cells ready to mature than we have just at that time small colonies to take them, there would be little more use for the queen-nursery than for the lamp-nursery.

For these reasons, however, I think that the use of the queen-nursery pays well; for it allows us to sort the Virgin Queens in sending them off, so we need send only the best; while it many times enables us to save a nice lot of Queens, which we would otherwise lose.

In using it, I put it in the upper story of any colony having a queen-excluder between it and the lower story; often putting it in the same apartment where the bees are rearing Queens, as it makes no difference to the bees as regards cell-building, the work going on just the same if there are 50 Virgin Queens caged in the same apartment. At first I feared that cell-building would stop, or the cells already built would be destroyed, but after testing the matter, I find that Queens thus caged have no influence along these lines whatever.

The kind of nursery that I prefer, is like the one [Fig. 10] invented by Mr. Alley, and described in his book; but if my memory serves me rightly, Dr. Jewell Davis was the first one to bring the queen-nursery before the public.

I make the queen-nursery as follows: Sixteen blocks are gotten out, 2 5/8 x 2 5/8 x 7/8 of an inch, which exactly fill one of my frames. A one-and-one-half inch hole is bored in the centre of each of these blocks, over which is tacked a piece of wire-cloth having 12 to 16 meshes to the inch, and being two inches square.

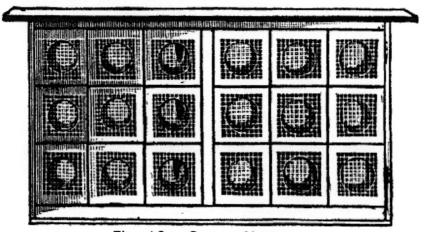

Fig. 10.—Queen Nursery.

Before tacking on the wire-cloth, I bore in the edge of the block (which is designated for the top after the block is put in the frame) a three-fourths inch hole, boring down to within one-eighth of an inch of the one-and-one-half inch hole. I now finish boring the hole with a one-half inch bit. This hole is for the queen-cell to be placed in, and the reason for the two sizes of holes is to give a shoulder, so that the queen-cell can hang in the block, the same as it does on the comb, and still be in no danger of slipping through into the block. This hole is bored a little to one side of the centre, so as to allow room for a one-half inch hole on the other side, which is to receive the candy; the latter hole is so bored that it comes out near one side of the one-and-one-half inch hole, and when it is deep enough, so that a hole large enough for the Queen to enter, is made, I stop boring, for a shoulder at the bottom is needed to keep the candy in place [Fig. 11]. Now fill the hole with candy, packing it in with a plunger made to fit the hole loosely, and tack on the wire-cloth. The blocks can be made so that a given number will fit any frame in use. I only gave this description as the right size to use in the Gallup frame.

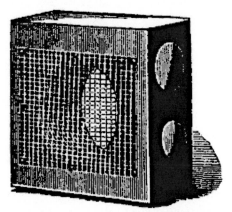

Fig. 11.—Nursery Cage.

"Good" Candy for Provision

Having the cages provisioned with the "Good" candy made of granulated sugar and honey (granulated sugar is preferable to the powdered, to use in these cages), the cells are taken off the frames as before, when, a little honey is put around the end, just where the Queen will bite off the cover in hatching, so that she can feed herself before coming out of the cell, the same as the bees feed her, when she hatches in the hive. When first using the nursery I did not know of this "trick," consequently I lost many Queens by their not having strength to get out of their cells; but when I saw the bees feeding a Queen one day, through an opening in the cell, I took the hint, and have had no trouble since.

After the cell has the honey on it (do not put on enough so that it will daub the Queen), the cell is put in the hole made for it, and the cage put in the nursery-frame, and so on till all are in, when the nursery is taken to the hive and put where it will be kept warm by the bees.

If in early spring or late in the fall, it is put in the lamp-nursery, which is then kept running, as the heat in the hive is not great enough at this season of the year to hatch Queens where the cells are in a nursery. I have lost many Queens, even after they had hatched, when trying to keep them in the nursery, placed in a colony, after the bees had become partially inactive in the fall; so the queen-nursery is of no value in this locality after Sept. 15, unless it is used in connection with a lamp-nursery. If all has been done as it should be, there will be a splendid Queen in each cage, in due time, ready to use when wanted.

In all of this work with cells, they should, as far as possible, be kept in the same position as they are in the hive, and if laid down, it should be done very gently, so as not to injure the young Queen within.

Before putting in another lot of cells in the cages, place a drop of honey on top of the food in the candy-hole, instead of renewing the candy; for this honey will work its way down through the sugar, so as to be where the Queen can get it when she hatches, which is as good as to renew the candy each time.

CHAPTER X. THE QUEEN-CELL PROTECTORS.

Caging Queen-Cells

Some few years ago, after becoming disgusted by losing many queen-cells in trying to get the bees of a colony to accept such, as soon as a laying Queen had been taken away (as some claim can be done), I resolved that I never would again lose a nice lot of queen-cells, by trying to save from 24 to 48 hours time to my nuclei, as I had so often done in trying what proved to be, with me, an impossibility. In so deciding, I felt somewhat sad, for "time," with the Queen-breeder, "is money," in the summer months. However, the loss of queen-cells was of more money value than the loss of time to the nuclei, hence I was driven to the above decision, although with great reluctance.

That afternoon Mrs. Doolittle was away from home, not expecting to be back till in the evening; so when it grew so dark that I could see to work no longer, I went into the house, and, being rather dissatisfied with myself, in having to own up as being entirely beaten by the bees, on account of the above, I threw myself on the couch, instead of reading or answering correspondents, as is my custom generally before retiring. Being fatigued, I soon fell asleep, and slept about an hour.

Upon waking, the first thought that came to me was, why not make a cage of wire-cloth to protect the queen-cell, so that the bees could not get at the side of the cell until they became aware that they were queen-less, which would be at about the time the cell would hatch; and in this way be victor, instead of owning up beaten, as I did a little while ago? Very soon the picture of a cage and all how to use it, stood out before me, so plainly that I could see it with my mind as distinctly as I ever saw any picture with my eyes.

The idea of caging queen-cells was old, and cages especially adapted to this purpose had been advertised for a good many years; but the ideal cage presented before me at this time, was for the special purpose of allowing the safe introduction of a queen-cell, nearly mature, to a colony at the time of taking away a laying Queen, the cage being so constructed that the Queen could hatch and walk right out among the bees, the same as if no cage was there; while at the same time the cell was safely secured against the bees, so that they would not destroy it.

All are aware that when bees destroy a cell, they bite into the side or base of it, and never at the point, on account of the cocoon being so thick at the point end, that they cannot make much headway in trying to get a hole in there. Well, the cage I saw on awaking, was to be made so as to protect all parts of the cell from the bees except the point, and this was not easy for them to get at, even if they did try to bite through it at this place.

The cage was made by rolling a small piece of wire-cloth around a Y-shaped stick, so that a small but not very flaring funnel was made, the hole in the small end being as large as an ordinary lead-pencil. After making the cage, I cut off a piece of five-eighths inch cork for a stopper, put in a nearly mature queen-cell, with the point down into the lead-pencil hole as far as it would go, when the piece of cork was put in so that the bees could not get in at the base [Fig. 5].

I now took a fine wire and run it through the meshes of the wire-cloth just above the cork, so as to keep the cork in place, while the other end of the wire was bent so that it would hold on to the top of the frames, in order to keep the cage in the position I desired it, between the combs. These caged cells were hung in the hives at the time the Queens were removed, and in from 24 to 48 hours, according to the age of the cell, I had a nice Virgin Queen in the hive wherever a caged cell was hung.

As soon as I found it was a success, I made many more cages, so that now I have no more trouble with bees destroying cells, not having over two per cent, destroyed out of the thousands so caged during the past three years. The cage protects the cell everywhere except at the point, but allows the bees to get accustomed to the presence of it, the same as if the cage was not there. The lead-pencil hole allows the Queen to hatch the same as if the cell were not caged, while the bees can feed the Queen and hold her in the cell as long as they please. The few cases where Queens are killed, seem to come about by the bees destroying them after they have been let out of the cell; yet I have thought that in a very few instances, the bees had torn away the extreme end of the cell and dragged the Queen out; but the instances of Queens being destroyed where cells are so caged, are so few, that the plan can justly be called a success.

Instead of using cork for stoppers, I now use a piece of corn-cob, which will fit nicely in the cage, above the base of the cell; and instead of hanging the cage in the hive with a wire, I press the cage into the comb, the same as I have told of doing with an uncaged cell. In pressing the cage into the comb, bear on the cob, for in this way there is no danger of pressing the meshes of the wire-cloth into the cell; and only press it far enough so that the cage is secure to the comb. In this way the bees will not gnaw through the septum to the comb, so as to injure it, if the cage is left in the hive for several days after the Queen hatches, as they are sure to do if the cage is pressed down till it strikes the center of the comb.

When I have many Queens to send off, in days when there is danger of over-heating the cells, or chilling them, I put up the Queens first, then take the caged cells in the basket or heated box and put them in afterward, the same as is done with uncaged cells.

Some who have tried this plan of caging cells, have complained that they had to trim the cell so much to get it in the cage, that it made lots of work, which is the truth, where the old plans of cell-building are used; but with the plan of wax-cells, as here given, all of this trimming part is done away with. This also accounts for the reason that some have not succeeded with them, for the cells built by the old plans would not fit the cages so well but that the bees could get at them to bite in at the sides of the cell.

This caging of cells is of great value, where Queens are placed in the upper story of full colonies, for fertilization, as is soon to be described; for in this case I have never had one destroyed, even when put in but a few hours previous to their hatching.

Borage.

CHAPTER XI. HOW TO FORM NUCLEI.

There are nearly as many ways of forming nuclei as there are different individuals who make them; yet that does not alter the old saying that, "there is a right and a wrong way" to do almost anything. Many of the plans employed are very poor, to say the least, as I have proven by trying nearly all of the methods given. Of these poor methods I will not speak, for there are plenty of plans which are fairly good, hence I can see no reason for any one using a poor method.

The plan in most general use, is to go to any colony which can spare the bees-between the hours of 10 a.m. and 2 p.m., when the bees are flying briskly, so that most of the old bees are from home-and take a frame of brood and one of honey, together with all of the adhering bees, being careful not to get the old Queen, and put the frames into a hive where you wish the nucleus to stand. The bees are now to be confined to the hive for 48 hours, so that they will not go back to their former location, and, just at night, they are to be set at liberty, and a queen-cell given them, when in a week or so they will have a laying Queen.

The frame of brood should have plenty of young bees just hatching from the cells, and to make sure that you do not have the old Queen, she should be found, and the frame which she is on, put outside of the hive while you take the other two frames out. This plan does very well where there are but few nuclei to be made, and has the advantage of being so simple that even those most unaccustomed to bee-keeping can understand it.

A plan fully as simple, and one which I like better (inasmuch as the bees need no confining), is to make a colony queenless, thus causing it to rear Queens, as given in Chapter VI, only using the wax-cells with royal jelly in them. As soon as the cells are sealed over, give the colony all of the frames of hatching brood that they can care for, so as to make it a powerful colony; when, two days before the Queens should hatch, a cell is to be transferred to each of the combs, by pushing the base of the cell into the comb, as I have explained.

Nuclei for Virgin Queens

On the next day, each frame having a cell on it, is to be carried to a hive where you want a nucleus to be, and after placing it inside the same, a frame of honey is to be given, drawing up the division-board so as to adjust the hive to the size of the colony. Now from any hive take a frame of brood, and after shaking the bees off from it, carry it to the now combless hive whose colony reared the queen-cells, fasten a cell to it (which should have been reserved for this purpose), and leave the frame in this hive to be taken care of by the bees that were not taken with the removed combs, and those which will return from the fields. In this way, from 5 to 10 nuclei can be made from one colony, according to the time of year when the plan is practiced.

Both of these plans are to be employed only where cells are given, for the first will not receive a Virgin Queen, and the last will not stay with a strange Virgin Queen as well as they will with one of their own cells.

However, it often happens that we have more young Queens than we have nuclei to care for them, and as we wish to save these, and also get along just as fast as possible, I experimented until I found a plan whereby young Queens from one to five days old could, be introduced, and nuclei formed at the same time, so that the day after the bees were liberated, they would have a fertilized Queen.

Wire-Cloth Cage for Frame

The way of doing this, is to make a wire-cloth frame that will fit into a hive, and yet admit one of the frames from the hive inside of it, which should also have a cover fitted over the top, so that no bee can get out. Now take a frame having much brood about to hatch, and hang it in the wire-cloth cage, when one of the young Queens that you have on hand is to be allowed to run down inside the cage. Next, put on the cover, and hang the cage in any hive where the colony-is strong enough to keep up the desired temperature. If you have an upper story on any hive, this is the best place to put it, for it requires more room for the cage and frame, than it does for the frame without the cage, as there must be from one-fourth to three-eighths of an inch between the wire-cloth and the comb. If you use a lamp-nursery, and can control the temperature perfectly, this caged frame may advantageously be put in that.

In four or five days from the time the frame was put in the cage, take the cage from the hive (when you will find it pretty well filled with bees, providing that you made a good choice in the selection of the frame of hatching brood), and carry it to the hive where you wish it to remain. Now put the cage down near the entrance of the hive, take off the cover, and after having removed the frame from the cage, put it in the hive, placing a frame of honey beside it; adjust the division-board, and after closing the hive, shake the bees out of the cage near the entrance, when some of them will run in, with the call of "home is found."

In making this change, probably many bees will fly in the air; but if the cage was left at the entrance, as it should be, all will enter the hive when through their play-spell.

In one or two cases the Queen was fertilized on the day they were put in the hive, but usually not until the next day.

This plan has the advantage over the others, inasmuch as there is no danger of any bees leaving the nucleus; and also, that about as soon as formed, they have a laying Queen. After having all in readiness, nuclei can be made in this way with very little trouble, and I like the method very much.

There is also another way of making a nucleus, and introducing a Virgin Queen of any age, which I use more often at the present time than any other; and as a part of the process applies to the safe introduction of Queens received through the mails when we do not expect them, I will give the plan at length.

Making a Nucleus-Box

First, make what I have termed a "nucleus box," as follows: Take two pieces of wood, six inches long by six wide, and three-fourths of an inch thick; to them nail two pieces twelve inches long by six wide, and one-fourth of an inch thick, so as to make a box ten-and-one-half inches long by six wide, and six inches deep, inside measure, without sides.

I next take two pieces of wire-cloth, twelve inches long by six-and-one-half inches wide, one of which is permanently nailed to one side of the box, while the other is nailed to four strips which are one-fourth of an inch square, these strips going around the outer edge of the wire-cloth. Through the center of these strips is driven a five-eighths inch wire-nail, so as to fasten this wire-cloth to the opposite side of the box in such a way as to make it readily removable at any time, with an ordinary pocket-knife. In the centre of the top of the box, bore a hole of the right size to admit the small end of a funnel, to be made of tin, and large enough so that you can shake the bees off the frames into it, the same as is done in putting up bees by the pound. Over this hole fix a slide, button, or something of the kind, so that the hole can be closed quickly, when you wish to do so, when taking out the funnel, or putting in a Queen.

Having the box and funnel made, go to any hive having an upper story with a queen-excluder under it (so that there will be no danger of getting the old Queen), and take out two combs well covered with bees, if you want a good-sized nucleus; or only one comb if you desire a small nucleus. Put these combs on the ground, resting them against the hive, then jar them a little, by rapping on the comb with a stick, knife-handle, or the thumb-nail, so as to cause the bees to fill themselves with honey. As soon as the bees are filled, shake them off the combs into the funnel, and they will roll down through the hole into the box.

Take the funnel out of the box, and close the hole, when, after putting the combs in the hive and closing it, take the box to the bee-cellar, or any dark, cool place, and leave it.

If more than one nucleus is to be made, I go to other hives, and take out frames, the same as I did at first, thus keeping at work while the bees are filling themselves with honey; in this way as many nuclei can be made as we have boxes, doing the work of getting them into the boxes as soon as the bees are filled with honey.

To secure the best results, and have the small colonies so that they can be cared for at about sunset, they should be made between the hours of 9 and 11 a.m. If it is not handy to carry the boxes of bees to a room or cellar, I simply put them on the shady side of the hive, and place an empty hive or hive-cover over them, leaving them where they are.

Queenless Colonies

In less than an hour these bees will almost "cry" for a Queen of some kind; but as they will sometimes cluster (or ball) a Queen if given too soon, especially where Virgin Queens are used, I wait from 3 to 4 hours, when they will fairly "beg" for a Queen. I now get as many Virgin Queens, from 4 to 8 days old, as I have boxes of bees, and put them in with the bees. All I do in putting them in, is to set the box down suddenly, thus jarring the bees all to the bottom, so that they will be out of the way while I have the hole open, then put the Queen down through the hole.

The box is now left until nearly sunset, when the bees will be found all quiet, and clustered like a swarm. To have it so that the movable side of wire-cloth can be taken off without disturbing the cluster, I incline the box when leaving it, after putting the Queen in, so that the bees will cluster away from this side.

They are now hived in a hive, having in it a frame of honey, and a frame having some sealed or hatching brood in it. If possible, do not give them any unsealed brood, for if you do, they will sometimes kill the Queen, and rear cells from the brood given. It is not natural for a colony to have an oldish Virgin Queen at the same time that they have eggs and larvae; for in Nature, all brood would be sealed before the young Queens were three days old.

To hive them, I put the frame of brood and honey on one side of the hive, together with the division-board, when, with a quick jar, I dislodge all of the bees from the box, on the bottom of the hive, near the opposite side. I now quickly slide the combs and the division-board across the rabbets to this side of the hive, when the bees are immediately on them. In three or four days more, this young Queen is fertilized and laying, when the nucleus is ready for sending off Queens at once, with less time and labor than it takes to get a laying Queen in any other way.

Nuclei for Virgin Queens

All nuclei, in whatever manner made, should be put close to one side of the hive, having a division-board drawn up to the combs which the bees occupy; while the entrance should be on the opposite end of the hive, at the bottom, else the nuclei may suffer from being robbed by strong colonies in times of honey-dearth. Since I began to thus arrange my nuclei in hives, I have not had one robbed; while previous to this, I was much annoyed from this source.

In these four methods, I have given the reader the "cream" of all the plans now in use, without their having to go over scores of methods of making nuclei, as I was obliged to do. At all times when bees are working in upper stories, I consider the last plan the best yet known.

After each Queen has commenced to lay, she should be allowed to stay in the hive a day or two before sending her off, so as to supply the combs with eggs, thus keeping up the strength of the nucleus. If she does as well as she ought to, and if we get a laying Queen as often as we should, one-half or more of the nuclei will begin to get stronger than they need to be, to secure the best results; for it is generally known that Queens will commence to lay sooner, and be received much quicker, in rather weak nuclei, than they will in those which are very strong.

For this reason, and also not wanting to take any more bees from my full colonies than I could possibly avoid, I have been in the habit of "multiplying nuclei," as I term it, every time that I came across a strong nucleus, where there were many young bees hatching from the combs; and doing this as long as I want an increase of such small colonies.

As already stated in a former chapter, bees will adhere to combs on which is a nearly-mature queen-cell, when carried to a new location, much more closely than they will adhere to frames of brood; but nothing will hold bees to a new location equal to a Queen that has just commenced to lay.

Taking advantage of this fact, I place an extra comb in all nuclei which I think will be strong enough to spare one from it, by the time the Queen commences to lay. After she begins to lay, she is left till there are eggs in the two combs, when the comb having the most mature brood in it (together with the Queen and all adhering bees) is taken to a new hive, thus starting up a new colony. The two combs left in the old hive are now brought together, a caged queen-cell given to the remaining bees, and the hive closed. A frame of honey is given to the bees and Queen in the new hive, the division-board adjusted, and that hive is also closed. They are now left in this way for three days, when the Queen is sent off, and the bees are given a caged cell.

Where Queens are reared in large quantities for market, this plan of multiplying nuclei is of great advantage, and I have used it very largely during previous years. A little practice will enable any one to know when these small colonies are strong enough to be divided, and convince him that this is the quickest and simplest way of making nuclei now known.

When through with these nuclei in the fall, enough of them are put together, or united, to form a good colony for wintering. To do this I send off Queens enough on one day, or otherwise dispose of them, so that the number of nuclei un-queened will make a good colony. I now proceed to cause the bees in each to fill themselves with honey, when they are shaken down through the funnel into the box used in forming nuclei, and treated in the same way, by giving them a Queen, hiving them, etc.

I am glad, however, to tell the reader that all this work of forming nuclei and uniting them, will soon be a thing of the past; for in the next chapter you will find a more excellent way of getting Queens fertilized, than by the use of nuclei. The near future will see all Virgin Queens fertilized in full colonies, each colony having a laying Queen.

CHAPTER XIII. QUEENS FERTILIZED IN THE SAME HIVE WHERE THERE IS A LAYING QUEEN.

QUEENS FERTILIZED IN THE SAME HIVE WHERE THERE IS A LAYING QUEEN. (Also see APPENDIX for this subject)

To secure the fertilization of Queens, without having to form nuclei, has been a hobby of mine for years. By this I do not mean what is called "fertilization in confinement," as my faith in this has always been small, although I have spent much time in this direction; and right here I wish to say, that I cannot help thinking that in all cases of success reported, there must have been a mistake somewhere; else why could not such men as Prof. A. J. Cook, L. C. Root, and hosts of other practical bee-men, who have tried all of the plans ever given to the public, have had *just one* case of success? As near as I am able to learn, all who have reported success, were either novices in the business, or were those who conducted their experiments so loosely, that there was a chance that their Queens became fertilized by the ordinary way, at some time when the experimenter might have been found "napping."

My hobby has been that of letting the Queens fly out to meet the drones, the same as they always do, yet without despoiling colonies, by making nuclei to keep them in, from the time they were hatched till they commenced to lay. My first plan was to take Virgin Queens, from eight to ten days old, into the fields to places where I believed that drones congregated, by the loud roaring which I heard in high altitudes, between the hours of 1 and 3 o'clock p.m.

I would then let them out of the wire-cloth cages which I had carried them in, leaving each one in a separate place, near some old stump or stone, from which they could mark the location of their cage. The Queens would mark the place from which they went, the same as they would when coming from a hive, circling farther and farther, till lost from sight, some of them being gone a long time (long enough to meet a drone), when they would return and re-enter the cage, and if I was on hand, they could be easily secured again; but I have to report only failure along this line. If allowed, to do as they pleased, after returning, they would fly out again and again, till they would finally go off, never to return.

My next plan was to take a very few young bees and a little piece of comb in these cages, but with this I was no more successful. Why no Queen should ever come back under such circumstances, bearing the marks of fertilization, is more than I can understand, yet such has always been the case.

Through the suggestion of Mr. D. A. Jones, I next tried putting the Queen over a hive of bees, keeping her in a double wire-cloth cage, the wire-cloth being so far apart that the bees from the hive below could not reach her, while an entrance was made from the cage to the outside of the hive through a tube. Here the Queen would stay, with no apparent desire to go out, any more than she would have, if she were kept in a queen-nursery till she was too old to become fertilized.

I tried putting a few young bees and a little piece of brood in with them, but in a little while the Queen, bees and all, would be gone, only to appear, perhaps, where they would do lots of damage by entering a hive having queen-cells in it, or one having a valuable Queen. Then the ants were determined to reside with these isolated Queens, as the few bees with them would not keep the ants away; so that, on the whole, the failure in this was even greater than the first, for all were eventually lost.

My next plan was to take the few sections from the hives, which I sometimes found with brood in them, taking bees and all, after which glass sides were put on so as to keep the bees from getting out. They were then carried to the bee-cellar, where they were left one day, when a Queen was given them. They were then left till the Queen was old enough to become fertilized, when they were put upon a shelf on the inside of the shop, near a hole which had been made for them to go out through-a hole being cut through the section to match. In this way I succeeded in getting a number of laying Queens; but as these were only nuclei on a very small scale, and as the bees bothered me by going out with the Queen, the whole thus becoming lost, I gave it up as a thing not worthy of pursuing farther.

Two Queens in a Hive

At about this time, I saw in some of the bee-papers, that, by accident, a Queen had become fertilized, in an upper story of a hive worked for extracted honey, the same having a laying Queen below, with a queen-excluding honey-board between the upper and lower story, the Queen having gone out to meet the drone through an opening which had been left between the upper hive and the queen-excluder. I was not long in seeing where my hobby might now be brought to the desired consummation, so I began experimenting.

I first tried to see if I could get a Queen to laying in an upper story, the same as I had read about; so I put some brood "up-stairs," and the next day I gave a queen-cell nearly ready to hatch, the same as I would have done had it been a queenless colony or a nucleus.

In a day or two afterward, I examined the upper story, and found that the Queen had hatched, and apparently as much at home as if she had been in any ordinary colony.

In four days more, I bored a three-fourths inch hole in the back part of this upper hive, which was left open till the Queen commenced to lay, being about the usual time, taking Queens as they average.

I expected that the bees would use this hole for an entrance, to some extent at least; but in this I was mistaken, for scarcely a bee was ever seen at the hole, although a few came out on account of the disturbance, when the hole was first put through.

This was late in the fall, but so confident was I of continued success, that during the winter I prepared several hives, so that I could slide down a sheet of queen-excluding metal, three and one-half inches from either side, at any time that I wished during the next season. This space gave ample room for handling the two frames that it was designed to contain, manipulating them the same as I would in a nucleus hive,

Tiering-Up

When the season for tiering-up arrived, the next year, these hives were put on as upper stories, over queen-excluding honey-boards; and when the colonies became strong enough to fully occupy them, a frame having a little brood was substituted for one of the combs at each end, and the queen-excluding metal placed in the grooves made for it.

I now had in each end of these upper hives, one comb like the rest in the upper story, and one having some brood in it, to which the Queen after hatching would be confined, while the bees were at liberty to roam over the whole of both stories of the hive at pleasure.

Into each end of the hive I then placed a queen-cell nearly ready to hatch, pressing it on the comb the same as I have spoken of in a preceding chapter; then I awaited results. The Queens in the lower part of these hives were very prolific, the same being selected on purpose, as I desired to try the plan under the most unfavorable circumstances, so as to know if there was any chance of a failure. An examination, two days later, revealed that all of the Queens had hatched and were perfectly at home.

Four days later, a three-fourths inch hole was bored through the back part of the hive near each end, so as to come into the apartment where the Queens were confined; while a button made of inch stuff, was fixed to turn on a screw, so that when the hole was open, the button formed a little alighting-board, immediately underneath the hole, and when turned, it closed the hole entirely, leaving the hive as tight as it was before. The holes were left open for the next four days, when an examination showed that the Queens had commenced to lay; and they were as nice, large Queens as I ever saw, when at this stage of their existence.

Rearing Queens for Market

The buttons were now turned, and the Queens left for two days, when they had filled every available cell with eggs- probably to the amount of three-fourths of a frame, as there was considerable honey in the combs. They were now taken out and sent to customers, or used in the apiary, according as I had place for them, when more cells were placed on the combs, and the buttons turned open again six days afterward. In due time I had more laying Queens ready to use, and that without hindering the work of the hive a particle, the bees working right along, and the old Queen doing duty below, the same as if she was the only Queen in the hive. More cells were given again, and so on during the season, success attending every effort.

As hinted at in a former chapter, Queens can be fertilized from sections in the same way, by having a little brood in one of them, and enclosing those which the Queen is allowed to occupy, with perforated-zinc; but, as I said before, I do not recommend the plan to be used that way. It will work equally well where using half-depth frames, as many do when producing extracted honey; only, to be sure of success, there should be a little brood in one frame or comb.

When I found that my "hobby" was really an actual fact, I felt to rejoice, I assure you; and had it not been that I had resolved to give the matter in book form, these facts (together with how to get the queen-cells built, just when and where they were desired), would have been given to the public long ago.

Fertilizing Virgin Queens

I find that to get the best results, the holes through which the Queens pass, when going out to be fertilized, must be on the back part of the hive, or on the opposite side from the entrance; for if on the sides of the hive, or in front, now and then a Queen will go to the entrance, upon returning from her wedding-tour, and, as the bees are all of the same family, this young Queen will be allowed to go in and kill the one reigning below. While experimenting to find out where these queen-entrances should be, I had three Queens killed in the lower part of the hive during one season; the last two of which were young Queens, having been laying only a month or two.

This is a singular freak, and one which I do not know how to account for; but I do know, that so far, every Virgin Queen that has succeeded in getting from the upper story into the lower one, has superseded the Queen reigning there, whether that Queen was young or old. Why they should think more of a Virgin Queen than of a laying one, under these circumstances, is the mystery; for in all other cases, it is almost impossible to get a colony, having a laying Queen, to accept of a Virgin, as thousands of beekeepers are ready to testify.

If it is desired to have more than two Queens ferti- lized from one upper story, it can be done by making more Queen apartments with the perforated-zinc, and inserting the cells so that they will hatch at different times, when, by keeping the buttons over the holes where the Queens are too young to be fertilized, several can be allowed to go out on the back part of the hive, as they .are ready to mate. If many upper stories are used in the apiary, probably the plan as I have given it, will yield all of the Queens required, except for those doing a large business at Queen-Rearing.

These holes in the upper hive do not materially in- jure the same, for, if at any time they are wanted to be closed permanently, all we have to do is to cut some plugs of the right size, with a plug-cutter (such as is used by wagon-makers, in cutting plugs to put over the heads of screws), and put them in the holes, when one or two coats of paint will make the hive as good as ever.

Rearing Queens for Market

By using the above plan, nuclei never need be formed, except by those who want to rear early Queens for market, or by those who rear Queens by the thousand for sale; in which case more or less nuclei would doubtless have to be made; for we could not get our colonies strong enough for the upper stories, very early in the season, and unless the apiary was a very large one, there might be a limit to the upper stories, in which to have Queens fertilized.

I think that no one will deny that the plan as given in this book, of rearing Queens at pleasure, and having them fertilized in the same hive with a laying Queen, is quite a step in advance of what we were 25 years ago, in this part of our beloved pursuit. The doing of this, without in the least interfering with the working of any colony, must, it seems to me, commend itself to every apiarist.

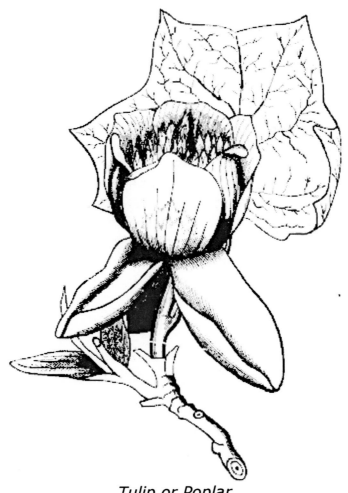

Tulip or Poplar.

CHAPTER XIV. BEE-FEEDERS AND BEE-FEEDING.

In all of the work of Queen-Rearing it is essential, if we would produce good Queens, to feed the Queen-Rearing colonies when honey is not coming in from the fields, whether the Queens are reared in upper stories or in queen-less colonies; even where a frame of prepared cells is placed in a colony which is preparing to swarm (as I frequently do in the swarming season, when conducting new experiments), the bees will do nothing with them, unless they are getting sweets from some source.

Many, undoubtedly, have bee-feeders of their own, which they think work well, and no doubt they do in general feeding, for at such times almost any feeder will answer; but for feeding during Queen-Rearing, I am satisfied that there is no feeder that will compare with the division-board feeder, after having tried nearly every feeder in use. By using this feeder, the food is near the bees, being in the same department with them, and in such way, that where even small nuclei are fed, there is no more danger of robbing, than there would be were so much stores in the comb, placed in the hive.

Division-Board Feeder

The engraving [Fig. 4] represents one of the division-board bee-feeders now in use, but I prefer one made as follows: Make a frame, just as you would make one of the frames for a hive, except that for the bottom-bar you are to use a piece of wood one inch thick, and of the same width as the side-bars. In making this feeder, it will be enough better to pay, if the joints are put together with white lead; for, by so doing, there is no danger of its ever leaking.

Having the frame made, nail on each side (nailing quite thickly with wire-nails three-fourths of an inch in length) a piece of one-eighth inch stuff, as wide as the frame is deep, lacking three-fourths of an inch. Having this nailed on, take off the top-bar (which should have been only slightly nailed on at first), and slip down in the centre of the feeder a side-bar of a frame, having previously bored some holes through it, so that the food may flow freely from the side where it is poured, into the opposite side, through this centre-piece. Now nail through the side of the feeder to this side-bar, thus fastening the thin sides to it, so that, should you ever wish to entirely fill the feeder, when doing rapid feeding, the sides would not spring out against the combs.

Next, heat some wax or paraffine quite hot (paraffine is preferred), so that it will penetrate the wood thoroughly, and pour it into the feeder till the same is nearly full, allowing it to remain for about one minute. By this time it will begin to come through the thin sides, thus showing when it should be poured out. In this way, the wood is so filled with paraffine, or wax, that the feeder will never soak up the food so as to become sour. After pouring out the paraffine, when the feeder becomes cold, nail on the top-bar, which should have a hole (of the right size to fit the funnel that you may have) bored through it near one end, to place the funnel in, when pouring in the feed.

Now hang the feeder in the hive the same as you would any frame, only let it be next to one side or the other of the hive, touching the same, so that the bees cannot go on the rear side of it. Here it can be always left, unless you want it out, for some reason. The reason for placing it close to the side of the hive is, that less heat will be wasted in this way than otherwise, and that the bees will have no loafing-place behind if, should you ever want to leave it in full colonies.

Whatever is used to cover the hive (whether honey-board enameled cloth, or a quilt), should have a hole in it to match the hole in the feeder. I prefer to use enameled-cloth, as it is always removed easily, during the many manipulations which must be performed, with any colony that is rearing Queens. Then all we have to do to feed, when such cloth is used, is to cut a slit in it over the hole in the top-bar of the feeder, through which the end of the funnel can be inserted. When the funnel is removed, the hole closes up, so as to exclude all bees, besides keeping the warmth in, which is also quite an advantage. If a honey-board is used, then a little block must be provided to place over the hole.

In feeding, I use a common watering-pot, minus the "rose" (such as is used in watering plants), to carry the food in. This will hold about 25 pounds of food, and is fixed so that I may know just how many pounds I feed each colony at one time, and so that I can feed as little, or as much, as I choose. To do this, I pour the food into the vessel on the scales, and when the scales indicate one pound in it, I stop pouring, and mark the vessel in three equidistant places, just at the top of the food; when another pound is poured in, and the vessel marked as before. In this way I keep on until the top is reached, when the food is poured out, the vessel washed and dried, after which a fine line of paint is put over each mark on the tin. At the end of the first line at the bottom, the figure one is placed; at the next, the figure two, and so on, placing the figure at the end of each mark, till I reach the top.

I let this paint become thoroughly dry, when I .have something that will be a permanent register of the number of pounds of food that I have in the vessel at any time, when the vessel stands upright.

I now put in 10, 15, 20 or more pounds of food in the vessel, according as I want to feed, and go to the bee-yard. If I desire to feed a colony one pound, I notice where the syrup stands, and, after inserting the funnel in the feeder, I pour in what I judge to be a pound, when the vessel is allowed to hang in an upright position, and at a glance I can tell how nearly right I was.

After a little practice one becomes so accustomed to this matter, that it will be a rare thing that a second pouring is needed. If I want to feed but one-half pound, I can arrive at this close enough, by dividing the space between the marks, with the eye; if more than a pound, count off the marks, pour in the food till it comes to the mark desired, and the work is done, without any fussing or guessing about the amount that has been fed.

As bees do better work at cell-building when the colony building them is getting honey liberally, I prefer to feed these colonies quite heavily at times when no honey is coming in from the fields; and when I find a nucleus that needs feeding, I do this by exchanging one of its combs for a full one, from one of these Queen-Rearing colonies. In this way I keep the food out of the way of these colonies, feed them as they should be fed while cell-building, and feed the nuclei, all at the same time.

As the feeder is always in the hive handy, and the scale which marks off the food is always with me when I go to feed, I think that it is the easiest and best way that feeding can be done.

It will also be seen that if a robber-bee tries to procure food out of this feeder, it must pass up through the cluster of bees, to the top of the hive, then down into the feeder, before it can get to the food, which practically excludes robbing from this source.

When Queen-Rearing is carried on after the honey-harvest is over in the fall, Queens are slow to become fertilized, and the later in the season it is, the more loath to go out they become. In such cases it is necessary to feed the nuclei a little, so as to stimulate the bees into activity, which will cause the Queens to fly when they otherwise would not. One-fourth pound of food, poured into the feeders of nuclei having Queens old enough to be fertilized, will bring them out every time, if poured in at about 11 a.m. On all other occasions, I advise feeding just at night, but it will do to feed in the forenoon in this case; for, let the little colonies get excited as much as they will, there are not enough bees in each hive to get on a rampage, as does a strong colony.

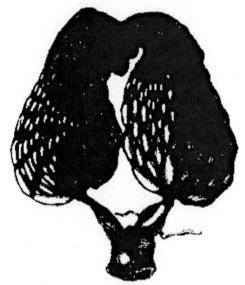

Ovaries of the Queen, greatly magnified.

CHAPTER XV. SECURING GOOD DRONES.

It is my belief that we, as apiarists of the Nineteenth Century, do not look to the high qualities of our drones as much as we ought, or as much as we do to these qualities in our Queens. To me, it seems that the matter of good drones is of greater value, if possible, than is that of good Queens; for I believe that the father has as much, or more, to do with the impress left on the offspring, than does the mother. We select our Queens with great care, but leave them to mate with drones of a promiscuous rearing from all of the colonies in our bee-yard, as well as with the "scrubs" reared by our neighbors, or from such swarms as may be in the woods near us. Now this ought not so to be; for if we would have the best of bees, our Queens must mate with the best of drones.

Fertilizing Queens in Confinement

To this end, it seemed to me that one of the most desirable things possible about Queen-Rearing, would be the fertilizing of Queens in confinement. For this reason I have tried every plan given to the public, for the accomplishment of this object, but, as I said in a previous chapter, I have so far nothing to record but failures. I would willingly give $500 for a plan by which I could mate the Queens that I rear, to selected drones as I wish, and do this with the same ease and assurance that our other work about the apiary is carried on.

As we cannot as yet, do this, I find that the next best thing that I can do, is to set apart two or three of my very best Queens for drone-rearing, causing them, as far as may be, to rear all of the drones in the apiary. I do this by giving these colonies a large amount of drone-comb, and keeping up their strength, if need be, by giving them worker-brood from other colonies.

Rearing Good Drones

The other colonies are largely kept from rearing drones, by allowing only worker-combs in their hives, and by giving them a comb of drone-brood occasionally from one of the colonies rearing drones, just when they want drones the most; for if this is not done, they will have drones anyway, even if they have to tear down worker-comb to build such as is needed to rear them in. As soon as the major part of the drones from this comb have hatched, it is taken away, before the inferior drone-brood (if any is placed in the comb) has time to mature. In this way I get all the drones reared from my best Queens, and only fail in not being able to sort out the weak and feeble ones, or, in not being able to select the most robust drones for the Queens.

To be sure, we can use the drone-traps now before the public, to keep the drones of the poorer colonies from flying; but to me, this causes more work and more disturbance with the bees, than the plan outlined.

Again, the rearing of drones, causes a great consumption of honey, and it seems foolish to be wasting honey in rearing drones, only for the sake of killing them after we have them reared.

Beside knowing how to rear mostly good drones, we want to know how to get them early in the spring. This is something not often spoken of, but it is one of the things which must be done by the Queen-Rearer who would please his customers. To do this, I place drone-comb in the center of the hives having my drone-rearing Queens, doing this in the fall, so that whenever the bees have any desire for drones, such comb will be handy for the Queen.

If these colonies are not very strong in the spring, I make them so, by giving hatching brood from other colonies, till they are running over with bees, while in addition to this, I often insert a drone-comb full of honey, right in the center of the brood-nest; for in the removing of a part of all this honey, the bees coax the Queen to lay in this drone-comb, to a degree that otherwise could not be attained. In this way I usually succeed in getting drones from one to three weeks earlier than I otherwise would.

Drones Late in the Fall

To keep drones late in the fall, I make a strong colony queenless, at the close of the honey-harvest, and in this colony I put all of the drone-brood that I can find in my drone-rearing colonies at this time. As much of this brood is in the egg and larval form, when given to the queenless colony, I have them hatching after all the other drones are killed off, for queenless colonies which are strong, are very choice of drone-brood. In this way I generally have a hive full of nice drones, as late as I desire to rear Queens, keeping them frequently into October.

As soon as I am through with such drones, I introduce a Queen to the colony, when the bees will destroy them at once, if feeding is withheld. I always feed a colony keeping drones when honey is not coming in, for they need much food to make them fly freely, and that is what we want them to do, on every warm day at that season of the year.

One other item that I wish to notice at some length, before closing this chapter on drones, is this: From the fact that worker bees can lay eggs that will hatch drones, and that Virgin Queens can also lay eggs which will also produce drones, the theory has obtained very largely among bee-keepers, that the drones from a fertile Queen must of necessity be of the same blood, as they would have been had this Queen produced drones before she was fertilized. In nearly every book written on bees, that I have read, where this subject is touched upon, we find words to the effect that, "a pure Queen, however mated, must produce a pure drone of her own variety." Mr. Alley's "Queen-Rearing" is an exception to this, I am happy to note.

Contamination

Now I am not prepared to say how, nor wherein, the drones are changed by the mating of the Queen; but this I do know, that drones are contaminated, to a certain extent, by the mating of a Queen of one blood, with a drone of another blood. Any one can prove this, for in four generations, by mating the Queen each time to these pure (?) drones, a bee can be produced which no man can tell from a hybrid. That this contamination does not show in the first cross, is the reason, I believe, that the theory has been accepted, by nearly all, as the truth.

To illustrate: Take a pure black Queen, and after she has mated with a fine, yellow Italian drone, let her rear all of the drones produced in an apiary containing only black bees. Of course, the drones from this Queen will all be black to look at, the same as they would have been had she mated with a drone of the same blood as herself. Now rear Queens in this apiary, from any of the pure black mothers in it, and these young Queens will mate with the drones from this mismated Queen. These young Queens will apparently produce all black workers and drones, the same as they would have done had these drones come from a pure black mother, mated with a pure black drone; but when we rear Queens from these young mothers, now and then one will show a little yellow, which would not have been seen, had not the drones from this mismated Queen been the least bit contaminated, To detect any slight contamination of blood in our bees, we must always look to the Queen progeny, for the Queen is the typical bee of the hive; hence they will show an impurity where the workers and drones would not,

Now, take one of these young Virgin Queens showing a little yellow, and have her mated with a pure yellow Italian drone-the same as was done with the first Queen. From this one rear all of your drones again, while you rear Queens from her mother, which young Queens would be sisters to the one now producing drones. Having one of these last young Queens fertilized by the desired drones, next rear Queens from her, and you will find that some of these Queens will show quite a little yellow on them; yet so far the drones and workers show little if any difference.

Take one of the yellowiest Queens from this last lot, and have her mated with a yellow drone again, going over the same process of mating as before, and you will get Queens in this third generation, which will (many of them) be quite yellow; while the workers and drones will show "yellow blood" about them, by occasional "splotches" of that color. Now follow out the same line of breeding once more, and you will get both workers and drones, which any Queen-breeder in the land will call hybrids-calling them rightly so, too. These hybrids could not possibly come about by this way of breeding, only as drones from a mismated Queen are contaminated; for so far we have used no drones except those which were pure black, according to the parthenogenesis theory, yet we have a hybrid bee as the result. Worker bees and drones do not show a little variation of purity, as much as does the Queen, hence if we would know of the stock which we have, we must rear Queens from them. Failing to do this, we often decide that we have pure drones for breeding purposes, because these same drones look all right.

If I have made this matter plain, and I think that I have, it will be seen how much value it would be to the scientific breeder of Queens, if he could select just the drone he wanted, and then have a valuable Virgin Queen mated with that drone. In this way we could accomplish as much in securing the "coming bee," in two years, as we now accomplish in a life-time.

Let no one be longer deceived about pure drones from a mismated Queen; for if such drones are allowed to fly in your yard, you cannot expect any satisfactory degree of purity from Queens reared therein. I have been forced to this conclusion by many carefully-conducted experiments as already described.

Teasel

CHAPTER XVI. THE INTRODUCTION OF QUEENS.

Perhaps there is no one subject connected with bee-keeping that has received so much notice in our bee-papers and elsewhere, as has the introduction of Queens; yet all who have read the methods and discussions given, must have plainly seen that success does not always attend the efforts in this direction. On the contrary, many losses have been reported, and these losses are not confined to the inexperienced altogether, for we often hear of our most practical apiarists occasionally losing a Queen.

The reason for so many losses, it seems to me, arises from the fact that bee-keepers in general do not understand that a discrimination should be made between Queens taken from one hive and placed in another, and those which have come long distances by mail or express. In introducing Queens, it should always be borne in mind that a Queen taken from one hive in the apiary, and introducing into another, does not require one-half the care that must be given to a Queen coming from a distance. The reason for this seems to be, that a Queen taken from a hive in the same yard, is still heavy with eggs, and will not run around, provoking the bees to chase her, as will a Queen after having had a long journey.

In introducing all ordinary Queens coming from my own apiary, I generally adopt one of the two following plans: The first is, to go to a nucleus or other hive from which I wish to get a Queen, and when she is found, I take the frame of brood she is on, bees and all, together with another frame from the same hive, carrying them to the hive from which I am to take the superannuated

Queen, when they are left with the Queen between the two combs, while I secure the poor Queen and dispose of her; then I take out two frames from this hive, and place the two frames, brought from the nucleus, in their places, and close the hive. I now shake off the bees from the two frames in front of their own hive, carrying the combs to the nucleus; or if the nucleus will be too weak, I carry bees and all to it.

The object in taking the two frames with the Queen, is so that while waiting outside of the hive, she and the most of the bees may cluster between them, thus becoming quiet. When placed in the hive, both are put in together, thus leaving the Queen quiet among her own bees. In this way I do not lose one Queen out of fifty, and as the operation is so simple, and the Queen so quickly installed, the advantages more than over-balance so small a loss.

The second plan, is to go to any nucleus and get the young Queen in a round wire-cloth cage (such as all beekeepers have in their apiaries) before looking for the Queen to be superseded. After she is in the cage, I place her in my pocket, and close the hive that I took her from, and look for the Queen that I wish to remove; having found her, she is killed or otherwise taken care of, and this hive is also closed. I next blow in at the entrance enough smoke to alarm the whole colony, pounding with my fist on top of the hive until I hear a loud roaring inside, which shows that the bees are filling themselves with honey. I now let the Queen that I have in the cage, run in at the entrance, smoking her as she goes in, while I still keep pounding on the hive. In doing this, nothing but wood-smoke should be used, for if tobacco-smoke were used, many of the bees would be suffocated.

If this is done when there is danger of robbing, I wait till just at night, about the operation. If more convenient, the Queen can be taken out of the hive at any time during the day, and the operation of putting in the new Queen done just at night. Some seem to think that the operation will be more successful if done in this way, but so far I fail to see any difference as to results. The idea is to cause the bees to fill themselves with honey, at the same time smoking them so that the Queen and bees smell alike. This plan is as free from loss as the other, still it is not quite so simple as the first-I only adopt it where it is not handy to use the former.

Queenless Colonies

Where any colony has been queenless from three to five days, a Queen can generally be successfully introduced by dropping her in honey, and rolling her over in the same, till she is thoroughly daubed with it, when the cover to the hive is lifted, and the Queen dropped from a spoon right down among the bees. This is equally successful with the others, but I do not like the plan, on account of having to keep the colony queenless so long. Even a Queen coming from a distance, can generally be safely introduced by this plan.

Introduction of Valuable Queen, Push in Cage

To introduce a Queen that has come to me from abroad, or one which I consider of more than ordinary value from my own apiary, I proceed as follows: First, I take the cage containing the Queen and her escort of bees, to the little room where I handle queen-cells, and open the cage before the window, so that if the Queen takes wing, she will not be lost. I then catch the Queen and clip her wings (as given in the chapter on that subject), when she is placed in a round, wire-cloth cage; but I allow none of her escort to go with her, as I consider such bees when left with a Queen one of the prime causes of the many losses which occur to the purchaser of Queens.

Having the Queen's wings clipped, and in the cage, I next take a piece of wire-cloth, containing 14 or 16 meshes to the inch, and cut it four-and-one-half by eight-and-one-half inches in size. Now cut a piece three-fourths of any inch square out of each corner, and bend the four sides at right angles, so as to make a box, as it were, three inches wide by seven inches long, and three-fourths of an inch deep. Next, unravel the edges down one-half way, so that the points can be pressed into the combs, and if the corners do not stay together as they should, they can be sowed together with one of the wires which were unraveled [Fig. 5].

Having the cage ready, and the Queen to be introduced, in your pocket, proceed to look for the Queen to be replaced, and after removing her, examine the combs until you find one from which the bees are just hatching, or where you can see them gnawing at the cappings of the cells, which comb should also have some honey along the top-bar of the frame above the hatching brood.

Now shake and brush every bee off this comb, and place the Queen that you have in your pocket on it, by putting the open end of the cage near the comb over some cells of unsealed honey, when she will go to the comb, and as soon as she comes to the honey, she will begin eating. While she is doing this, put the large cage over her and the hatching brood, as you wish, taking all of the time that is needed, for as long as she continues eating, she will not go away, nor be disturbed by any of your motions.

Having honey in the cage is necessary, for the bees outside of the cage cannot be depended upon to feed a Queen when she is being introduced. Some claim that if the cage is made of wire-cloth having large meshes, the bees will feed them; but after losing many Queens by depending upon the bees to care for them, I say always provision your introducing-cage in some way, so that the Queen is not dependent upon the bees for her food while in the cage.

Queen-Nursery

Even when keeping Queens in the queen-nursery, where placed in queenless colonies, I find that the bees often refuse to feed them; so I now provision all cages of all kinds, notwithstanding the claim put forth by some of our best bee-keepers, that several caged fertile Queens will be fed by a colony having a laying Queen, if they are put between the combs in a hive having such, for safe keeping. Finding a whole queen-nursery full of dead Queens, after trusting them to the care of a colony of bees having a laying Queen, is generally more convincing than many words given in support of an untruthful theory.

But to return: Fit the cage so that it comes over three or four square inches of honey, and as much of the hatching brood as possible; for these hatching bees have much to do with the speedy introduction of the Queen. Having all fixed, leaved the hive from 12 to 48 hours, according as your other work will allow you, when the hive is to be opened, and the cage examined.

If all has worked as it usually does, the bees will be found spread out evenly over the cage, the same as they are on any of the rest of the combs. When such is the case, the cage is to be carefully lifted from over the Queen, letting her and the young bees that have hatched during her confinement, go where they please, keeping watch all the while to see that the bees treat her kindly; if they do this (as they will, nineteen times out of twenty), the comb is to be placed in the hive; if not, she is caged again. In from one-half to one hour after liberating her, look at the Queen again, and if she is now treated as their old Queen was before her removal, the hive is closed, and she is considered safely introduced.

If, on the contrary, the bees are found clustered thickly on the cage, biting the wire-cloth and showing signs of anger, the frame is to be placed back in the hive and left till the next day, when, if they still show the same symptoms, you must wait until they are scattered over the cage, as spoken of at first, before letting her out.

Releasing Queens after Shipment

I often release a Queen in 12 hours, and find that she is all right; and I rarely have to wait to let her out, more than 24 hours. Still, in extreme cases, I have been obliged to keep them caged nearly or quite ten days.

Balling a Queen

Do not be afraid of the Queen dying in the cage; for if she is placed over honey, as I have advised, she will live a month, and there is no need of losing any Queen if there is not too much haste used, in letting her out. Even then, there should be no danger, if the apiarist is on hand to release the Queen from the bees which cluster (or ball) her, as they always do a Queen for sometime before they kill her. Such clustered Queens can easily be released, by smoking the bees till they free her.

In liberating a Queen from a "ball" of infuriated bees, she is liable to take wing and fly away, thus losing her in that way. To guard against this, I either clip her wings before trying to introduce her, or take the "ball" of bees into a room while smoking them apart. Again, there is some danger that after the Queen is free, a bee from the cluster will sting her, if this bee gets to the Queen singly; and for this reason, I always secure the Queen in a wire-cloth cage as soon as the last bee has let go of her.

If the bees of any hive have once clustered a Queen, I find that it is very hard work to get them to accept the same one afterward; for this reason, I generally take a Queen that has been clustered, to some other hive and introduce her there, giving the infuriated colony another Queen or a queen-cell.

However, not one Queen in one hundred is treated in this way when using the above plan, for, as a rule, I find that the young bees that have hatched in the cage with her, have accepted this Queen as their mother; thus the news is conveyed from them to the rest of the bees in the hive, so that she is fed by "all hands," which causes her to keep the cells enclosed by the cage, from which the young bees have hatched, well supplied with eggs.

After the Queen has been laying eggs for one or two days, she is as safe as if she had been reared in the hive; and for this reason, I do not liberate the Queen till I see eggs in the cells enclosed by the cage, unless it is in the fall of the year, after Queens have ceased laying.

At this time of the year (fall) I am in no hurry to liberate a Queen, for she is of no especial use out among the bees when she is not laying eggs, hence I generally leave them in the cage for a week or two, until I know that the bees will accept of the Queen after I take the cage from over her, without further trouble. Now there are no bees hatching from the combs, so in caging the Queen I only see that she is in the center of the cluster, and has plenty of honey to eat inside of the cage; for when a Queen is not laying, she has to help herself to honey, the same as any other bee.

In using these cages, the comb next to them should be a left a bee-space from the cage, so that the bees can go all around it, thus getting acquainted with the new Queen much more quickly than they otherwise would. If this space cannot be procured in any other way, one frame should be left out of the hive for the time being.

The advantage that this plan has over any other where the Queen is to be caged in the hive, is in the young bees hatching out in the cage with the Queen; and as they have known no other mother, they accept her at once, thus forming an escort which the older bees, sooner or later, are obliged to accept, as being a part and parcel of the colony.

By any of the above plans, there is very little danger of losing a Queen, yet none of them are absolutely safe; nor would I use them were I to receive a very valuable Queen, say one worth $10, for with such Queens we do not wish to take a particle of risk.

Caging the Bees

After studying on the method of forming nuclei by the "caged bees" plan, as I gave in the chapter on that subject, I saw that by using that process, I had an absolutely safe plan of introducing a laying Queen, even were she worth $100. I have used this plan with all the valuable Queens for several years, and have not lost a single Queen, nor do I believe that I ever shall lose one by it, unless she should happen to fly away in putting her in the box with the bees; nor will she do this, as long as I clip all of my Queens' wings.

My usual method of using this plan, is to get bees enough from the upper stories of different hives to form a good, strong colony, doing it just the same as I gave in the chapter on forming nuclei, only I take the bees out of four or five different hives, and off from 10 to 15 combs, according to the strength that I want the colony. After having the bees in the box, they are treated just the same as there described, giving them the valuable Queen, in the same way that the Virgin Queen was given.

In hiving them, give as many empty combs, or combs of honey, as you choose, but do not give any more brood at this time than you did to the nucleus; for if more brood is given, the bees sometimes will swarm out with the Queen in a few days, where made so strong, the same as a natural swarm. If you desire to give brood, do it by giving a frame or two at one time every few days, after waiting four or five days from the time of hiving, before giving the first frames.

If you do not have bees in upper stories having a queen-excluder under them, then go to two or three colonies in ordinary hives, look for the Queens, and as fast as they are found, put the frames that they are on, outside of the hives. Now smoke and jar the bees on two or three frames from each hive, till they fill themselves with honey, when you are to shake as many bees down through the funnel into the box, as you want in your colony, and proceed as before.

If you desire to introduce the Queen to a certain colony, (the same as we have been doing by the other plans given), kill or take away the old Queen, and cause the bees to fill themselves with honey, the same as in the last instance; when all the bees that you can get, are to be shaken off the combs through the funnel into the box.

Having all of the bees in the box that you can possibly obtain, treat them the same as before, until you are ready to hive them. After they are placed in the cellar or other cool place, take all of the combs having brood in them, and give them to the other colonies, leaving one or two frames of honey in the hive, to hold till night, the bees which you did not succeed in getting into the box, and those returning from the fields. These combs should be put in the centre of the hive, so that when night comes the bees will be mostly clustered on them, instead of about the side of the hive, as would be the case if they were left next to one side of the hive. When you hive the bees having the new Queen with them, take these two combs with the bees out of the hive, putting in other combs as before, using only one having a little brood in it, and that taken from another hive, so that they are not given their own brood.

Having all prepared, proceed to hive the bees as was done with the nucleus; or, if preferred, the bees can be shaken down at the entrance, for, as this is their old home, they can go nowhere else, even should they try to do so. After the larger part of the bees are in the hive, shake the bees off from the two combs, and let them run in with the others. In five days, commence to give the brood back again, and keep on doing so occasionally until all is back in the hive, as it was before.

The above I believe to be an infallible plan for introducing Queens, and well pays for the time and trouble, when we have a very valuable Queen coming from a distance, which we would not lose on any account; yet it will hardly pay to spend so much time on ordinary Queens, except by way of experiment, or when desiring to make new colonies in addition to introducing Queens. Where a Queen comes to me very unexpectedly, I always use this plan, taking the bees from an upper story or two, thus forming a small colony with the Queen, which colony is built up later on, by giving frames of hatching brood. Using it in this way, it always gives me the assurance of success, in any case of emergency. .

CHAPTER XVII. INTRODUCING VIRGIN QUEENS.

That just-hatched Virgin Queens, which are so young as to be white, weak and fuzzy, can be introduced to any colony that will accept a sealed queen-cell, is a fact generally known to all; and if there was no need of ever introducing Virgin Queens older than these, this chapter would never have been written.

However, in these days of progress, and of close competition in the Queen traffic, it is very desirable to have some plan whereby we can introduce a Virgin Queen from 5 to 8 days old, to a nucleus, as soon as a laying Queen is taken away from it; as well as to introduce one into any other colony where we wish to place a Virgin Queen coming to us from a distance, which we have ordered to improve our stock, by a direct cross between her and one of our drones. From the fact that not one colony in 500 will take such a Virgin Queen, when giving her at the time of taking away the laying one, comes the reason that such a plan of safe introduction will be of greater value to us, than it would be could we succeed in introducing these Queens as well as we can a laying Queen.

On no one thing in bee-keeping have I spent so much thought, as on how to successfully introduce Virgin Queens, from 4 to 10 days old; and I am happy to say that I am master of the situation; not that I have dug it out all alone, for I have not. I have picked up little things here and there for several years, and by saving every little item that proved to be in advance of what I already had, and applying them, together with what I could study out myself, eventually gave me success.

As I said in the preface, I cannot give credit to all from whom I have gained knowledge, for I have not tried to keep the authors of all these things in mind; besides, there has been scarcely a writer in the past, who has written for our bee-papers, from whom I have not gained some light; so if I were to single out some, I would do injustice to others. I claim very little as original with myself, and I am glad to know, that it is the "littles" of the past, coming from the thousands who have engaged in our pursuit, that have made the "mickle" of the present; hence very few are able to say, "I am more holy than thou."

My first ideas in this matter, came from the need of procuring a laying Queen from a nucleus, more often than could possibly be done, by the old plan of giving the nucleus a queen-cell 24 hours after taking a laying Queen away from it, in order to overcome the low prices to which Queens had fallen, owing to the close competition in this branch of our industry. If a five-days-old Virgin Queen could be introduced into a nucleus so that she would commence to lay in five days from the time the other was removed, two Queens could be taken from one nucleus during the same time we had formerly taken one. All know that by the old plans, a laying Queen cannot be taken from a nucleus oftener than once in 10 or 12 days.

Virgin Queen Traffic

This one item alone I considered worth striving for; but when it came to be fully understood that it was an object for us as apiarists, to change the blood in our stock by a direct cross, as often as possible, so as to give greater life and vigor to our bees, then such introduction of oldish Virgin Queens became almost a necessity. Since this idea was first advanced, it has gained ground rapidly in the minds of our best bee-keepers; and I believe that the day is not far distant, when the traffic in Virgin Queens will assume greater proportions than at the present. A Virgin Queen is not fit to start on a journey until she is at least 24 hours old; and as from 2 to 4 days must be required in her transit, none of the plans of introducing young Virgin Queens would work in this ease.

Without taking the reader over much of the ground which lead to the discovery of a plan for the safe introduction of Virgin Queens, I will give the three plans which I employ- using them according to the circumstances which I am placed under, as to the number of Virgin Queens on hand, length of time the nucleus or colony has been queenless, etc.

Some 10 or 12 years ago, I had a colony rearing Queens, that had a nice lot of queen-cells just sealed, when one day a Virgin Queen escaped from me and flew out of sight. I waited for her to come back, but as she did not, I concluded that she was lost. Upon going to get my queen-cells when it was time for them to hatch, I found the cells all torn down, and the Queen that I had lost was in the hive just commencing to lay.

Here I was shown that a colony that had been queenless long enough to have their queen-cells capped, would accept a Virgin Queen under almost any condition. In fact, I had read of this before, but nothing convinces us as does sortie-thing which comes close at home, to ourselves or our family. Through this loss of cells, which occurred just when I needed them very much, came something of great value to me, which I might not have *fully* known had I not lost them.

From this, I found that whenever I came across a nucleus or colony having queen-cells sealed, all that I had to do to introduce a Queen was, to go to my queen-nursery and pick out a nice Virgin Queen, and drop her in some honey; when, after pouring some of the honey out of a tea-spoon on her back, and rolling her about in it until she was thoroughly daubed, the quilt was raised from over the frames, and after scooping her up together with some of the honey, I turned the whole down among the bees between the combs. The hive was then closed, and I would usually have a laying Queen in three or four days. To prevent the Queen from flying, when introducing her in this way, I held the mouth of the cage close down to the honey (which I generally take in a tea-cup), when, by a sudden jar, caused by striking the cage, she was thrown down into the honey, thus daubing her wings, after which there was no further danger.

This plan I also use when receiving a Virgin Queen from abroad, if I have a colony that has been queenless long enough to have cells sealed. Of course we do not expect many colonies in this condition, but all Queen-Rearers, as well as apiarists in general, have more or less of them coming from an unexpected loss of Queens.

The second plan is one that I use with younger Virgin Queens-say those from one to three days old-and in all cases where it is not convenient to use either the first or third. It is as follows:

Introducing-Cages

Make a round wire-cloth cage, about an inch in diameter and three-and-one-half inches long. Into one end of this fit a permanent stopper, and for the other saw off a piece of an old, soft-wood broom-handle, five inches long. Whittle one end so that it will go into the cage one-half an inch, when a five-eighths inch hole is to be bored through it lengthwise. Now fill this hole with "Good" candy, made of granulated sugar and honey, packing it in with a plunger quite tightly. Next, put the Virgin Queen into this cage, and put in the provisioned stopper.

When you go to remove the laying Queen, take the cage along with you, and after having removed her and replaced the frames in the hive, lay the cage lengthwise between the top-bars of the two frames having the most brood in them. Put a quilt over all, and close the hive.

As it takes the bees about four days to burrow through, or dig out, the five inches of candy, they become pretty well acquainted with their loss, and the existing state of affairs; hence they are ready to accept the Queen when she is set at liberty, by the removal of the candy. In about eight days time (counting from when the cage was laid on the frames), I generally find this Queen laying, without having to open the hive, except as I do it to take out laying Queens.

Right here I wish to say, that the cage here de-scribed is just such an one as I use about the apiary for all general purposes, except that when so using it, I put in a piece of corn-cob for a stopper, instead of the one filled with candy.

The third plan, and the only one that I know of that is absolutely safe at all times (for I sometimes have a loss with either of the other two) is as follows:

Get out a little block, two inches long by one inch square, through which is to be bored a five-eighths inch hole, boring the same through the block lengthwise. This is to be the base of the cage. Next bore a one-half inch hole through the center, so as to cross the five-eighths inch hole. Now get two pieces of frame-stuff, four inches long by one inch wide, and one-fourth of an inch thick, boring a five-eighths inch hole in each, near one end, to correspond with the five-eighths inch hole which was bored lengthwise through the little block. Having these ready, nail one to each end of the little block, so that the holes bored in them will match the hole in the block, thus making one continuous hole straight through.

Next get a piece of wire-cloth, eight inches long by two-and-one-half inches wide, and nail it to the frame-stuff and lower edge of the block, so as to form a cage three inches deep by two inches wide and one inch thick, through which the bees can become acquainted with the Queen. Now drive two three-fourths inch wire-nails into the edges of the frame-stuff, driving one into each piece and letting it project one-fourth of an inch, or, in other words, do not drive them up to within one-fourth of an inch. With a pair of cutting pliers, cut off the heads of each nail, and file them to a sharp point, so that you can fasten the cage on the side of the hive, or to whatever you like, by simply pressing these points into the wood.

The cage-is now ready for use. To use it, first put the Virgin Queen into the cage by letting her run through the half-inch hole down into it, when a long stopper is put into the hole to keep her from returning. Now proceed to fill the five-eighths inch hole with the "Good" candy, as used in the shipping-cages (this made of powdered sugar instead of granulated, as will be explained farther on), putting it in at both ends, and pressing it around the long stopper down in the center. When this is done, remove the stopper, and fill the place where it came from with more of the candy, when the cage and Queen is ready for the hive.

Next, take a frame having only a starter of foundation in it and the caged Queen, and proceed to the hive where you wish to take away a laying Queen; after having caught her, take all of the frames out of the hive, and stick the cage on that side of the hive where you want it, by pressing it against the wood. Now put in the frame with the foundation-starter and adjust the division board, closing the hive. Next, shake all of the bees off the combs near the entrance, letting them run in, and give these combs to another colony to care for. They are now to be left for four or five days, when you will find a laying Queen (providing that the Queen was four or five days-old when put into the cage), and also the frame partly-filled with worker-comb, in which the Queen has laid.

At times when no honey is coming in, the bees are to be fed what they need every night, so as to place them in the same condition that they would be, were honey coming in from the fields. To thus feed, I place a division-board feeder in the hive, fastening the caged Queen to that instead of to the hive; or, if preferred, the cage can be fastened to the division-board.

The candy is placed in the block, and the different holes made in it, so that the bees may be good-natured when coming to the Queen; and also to keep the bees from liberating the Queen till they have given up all hopes of getting their laying Queen or brood back again; for they would kill her at once if she was liberated sooner.

It generally takes the bees from 8 to 12 hours to eat out the candy, being about the time needed to get them reconciled to their new situation. If preferred, the cage described in the second plan can be used instead of this one, by cutting off the provisioned stopper to one inch in length; but the last gives a little better satisfaction, inasmuch as the bees have a larger surface of wire-cloth to cluster upon, and they can be eating candy at several places at once, hence they do not rush into the cage so fast when an opening is made into it.

If the Queen was not given to the bees when the combs were taken away, many, if not all of them, would go to other hives; for although the bees do not like her at first, yet she holds them where they belong, as they consider her better than nothing. Do not give them back their brood till the Queen begins laying, for if you do, they will at once kill the Queen, or "hug" her till she is nearly or quite spoiled; for in nothing are bees so determined as they are not to accept of a Virgin Queen, five or more days old, immediately after having their mother taken from them.

When the Queen commences to lay, take out the partly-filled frame, and give back the combs that you took away from the bees at first, allowing this Queen to stay in the hive a few days before you try the operation with the same colony again; for if you keep right on giving Virgin Queens and taking laying ones out, the colony will soon decline, on account of no young bees hatching to take the place of the old ones, which are dying all the while. Young bees could be shaken in the hive every little while, if it was preferred to giving back the combs, and in this way a laying Queen could be taken from a hive every five days. Of course, where this plan is used with Virgin Queens coming from abroad, the brood will be put back to remain, as we will desire to keep the Queen.

If the colony is other than a nucleus, we shall want to give two or three frames with starters of foundation, so as to give the colony the room they need. These partly-filled frames are used to advantage during the swarming season for new swarms, so that a colony treated in this way, is doing valuable work all this time, besides getting our Virgin Queens fertilized.

I have never lost a Queen in this way, no matter if she was 12 days old when placed in the cage; and I consider it an absolutely safe plan for introducing Virgin Queens, and one of great benefit to those who desire to improve their stock, by a direct crossing of Queens and drones; but for the purpose of getting Queens fertilized in nuclei, oftener than by the old plans, I doubt if it pays, on account of the large amount of work which it requires.

Before leaving the subject of introducing Queens I wish to say, that where any plan of introduction is used, by which the bees are liable to start queen-cells from their own brood, before the introduced Queen is liberated, I think that the idea which prevails-that the bee-keeper should look over the combs and destroy all of these cells-is fallacious. All of my experience in this matter, proves that a Queen will be as quickly accepted, when such cells are allowed to remain, and when so accepted, the bees themselves will destroy the cells. Where I find queen-cells sealed in any colony, I always roll a Queen in honey and drop her into the hive, letting the bees attend to the queen-cells when they get ready; and it is a rare thing that I lose a Queen by this process, even when such Queen is a Virgin.

Sour Wood

CHAPTER XVIII. KEEPING A RECORD OF CELLS, QUEENS, ETC.

When I kept but few bees, or reared but few extra Queens, I had no trouble in keeping track of what each hive contained; and even where I am working 100 small colonies for Queen-Rearing, I know what is in each hive in a kind of general way, but not enough so that, should I trust to memory, many blunders would be made. For this reason, when I began Queen-Rearing as a business, I found that I must have some way of knowing precisely what was in each hive, so I adopted something to help me in this matter.

The first thing that I used was small, flat stones, four of which were placed under the bottom-board of each hive; so that when an operation was performed with any hive, these stones could be made to tell me very nearly what I wanted to know, from just glancing over the tops of the hives, as some of them were placed in different positions, each time that I worked at the hive, to denote what had been done with it. These stones, together with a piece of section to keep the dates on (the piece of section being placed under the cover of the hive, to keep it from getting wet), does very well, where not over 10 to 20 nuclei are worked.

To use the stones intelligently, we must jot down somewhere what the different positions which the stones occupy indicate, until we get so accustomed to it that our memory is always posted in this matter. For instance: If I look over a hive on June 1st, and give the bees a queen-cell, I place one of the stones on the right-hand front corner of the hive, and put that date on the section. A glance over the yard shows me that all the hives having a stone on that corner, had queen-cells placed in them the last thing which I did; and the strip of section will tell me when that was. In the same way when I take out a Queen, I put this stone on the left-hand front corner, which indicates that the Queen from that hive is missing; and when I find a queen-cell hatched, or a young Queen in the hive, this stone is placed on the left-hand back corner, while for a laying Queen, it is placed on the right-hand back corner, the date being put on the piece of section each time, so that the last date shows when the stone was changed, the two together thus telling me all that I wish to know. The main trouble with this plan is, that it requires the lifting of the cover to find the date; but as I said at starting, this will answer pretty well where but few colonies are worked for Queens.

More of the stones are used to indicate other things. For instance: A stone in the centre of the cover shows that the colony is short of stores, and must be fed; while a stone in the centre of the back part of the cover shows that the bees are crowded for room, and that another frame should be given. A stone in the centre of the front indicates that there are too few bees to do good work; and so on, for these stones can be made to tell a great variety of matters.

Again, I use them on all hives worked for honey, having them tell me when the sections were put on, when more room was given, and when taken off; also when the honey was extracted from certain hives, which hives are worked for extracted honey, and which for comb honey, etc. In fact they are really indispensable to me in working an apiary, either for Queens or honey, and are in constant use, even when using the cards which are about to be described.

When I first commenced bee-keeping, I had no idea of rearing Queens for sale, nor did I think of it until I was crowded into it; so when my first order for a Queen came, I took the same from a full colony. This Queen seemed to give satisfaction, and soon the neighbors of this customer sent for Queens, and so on, till I found that I must have a few nuclei to supply this demand for Queens, which had apparently sprung up of itself. The Queens that were reared in partnership, as spoken of in the first chapters, were taken to different apiaries, and introduced by myself to colonies, at the suggestion of the one who did the Rearing, rather than being sent off to customers through the mails.

Queen-Registering Cards

When the business grew, so that I could no longer keep track of it with the stones and pieces of sections to advantage, I secured a sample of Root's Queen-Registering Cards. These suited me exactly, and they were sold very cheaply. I procured a quantity of them, and have used them ever since. To show the reader what they are, I give a sample card on the next page:

It will be seen at a glance, that all we have to do after each manipulation with the different hives, is to turn the pins to where they tell us just how and when we left the hive when last looked at, which, together with the stones to tell us about honey, etc., tell all that we want to know.

I have watched carefully to see if anything better was brought before the public, but so far nothing of the kind has come to my knowledge. These cards are used on the upper stories, the same as on nuclei, using one on each end where two Queens are to be fertilized from one hive, and on all hives where a change of Queen is made often. No Queen-Rearing apiary can be complete without something of the kind.

QUEEN REGISTER.

EGGS.

No. ......

MISSING.   **BROOD.**

LAYING.   o   **CELL.**

APPROVED.   **HATCHED.**

NOT APPROVED.

DIRECTIONS.—Tack the Card on a conspicuous part of the Hive or Nucleus box ; then, with a pair of Pliers, force a common pin into the centre O of each circle, after it is bent in such a manner that the head will press securely on any figure or word.

MARCH.

OCT.   **APRIL.**

SEPT.   **O**   **MAY.**

AUG.   **JUNE.**

JULY.

Queen-Register Card

DIRECTIONS.--Tack the Card on a conspicuous part of the Hive or Nucleus box; then, with a pair of Pliers, force a common pin into the centre O of each circle, after it is bent in such a manner that the head will press securely on any figure or word.

CHAPTER XIX. CLIPPING THE QUEEN'S WINGS.

Probably there is no other item about bee-keeping, on which there is so much diversity of opinion, as there is regarding the clipping of Queens' wings. Many of our very best apiarists stand directly opposed to others, who are equally as good authority. Some claim that the Queen is injured by having her wings clipped, and for this reason many are superseded by the bees; while others are equally confident that it is impossible to injure the Queen in the least by clipping her wings, if the clipping is delayed, as it always should be, until after the Queen has commenced to lay. However, when I look the ground all over, I believe that the greatest number of the "dollar-and-cent apiarists" of our land, are on the side of clipping the Queen's wings; and as I stand on that side myself, I trust that I shall be excused, if I tell the reader in brief, some of my reasons for clipping the wings of my Queens.

The second year of my bee-keeping life, I lost a splendid swarm of bees, being the second swarm that issued from my then small apiary, for I only had one swarm the first year. I felt this loss very keenly, and then and there I resolved that this would be the last one that would ever "runaway." In accordance with this resolve, I clipped all of the Queens' wings in the yard, and have kept them so ever since, except those that I thought, of late years, I might sell; and although I now think that resolve a rash one, yet in all of my twenty years of bee-keeping, that one swarm has been the only one lost from this cause.

A person can hardly pick up a paper that treats on bees, but that he will find an account of swarms going into the woods; and there is no question but what hundreds of dollar's worth of property "took wings and flew away" in just this manner; while if the Queens' wings had been clipped, this loss might have been saved.

By having the wings of all Queens clipped, the bees are perfectly under the control of the apiarist, and he can handle them as he pleases, separating them with pleasure where two or more swarms cluster together, and hive them on the "returning plan" when they come singly. In using this plan, all that we have to do when a swarm issues, is to step to the entrance of the hive with a round wire-cloth cage (such as has been described), into which the Queen is allowed to run, when the cage is closed up and laid in front of the hive. The old hive is now moved to a new stand, and a hive prepared for a new colony put in its place. In a few minutes the bees miss their Queen and come back, running into the hive with fanning wings, when the Queen is liberated and goes in with them.

I have followed this plan of hiving bees for years, and I know it to be a good one, as a good yield of honey is generally the result. There is no climbing of trees, cutting off limbs, or lugging a cumbersome basket or swarming-box about. It is so straight-forward, too-remove the old hive to a new stand, put the new hive in its place, and the returning swarm hive themselves without trouble, except the releasing of the Queen.

Again, I clip off at least two-thirds of all of the wings of the Queen, so that she is always readily found. In making nuclei, changing frames of brood and bees, mak-ing swarms, extracting, etc., if you find the Queen, you can always know that she is just where she belongs, and not in some place where she ought not to be. By having her wings cut short, you can see her golden abdomen as soon as you lift the frame that she is on.

Then the clipping of Queens' wings does away with that expense to the apiarist-the fountain pump-or one of some other manufacture, which all the apiarists think a necessity where their Queens have their wings, so that by the use of it swarms may be kept from clustering together, where two or more come out at once, or if a swarm tries to "run away," so that it can be stopped.

Some claim that a Queen with clipped wings is more liable to fall off the combs, and get injured, than she would be if her wings were not clipped; but I cannot see how their wings should help them to hold on to the combs as long as that part is done with the feet.

Others claim that unless the apiarist is constantly on hand, during the swarming season, many Queens will be lost, by the bees swarming out and going back, while the Queen stays out in the grass, she going so far from the hive that she does not find her way back. If the apiarist is obliged to be away from home, let some one of the family get the Queen in the cage, and lay her at the entrance of the hive till the apiarist returns, when he can divide the colony, or let the Queen go back, when she will come out with the swarm again the next day. If all are obliged to be away from home, the Queens can be readily found upon the return of the apiarist, by his passing through the yard and looking for the Queens, which will be found by the little balls of bees from the size of a butternut up to that of an orange; for I have yet to see the Queen, thus left in the grass, which did not have this escort of bees with her.

To find the hive that this Queen came out of, take the Queen away from the bees towards night, when the most, of the other bees have stopped flying, and they will return to the hive from, which she came, setting up their fanning at the entrance. Now let the Queen go in with them, and the swarm will issue again the next day.

If I desire to be gone from home for two or three days together, with my family, I hire a man to stay with the bees, from 9 a.m. to 3 p.m., instructing him to cage the Queens as they come out with the swarms, and leave them on top of the hive, arranging them in such a way that the bees in the swarm can have access to the cages when they return. ^Any man can do this, or a boy even, who would not think of hiving swarms. On my return, I liberate these Queens, when the swarm will issue again in a day or two; or if it is preferred, these colonies can be divided.

Still others claim that the bees will swarm out from the new hive, with the Queen, immediately after she has gone, into the hive with the bees; but as far as I can judge, all of these reports come from those who are using so small hives that the bees are not contented with them. In any event, this can be easily overcome, by leaving the Queen in the cage at the entrance of the hive until the bees have all become quiet, when she is released, with no danger of their coming out in the air again, as has been spoken of.

As to the claim that Queens are injured by having their wings clipped, I can only think that such claims are entirely fallacious; for during the past five years, I have kept many of my Queens with their wings whole (where I thought there might be a call for such Queens, which had been wintered over), and not one of them proved in any way superior to those whose wings were clipped. Again, I have had Queens sent to me from those who never clip the wings of their Queens, and these have shown no superiority over those in my own yard that had their wings clipped.

The clipping of the Queen's wings, often seems like a serious job to the timid and inexperienced, but after a little practice it is no more of a job than any other work about the apiary. Some recommend scissors for clipping Queen's wings, hut I think that a Queen is much more liable to be injured in using them, by having a leg or two cut off, than where a knife is used.

My way of clipping the wings is as follows: After having found the Queen, catch her by the wings, getting all four of them if possible, by using the thumb and fore-finger of the left hand. Now take a jack-knife, which should have one of its blades very sharp, and place the sharpened blade on the wings of the Queen. Carefully lower both hands down within an inch or two of the top of the frames, so that the Queen will not be injured in fall-ing, when the knife is lightly drawn, the wings severed, and the Queen runs unharmed below. In doing this, place the knife so that it will cut off about two-thirds of the wings; for there is no more harm in cutting off this much than there is in cutting off one-half of one wing, as some recommend. By clipping the Queen's wings in this way, she is easily found at all times.

Some claim that this destroys the beauty of the Queen; but to me it causes her golden abdomen to show off to a much better advantage; and even if it did not, the ease with which they are always found afterward, more than compensates for the lack in looks, to those who reason in this way. Do not be afraid of cutting the fingers, for if you stop drawing the knife as soon as the Queen drops, you cannot do so.

The best time for clipping Queens' wings, is during fruit-bloom, when there are but few bees in the hive, compared with what there will be later on; doing the same when the bees are industriously working during the middle of the day, so that few are at home.

CHAPTER XX. SHIPPING, SHIPPING-CAGES, BEE-CANDY, ETC.

Prior to the advent of the Italian bee into this country, the shipping of Queens was comparatively unknown, while the sending of Queens in the mails is something scarcely a quarter of a century old. In Queen-Rearing and queen-shipping there have been mighty strides made during the last 25 years-strides, which had they been told to our fathers, would have seemed little less to them than miracles.

Instead of bees in hives now being carried on a pole between two men (as were those which I first saw brought to my home), we now transport them all over the world by mail and express, although, as yet, we can hardly say that we send a colony of bees by mail; still the essential part of a colony is thus sent, and I believe that the day is not far distant when enough bees will be sent with a Queen by mail, to start a colony of bees which will make a "hive" of it, if sent early in the season. This will carry our beloved pursuit even to the "uttermost parts of the earth," so that every one can have the privilege of eating honey of their own producing, "under his own vine and fig-tree."

Here, again, we see the working of many minds, for no one man has accomplished all this; but a little here and a little there, has wrought out most of this grand advance during the present generation. The cages first invented for shipping Queens, would seem bungling affairs to us to-day, yet they had their place in working out this problem-the shipping of Queens through the mails.

When Queens were first sent by mail, it was thought that the apartment made to contain them and their escort, must be roomy, so that they should not be cramped;

but as time wore on, it was found that very little room was needed, and I am convinced that most of the cages now in use, are much too large, where Queens are to be sent by mail with only 8 or 10 bees to accompany them.

Probably there is no cage in as general use as is the one which is called the Peet Shipping and Introducing Cage; yet I firmly believe that the apartment for the bees is much too large in this one. I have used large numbers of these cages, yet I consider them faulty in this respect; nor do I like them as introducing-cages. They are faulty as shipping-cages, in the size of the hole which holds the bees, inasmuch as when the mail-bag is thrown off a train at full speed (as is frequently done), or thrown from the train to the ground, or even from off a wagon, the sudden precipitation of the Queen from one side of the cage to the other, often causes an injury from which she never recovers.

The hole in any cage, calculated for holding no more than 8 or 10 attendant bees with the Queen, should not be larger than an inch across the farthest way, and if thus made, the wings and legs of the 10 bees will be so close together, that they will form springs, as it were, to deaden the effect of any sudden concussion. When 30 or 40 bees are placed in a Peet cage, then it answers the purpose of a shipping-cage very well, except that it takes twice the postage that a shipping-cage ought to require, and this matter of postage makes quite an item, as regards our profits, in these days of close competition, and where Queens are sent out by the thousand.

No shipping-cage which meets the requirements as I have set forth, can be a successful introducing-cage; for to meet with the greatest success in introducing, the cage should cover at least one-sixth of one side of a comb, so that hatching brood and some honey can be enclosed. In the hatching of this brood, to form an escort of bees for the Queen, and in her laying eggs in the cells enclosed by the cage, conies an assurance of safety, not found in any other item regarding cage-introduction of Queens. When these young bees, which hatch out with the Queen, become so attached to her that they accept her as their mother, it is not long before the bees outside of the cage fall into line. They now begin to feed her such food as is given for egg-production which means safety to any Queen. That the Peet cage will not allow of such hatching of bees, is wherein it is faulty as an introducing-cage. As the introducing-cage which I prefer has been described in the chapter on introducing laying Queens, I will not speak further of it here.

The shipping-cage which I prefer, is made as follows: Get out a block of wood, two-and-one-fourth inches long by one-and-one-eighth inches square. Near one end bore a seven-eighths hole, having the same one inch deep, and boring it across the grain of the wood. In the center of the opposite end, bore a one-half inch hole, boring it lengthwise of the grain of the wood, until it comes in contact with the seven-eighths hole which was bored before [Fig. 5]. This last hole is for the candy for the bees to live on during their journey, while the former is for the bees themselves. Next, get a piece of wire-cloth one inch square, and a piece of wood 2 1/4 x l 1/8 x 1/8 inch for a cover to go over the top of the cage after the bees are in and the wire-cloth is nailed on.

Bee-Candy

The next thing to be done is to prepare the candy for the bees. This is made by taking a quantity of powdered sugar, and putting it in any dish; although I prefer what is known as "Agate Iron-Ware," because in the kneading process, about to be described, the candy does not take on any foreign substance like lead or tin, as it does where a tinned dish is used. If you do not have the

Agate dish, an earthen one is equally as good, providing you are careful enough not to break it, thus causing trouble in the family.

Having the sugar in the dish, set the same on the stove or over a lamp, and put some nice, thick honey to heat also (such honey as will not granulate easily being preferred, for spring and fall use), letting both heat slowly till of about the warmth that you can conveniently hold your hand in, when they are to be taken off the fire and some of the honey poured into a little hollow made in the sugar. To get the sugar evenly warmed through, it may be necessary to stir it occasionally.

Having poured in the honey, take a little stick and stir sugar into it, by putting the sugar on top of the honey and rolling the whole around. When enough sugar is mixed with the honey so that it will not stick to the hands, when they are rubbed with a little of the sugar, proceed to knead it, the same as your wife or mother kneads bread, keeping this up as long as much sugar will be incorporated with the loaf, or until the loaf will not spread out or change its shape, if placed on a board.

You need not have any fears that you will get the candy too stiff, for, as a rule, more Queens are lost by the candy absorbing dampness, or being left too soft so as to daub the bees, than by all other losses put together. This is the reason for heating the honey and sugar, so as to get them of about the consistency they would he in a hot mail-bag during some of the warm weather that we have when shipping Queens. This candy is called the "Good" candy, although that as first made by Mr. I. R. Good, and given to the world, was made without heating, and contained granulated sugar and honey as its ingredients.

Having the candy ready, wet the forefinger of the left hand by touching it to the tongue, when it is to be placed over the one-half inch hole, where the hole terminates on the inside of the cage; when the hole is filled to within one-eighth of an inch with the candy, pressing it in with a plunger.

The wetting of the finger is done so that the candy will not stick to it as it otherwise would do thus pulling a part of the candy out of the hole, leaving it rough and uneven.

The candy being in place, take a plug-cutter made to cut a one-half inch plug, and cut one out of the one-eighth piece which is to go on the cage for a cover; cutting it out of one end so that the hole where the plug comes out will come over the center of the hole to be occupied by the bees, thus making the ventilating hole in the cover. Now take the plug thus cut, and drive it into the hole over the outer end of the candy, when the cage is ready for the bees, all but the wire-cloth.

Put this in place and drive a tack in one corner of it, leaving the tack a little out from the wood, so that the wire-cloth will turn on it, when it is to be turned so as to form an entrance for the Queen and attendant bees, which are now to be put in. Place the left hand thumb or fore-finger over this entrance, and with the right hand pick the Queen up by both wings and put her into the cage.

If you are not accustomed to this kind of work, it will seem very awkward to you. At least it seemed so to me when I first began, so much so that several Queens got away, instead of going into the cage. To succeed best, go slowly, and see that the Queen and bees get their feet hold of the wood, rather than on the wire-cloth, when it will be natural for them to run in, instead of backing out.

Catching Bees to Send with a Queen

Having the Queen in, close the entrance at once with the thumb, when a bee is to be caught by the wings in the same way and put in with her. Do not raise your thumb off the hole in putting the bees in, but rather give it a rocking motion. As the bee's head nears the hole, rock the thumb back a little, only opening it just enough for the bee to go in, and if the Queen attempts to come out, make this bee's head stop the hole at just that instant. Now catch another bee, putting this one in, in the same way, and so on until you have enough, when the wire-cloth is to be brought back in place and nailed.

After you get a little used to this work, you can put bees into a cage almost as fast as you could peas or beans. If you catch a bee by both wings, it is impossible for it to sting you, so that you need have no fears unless you happen, to push it against the thumb you have on the cage, in which case you will be quite liable to be stung.

In catching bees to send off with a Queen, select those which are from 6 to 10 days old, as nearly as you can get at it; for very young bees, or those that have never left the hive to void their feces, are unfit to send with a Queen that is going along distance, on account of their soiling the Queen and cage with the accumulation with which they are filled while passing through the larval and pupal state; while very old bees have not vitality enough to endure a long journey. By a careful watching on your part, as to the development of bees during the first sixteen days of their existence, you will soon know how old a bee is by its appearance,

If the frame the bees are on is jarred a little so as to cause them to fill themselves with honey, they will stand the journey better, and are more easily picked off the combs, when they have their heads in the cells with their wings standing out.

The bees being in, and the wire-cloth nailed down, next nail on the cover, having the ventilation hole over the wire-cloth; after which you will put on the directions, when it is ready for the mails.

If the bees are put up twelve hours before they are mailed, and left with the face side of the cage downward, but raised a little off the table, the Queen will rid herself of eggs, and thus better endure the sudden jars which she will be liable to get.

If the cage has been made according to the foregoing directions, and light, soft wood has been used, the postage required will be but one cent, as it should not weigh more than one ounce after the bees and candy are in. If you send out a thousand Queens during the season, the saving in this alone will be $10 over what it would be if your cage required a two-cent postage-stamp; and a saving of $20, if it required a three-cent stamp, as our cages did not long ago. This saving of postage is an item worth looking after, when such saving does not conflict with the safety of the Queens.

In all handling of Queens, great care should be used, not to injure their legs or abdomens. That all do not use this care, is evident from the number of Queens that I have received minus one or two legs, and often with dents in their abdomens. In putting up bees, don't get excited, and handle them as a "baggage smasher" would a trunk; but keep as cool as possible, and if you find that you are nervous and shaky, put off the caging of them until some other time. I realize that with some, I am urging a very difficult matter, for I once knew a man (who came to get some Queens that he had ordered) to get more nervous and excited in putting them up, than he would have been in fighting with a bear. If you are not used to putting up Queens, do not undertake the job when some one is looking on, but go at it alone, when you are in a quiet frame of mind.

CHAPTER XXI. INJURED IN SHIPPING.

Probably there are very few who have received Queens from a distance, through the mails and otherwise, that are not aware that some of these Queens did not come up to those which they already had, as to prolificness; for such is so common, that many of our best breeders have been censured and blamed for sending out poor Queens, when they were not to blame at all. Scores of these complaints came to me before I ever reared a Queen for sale, and the same has been so general, that even Mr. Alley occupies considerable space in his book on Queen-Rearing, regarding this matter.

Sudden Stop of Egg-Laying

Now, as a breeder of Queens, I suppose that I should let this pass, if I would consult my own interests; but I feel that both duty and truth require that I should not pass over the matter without mentioning it. Probably no man in the United States has more flattering testimonials, according to the number of Queens shipped, than I have; yet this does not prove that none of the Queens that I have sent out have never been injured by shipment. By shipment I include all of the necessary evils attending the removal of a Queen from her hive and home, and sending her to another hive and home, where she is obliged to suddenly stop a profuse egg-laying, and continue in this condition for from three days to three weeks.

Years ago my attention was called to this matter, by some writer of the past, who attributed the trouble to the rough usage to which the Queens were subjected in the mails; and gave as a remedy, that all Queens should be sent by express. In this I thought that I saw an explanation of the unsatisfactory results which I had I experienced with Queens which I had purchased; so for some time after that I ordered all of the Queens that I bought, sent by express. However, as I saw little difference in favor of those that came by express, over those which came by mail, I concluded that I must look elsewhere for the trouble.

In studying over the past, to ascertain if I could find out wherein the difficulty lay, I remembered that such a Queen, sent me by a noted breeder, had not laid eggs enough during two years, to amount to as many as one of my ordinary Queens would lay in two months; so I wrote him, asking if he remembered whether the Queen was prolific in his apiary or not. His reply was, that she was unusually so, and at the time he took her out of the hive, she was keeping ten Langstroth frames full of brood.

Later on, I received another Queen from another noted breeder, for which I paid a very high price, thinking to get the best there was in the country; yet, while she lived, she was about the poorest layer I ever had; still I was assured that she was an extra Queen when sent.

Soon after this, I commenced to send out Queens myself, and during my experience as a breeder and shipper of Queens, several instances have come under my notice of Queens which proved of no special value as to prolificness, after they were received by the purchasing party; while I know that they were among the best, if not the best Queens, that I ever owned.

Mr. Alley, in speaking of this matter in his book, attributes the cause to sending off a Queen immediately upon her removal from a full colony, while she was filled with eggs; in which state, he claims, she was not capable of enduring the rough usage which she would be subjected to during shipment, and advises that all Queens be kept in a nursery for a few days before sending them out. Others have advised leaving the Queens caged for a day or two before sending them off; and still others, keeping them in a nucleus for a week or so, before mailing them. All of these things show that nearly, if not quite all of our Queen-breeders acknowledge what a few say is not true, as some claim that a Queen cannot be injured by ordinary shipment.

While thinking of this matter one day, I resolved that I would find out the truth regarding it, if possible; so I caught some of my most prolific Queens and caged them, the same as I would for shipment, giving them the usual number of bees for an escort, and placed them in my shop. A part of these were thrown about the shop, and handled about as I thought they would be when shipped away, while others were handled very carefully or let alone entirely; all being kept from the hive from one to two weeks. Upon returning them as the heads of colonies again, some of them proved of little value, and, strange to say, a part of those that were of the least value, were among those treated the most carefully. I was now satisfied that the cause very largely lay where I mistrusted that it did-in the sudden stopping of a Queen from prolific egg-laying; for whenever a Queen expects to leave a hive with a swarm, she almost, or altogether, stops egg-laying preparatory to leaving, but doing the same gradually.

If I am correct in the above conclusion, and I believe I am, then the plan of keeping Queens out of colonies for a week or so before sending them out, can only remedy the matter as far as they are liable to being bruised is concerned; while it has really no bearing on the main cause of the trouble. The keeping of them in a nucleus for a few days, would come nearer to Nature's way of preparing the Queen to leave the hive, than any of the other plans; yet this will not fully accomplish the object, nor do I know of any that will.

Having solved the matter to my satisfaction, that Queens were mainly injured by suddenly stopping them from prolific egg-laying, and not finding any plan to fully overcome this difficulty, I next tried to find out if this unprolificness had any effect on the daughters from these once prolific Queens, but now almost valueless mothers. I am pleased to be able to go on record as saying, that, so far as I can see, such injured Queens produce just as prolific daughters, after their confinement, as they did before. For this reason, I would advise all who receive Queens, that do not seem as prolific as they would desire, to rear Queens from them immediately, or as soon as any of their brood is old enough for that purpose. In this way the buyer gets a fair return for his money, even if the Queen bought does not prove to be all that he had expected or desired.

Honey Locust Tree

CHAPTER XXII. QUALITY OF BEES AND COLOR OF QUEENS.

Had I thought that this book would have been considered complete without it, I should have preferred to leave this chapter out; for I am well aware that we do not all agree as to which is the best race of bees, how these bees should be marked, etc. However, as I thought that all would not consider it complete, and as I desire to injure no one's feelings, I will try in a mild, brief, impartial way, to tell what I believe to be the truth about them, as looked at from the stand-point of this locality-Central New York.

The black, or German bee, probably all are quite familiar with. All the really good qualities that I know of them, are their readiness to enter the sections and build comb, and smooth, white capping of the honey of the same. Their poor qualities, as I find them, are their inclination to rob, and willingness to be robbed; their running from the combs, and out of the hive, unless handled very carefully; they do not resist the wax-moths, are poor honey-gatherers, except in times of plenty; are inclined to sting with little provocation, and do not work in a business-like way.

This last particular, I do not know that I ever saw mentioned; and by it I mean, that they live only from "hand to mouth," as it were, calculating only a day or so in advance. They go into the sections to work, and build comb only so long as honey comes in plentifully. The least slack stops comb-building, only that the cells are lengthened on those that are already built, so that I have frequently found sections one-fourth full of comb, and that one-fourth lengthened out, filled, and capped over, without being attached to the sections except at the top. I never saw anything of the kind with any other race of bees, for they all start and build the sections full of comb, as if they calculated to do something business-like.

If another yield of honey comes in a few days, these bees start the comb down a little further, when it is again stubbed off if the flow slackens; and again and again do the same thing, until I have counted as many as five times in a single section, where they have started and stopped, making the face side of the comb resemble a wash-board.

It has been claimed that there is a difference in these bees, some saying that there is a large brown bee of superior merit; others claim great things for their grey bees, both of which varieties are said to be a great way ahead of the little black bee; but I wish to say, that, after getting Queens from several claiming to have these superior strains, and placing them beside the "little black bee" that our forefathers used to have, there is not a bit of difference in them, so far as I can see, or any of my bee-keeping friends to whom I have shown them.

The Queen-bee of the German race, seems to be the most constant in color of any of the bees that have come under my notice; all of which are of a very dark brown upon the upper side of the abdomen, while the under side of the same is of a yellowish brown. Out of scores of specimens which I have examined, I could not detect the least variation of color, so that in these bees we have Queens which will duplicate themselves as to color, if we do not have such in any other race.

Right here I would say, that, in speaking of markings, I shall notice only those which are fixed, or permanent, as are those colors on the horny scales, or segments of the abdomen; for nearly all other markings are of hair or fuzz and are soon worn off, so that an old bee does not look nearly as showy as a young one, when the color of the fuzz is new and bright. The head and thorax of all the races of bees are very much alike, except as the color of this fuzz gives them a lighter or darker appearance. To be sure, the Cyprians have a bright spot, or shield, as it is called, at the back of the thorax between the wings; but as I find this same spot on the best marked Syrians and Italians, I do not see how it can be used as a test of purity of the Cyprian race, as some claim for it. Hence the abdomen of the bee is the place we are to look for the markings of the different races.

Perhaps I ought not to say anything of the Carniolans, for the two Queens which I received that were said to be pure, were not at all alike as to their worker progeny. From these two Queens I decided that it was a mixed race, when I looked at the progeny of one Queen; and that it was only a peaceable strain of the black bee, when I handled those of the other Queen. My trial of these bees from these two Queens, agreed with the reports of the most of those at that time, in that they were not nearly as good as the Italians. As to the "steel blue" color claimed for them, I will say that the same will be seen on a lot of black bees, just hatched, if held so that the light strikes them just right. From the experience that I have had with these and their offspring, I concluded that I had no use for them, so I superseded the Queens. Of late they seem to be growing in favor, and I shall try to give them another trial in the near future.

The few Queens which I reared from these mothers, varied from a jet black to a light brown, one of which was fairly a shiny-black, like a crow, or what we term a "crow blackbird." There was no constancy of color in either the bees or Queens, I have thoroughly tried the Syrian bees, and for this locality I consider them the poorest of all bees yet brought to this country. The two great faults which make them thus are, first, not breeding when they should breed, and then breeding beyond measure when they ought to breed but little; which results in few laborers in the field in the honey harvest, and countless numbers of consumers after the harvest is past, to consume all that the few gathered. Consequence: *no profit*.

Second, the workers begin to lay eggs as soon as the Queen leaves the hive, whether by swarming or otherwise, so that the combs are filled with a multitude of dwarf-drones, to the disadvantage of the bees, combs, and owner. Laying workers are always present with these bees. At times they sting fearfully; at other times they are nearly as peaceable as Italians. However, they will not venture an attack unless the hive is disturbed, as do the black bees. A colony of Syrian or Cyprian bees will let me stand an hour at a time right in front of their entrance, turning out for me, and not offering to sting; while in less than ten minutes a black colony will resent such impudence to the score of hundreds of stings, if I do not leave.

The Cyprians I disliked to part with, for they were really good bees in all points but one; but that *one point* was altogether too sharp for me. Of all the bees to sting when provoked, these bees "beat all." In opening a hive, smoke does no good, while the least mishap will, without warning, send hundreds of hissing, angry, stinging bees all over one's person. They also have "a touch" of the laying-worker nuisance, but nothing nearly so bad as the Syrians.

With me, the markings of the Cyprian and Syrian Queens are very much alike, except the stripes or rings on the Cyprian Queens have more yellow on them than do the Syrians; and the yellow is of a bright orange-color, while that on the Syrians is less bright, and often dusky. Every segment of the abdomen has both yellow and black upon it, unless it be the last one at the tip, which generally is nearly or quite all black, or very dark brown. The Queens of these two races of bees are next in constancy of color to the German Queens.

Italian Bees are Best of All

Lastly, we have the Italians, and it is hardly necessary for me to say that they are my choice among all the bees that I have ever seen, either for comb honey or for extracted. Some claim that they will not work in boxes readily, while others think that they give the cappings of their honey a watery appearance. In neither of these points do I find any trouble with them; for if rightly managed, so that the hive is filled with brood when the sections are put on, as it should always be, they work in the sections on the first appearance of honey in the fields; while I have none of the watery-appearing honey from them, which is produced by both the Syrians and Cyprians. To be sure, they do not use as much wax on their combs as do the blacks, but they use enough, when we take all things into consideration, such as the cost of wax, toughness of comb, pleasure of eating, etc.

Especially am I pleased with these bees, when we have a light yield of honey, for at such times they work right on, untiringly, storing a little honey in the sections every day, at times when hybrids and other bees are scarcely getting a living. They will also work on the red clover more than any other bees, as I have proven during many seasons, storing nice, white honey at the same time the German and hybrid bees are gathering only that of dark color. This one quality alone would give them the preference over the other races, with me, had they not many other redeeming qualities besides.

Crossing for Improvement

The Queens are very inconstant in color, especially those from an imported mother, such varying from that of a German Queen to a bright, golden orange-color the whole length of the abdomen; some of the best specimens of my home-bred stock, not having even a particle of black on the extreme tip, or point. By crossing the best specimens of my home-bred stock, with similar specimens from different apiaries from 100 to 1,000 miles from me, I have succeeded in securing bees of the Italian race which are far more constant in color than any I could get ten years ago; while at the same time my bees have vastly improved as to their working-qualities. By this method of crossing, I believe it possible to get a bee of the highest type, as to working-qualities, as well as to produce the handsomest bees in the world.

Sacrificing Working-Qualities for Color

While I would by no means sacrifice working-quality for color, or anything else, yet when we can have a beautiful bee combined with one having the very best working-qualities, why not combine pleasure with profit? It is one of the "queer" things which a Queen-breeder meets with (as nearly all such breeders will bear me out in), that where a party orders several Queens, writing that he does not care for color, only give him good working-quality, he will, nine times out of ten, select the very Yellowiest one you would send him, to breed from, while his next order will call for all yellow ones.

Of hybrid bees I have little to say, for I believe that the crossing of any -of the races with those of the same race, procured from some apiary 200 or more miles away, will produce just as good results as to honey, as will the crossing of the different races.

CHAPTER XXIII. REARING A FEW QUEENS.

Buy or Rear Queens-Which?

No apiarist-no matter how few bees he may keep-should consider that he has done his duty by his "pets" until he has learned how to rear Queens. Not only is this a duty which he owes to himself, but in the doing of it he will find the most fascinating part of apiculture. I know of nothing so enticing, or of anything that will so completely absorb the mind, and get one out of that complaining mood which we sometimes fall into, as will the rearing of Queens. When at this work, minutes and hours fly away as though they were not, and even a whole day spent in the closest of this work is only considered a day of recreation. Here we can get away from self and the cares of life, and be led out along a higher plain of thought-thought which grasps, to some degree at least, the mind of our Creator, when He made so many things for the comfort and enjoyment of us, His children. In no one thing can the handiwork of God be seen more, than in this particular branch of our beloved pursuit.

Again, the rearing of Queens is a duty that we owe to our families, if they are in any way dependent upon us for their support. Many times I see it advised, that the bee-keeper should buy his Queens, as though that was the best and cheapest way to Italianize an apiary. While I have Queens to sell, yet I object to such advice, or any advice which compels the man starting in apiculture, or already in the same, to take his hard-earned pennies-often earned in some other calling in life-away from his family, and send them for the support of some other man's family, who may have many luxuries that his own does not have.

If any one has plenty of money that is hanging idly on his hands, then I have no objection to his sending it when and where he pleases; but I do claim that the average beekeeper has not the right to scrimp his family by buying Queens, or anything else, that he can rear or make just as well as not during his leisure moments; and by so doing, keep his money to cheer the hearts of his loved ones, and at the same time be growing intellectually in his chosen pursuit. Of course, it will be necessary for any one desiring a change of stock to get a Queen of the desired race, but to purchase Queens for the whole apiary, or by the score, as is frequently advised, is quite another thing. The object of this book is to tell all how they can rear one or more Queens with ease, and in such a way that their bees can be improving all the while, instead of retrograding, as was often the case where the old plans were used.

"But," says one, "How can I rear one or two Queens by your plan, without going through with all of the work that the plan, as a whole, requires?" If you do not want to rear enough Queens to pay for using a stick of 12 cells, proceed to make 2 or 3 wax-cups, as I told you how to do in Chapter VII; or, if this is too much work, use embryo queen-cells, as given in Chapter VI. A few days previous to doing this, tier up a hive for extracted honey, as all want at least one hive worked in that way; if not wanted for that, it will be needed for the purpose of securing extra combs of sealed honey, to be used in feeding the bees when they need feeding, putting a queen-excluding honey-board between the hives, and having one frame of unsealed brood in the upper hive with other combs.

In four or five days, look at this comb having the brood in it, and you will find one or more queen-cells started? From which you can get royal jelly with which to supply the wax-cups that you have made. Just before putting in the royal jelly, go to the hive having your best

Queen-best as regards color, work and every other quali-
ty-and get a piece of comb containing a few larvae. If the
day is not quite warm, take all to the kitchen, where it is
always warm just before the noon-day meal; and after
putting the royal jelly in the wax-cups, transfer some
larvae into each. Roll the cups in a warm cloth, if the day
is cool, and upon going back to where you got the jelly,
press down some of the cells a little where the comb is
empty, by laying the side and end of the little finger
against the comb, thereby forming a place into which the
wax-cup will fit nicely, thus holding it in place on the
comb with the open end down.

Now put the frame in place in the upper story again,
and close the hive. If other cells were started on this
comb, besides the ones which you destroyed by taking
out the jelly, they should be destroyed also. In case a
Queen is allowed to hatch from such cells before those
hatch which you have started, they will destroy those of a
better quality that you have worked for.

In ten days from the time you prepared the cells, go
to the hive and slip in a queen-excluding division-board
near the centre, if you want two Queens to hatch and
become fertile. If you want only one, you need not fix the
hive in this way at all; for the first Queen that hatches will
destroy all the rest. All you need to do in this case, is to
bore a half-inch or larger hole in the back part of the hive,
five days after the Queen is hatched, and close it again
after she begins to lay.

But as it will be natural for you to want two Queens
at least, we will suppose that you have the Queen-
excluding division-board in place, after which you are to
get a frame containing a little brood, from any hive in the
yard, and after shaking the bees off this comb in front of
their own hire, you are to take one of the nearly-mature
queen-cells off the comb that they are on, and stick it on
this comb, the same as you did the wax-cup; using this

way of fastening the cells to the comb until you get enough accustomed to the work so that you will not injure them by pressing them into the comb, as I advised in Chapter IX.

Now place this frame on one side of the hive, and the one on which the cells were built on, the other side, having the queen-excluding division-board between them. If you have more than two completed queen-cells, and you wish to save them, of course you will need more upper stories, or will form nuclei for them, this being written on the supposition that you are only desirous of rearing just two extra Queens.

In five days after the young Queens hatch, bore a hole from the back part of the hive, into each of the apartments having the Queens, leaving these holes open till the Queens begin to lay, when they are to be closed. You will now have two as nice Queens as you ever saw, reared without much trouble; and they can be kept where they are until you desire to use them without their interfering with the workings of the colony in the least, any more than they have done so far, which is none at all.

This rearing of Queens and having them fertilized in a hive having a laying Queen in it, without in the least interfering with the working of the bees or the hive, is something which holds me almost spell-bound when I think of it, and something that we have heretofore considered impossible.

Superseding Queens

Another point right here (and one which I consider worth much more to any one, than the price of this book; even though he may keep only two colonies of bees), and that is: If you desire to supersede any Queen in your yard, on account of her being too old to be of farther use; or if she is of another race of bees from what you desire;

all you have to do is to put on an upper story, with a queen-excluding honey-board under it, place a frame of brood with a queen-cell upon it, in this upper story, and after the young Queen has hatched, withdraw the queen-excluder, and your old Queen is superseded without your even having to find her, or having the least bit of time wasted to the colony.

In fact, the possibilities which this perforated-metal may bring, have only just begun to loom up before us, so that what the future may bring forth in this matter can hardly be conceived by any. Truly, our pursuit is one of the most fascinating of any of those that are engaged in by man; and I am thankful to Him who ruleth all things, that I have a part and a lot in this matter.

That all who read this book may try to carry out the thoughts herein advanced, to still greater perfection, and strive in the future to rear only Queens of superior value, so that we may soon be able to say,

"THE COMING BEE IS HERE,"

is the best wish of the author.

Since writing the preceding Chapters, I have been experimenting further along the ideas contained in Chapter XIII, as I found that owing to the conditions under which I had tested the thoughts and experiments therein contained, there was a possibility of a failure along that line, when the plans were used under other conditions than those which existed during the times when I had formerly used them. In previous years, owing to my selling nearly all my stronger colonies of bees to fill the many orders which I had for the same, I had no colonies of suitable strength to tier up early in the season, so that the plans then tried were used only after the basswood had blossomed, and later, in having queens fertilized over a queen-excluding honey board. During the present year (1889) having more strong colonies than usual, owing to fewer sales, and the bees wintering better, I tiered up several hives early in the season, and very much to my surprise, found that that which had previously worked to perfection, was a failure, as far as the fertilizing of the queens from these upper stores was concerned. The cells were allowed to hatch just the same as before, but when the queens came to the age of two or three days, the workers began to worry and tease them which resulted in their being killed sooner or later, while in one or two instances the result was a general row "upstairs," in which many bees were killed besides the queen. At this time the bees were only living from "hand to mouth," as it were, for the forepart of our season was the poorest I ever knew.

When the Basswood begin to yield honey, I again began to have the same success which I formerly had, either owing to the peculiarity of this locality, which brought about former conditions, or to some additions which I made to the hive, or perhaps to both. When I saw that what I supposed was the same plan that I had formerly used was failing, I began to study into the matter to see if I could not find a remedy, and about the first thing which appeared was that I did not have the hive fixed as I had previously, although I now had it arranged the same as I gave in Chapter XIII in this respect.

Some may think that it would be strange for such a thing to occur, and perhaps it was, yet it was one of the most natural things in the world, as you will soon see. As all of the older readers of the bee-papers are aware, when I commenced using the Gallup hive, I used it the same as Gallup recommended, using twelve frames in the hive. As the years passed by, I believed that twelve frames were too many for the brood-apartment, so I made dummies or division-boards to take the place of one or more frames, according to the time of the season, or as I wished to contract or expand the hive, my custom being to expand the hive during the forepart of the season, and contract it the latter part, or contract at the beginning of basswood bloom.

After a little thought along the line of what had caused the failure this season, when no failure had occurred before, it began to dawn upon me that in my former experiments I had contracted the lower hive down to eight frames, so as not to rear a large number of bees during the basswood bloom, to become consumers of the honey later on, as we have no fall flow of honey here; and in this contraction might be found a solution of the problem, for I now had both stories of the hive filled with combs, as it was the forepart of the season, the lower hive being now filled so as to rear workers for the harvest. In this latter case the brood came directly under that part of the queen-excluder running under the apartment petitioned off with the perforated-zinc division-board, so that when the young queen ran down on the zinc, she and the old queen could get their heads together and try to kill one another, which resulted in the bees worrying the young queen when she was old enough to be recognized as a queen, the same as bees always try to worry virgin queens in the queen-nursery after they are two or more days old, as they always do when such nursery is hung in a hive having a laying queen.

When younger than this, the bees do not seem to notice them in either place, nor does the young queen try to get below. Without intending it, I had so partitioned off the upper story in my previous experiments that the apartments the queens were in at each side of the hive came directly over the dummies, so that there was no temptation for the old queen to come out in the bee-space over and between the dummy and the queen-excluder gave the worker-bees free access up through the bottom of the apartment, as well as through the zinc division-board in the upper story.

When writing Chapter XII, I had not the remotest idea that these dummies played such an important part in the matter, nor am I now fully certain that they will make the plan a success always in all localities and at all times of the year, but I believe that they have much to do with the plan working so successfully in this locality; for nothing could work more perfectly than it has with me since the dummies were put in the lower story when fixing the hives for the basswood bloom.

Right here I would say what I forgot to say in the body of the book, which is that I tack the queen-excluder, used between the upper and lower stories, to the bottom of the upper hive, tacking it on lightly with small wire-nails. This makes it so that when I wish to get to the lower hive for any manipulation of the same, all I have to do is lift off the upper story, the same as would be done were there no queens above, or any queen-excluder used.

In this way there is no more danger to the young queens when the hive is off, than there is at any other time.

After finding what I believed to be a solution of former trouble, and knowing that all would not want to use dummies under these queen-rearing apartments, I began to experiment to see how the matter in regard to the young queens going down on the perforated metal, so as to cause trouble, might be obviated, and arrived at the following:

My queen-excluding honey-board is what is called the "wood-and-zinc" board, having a full bee-space on the upper side of it. On this upper side I tacked a string of wire-cloth of the right width to come out to the queen excluding division-board, tacking it on each edge of the wood which formed the bee-space, thus giving a bee-space between the honey-board below, and the wire-cloth, which entirely prevented the virgin queen from getting to that part of the queen-excluder immediately under her apartment, yet at the same time allowing the warm air from below to come up into the apartment, the same as it would were the wire-cloth not there.

With this I have been equally successful in having queens mated from these apartments, the same as I was where the dummies were used, and I believe the same will overcome nearly all of the difficulty which I experienced during the forepart of the season, although I cannot say positively at this date, as I have not had the chance to try it, except during the basswood bloom, and later. If it should not, my next plan would be to make the division, which forms the queen apartment, or wire-cloth, except say three or four rows of perforated metal at the top, so that all bees entering this apartment would be quite a distance from the reigning queen below, when entering this apartment, which I think would make the plan successful in localities where all else failed.

Now, as there seems to be a chance that a failure may possibly result in some localities, and at some seasons of the year, I would advice all to try only one or two colonies at first, to see if the plan will work in their locality; so that, should it not work, they will be but little labor and time out, in trying the experiment.

I still believe that there is a great future before us, along this plan of having queens fertilized from an upper story and as I have intimated in other parts of this work, it is my desire that the plan which I have here given may be so improved upon that there shall not be a doubt about this matter, and we as bee-keepers be led out to a wider plain than any heretofore enjoyed.

Already some are branching out along different lines, notably among which Dr. Tinker with his "Queen Rearing Chamber." There is little doubt but what his plan will work, but that "Chamber" seems to be more suitable to the large queen-breeder than to the rank-and-file of beekeepers; while my design was to bring out a plan that would be of benefit to all, from the person having but two colonies up to one who numbered his colonies by the thousand.

Some seem to feel (or act as if they so felt) that I was trying to crowd my plans upon them for some irritation has been shown by a few, since this work was published; but such is not the case. All are free to use, or refuse, these plans which I have outlined as they please. No, dear reader, I have not the least desire to crowd anything upon you. All I have done has been done with the hope that I might be of benefit to the world-- benefiting some one by smoothing over the rough places a little, the same as some of the writers of the past smoothed the way before my tender feet, when they were still youthful in the pursuit of apiculture.

As I have freely received of the good things in the bee-literature of the past, so I as freely give of the little I know, that I may, in a measure, pay the large debt I owe to those who have preceded me in the way of our delightful pursuit.

Borodino, N.Y., October 1, 1889

Yours Truly
G. M. Doolittle

About the Author

The late G .M. Doolittle preaching from the pulpit so long occupied by him.

Edited from his Obituary in July 1918 GLEANINGS IN BEE CULTURE pg 396

"Gilbert M. Doolittle born April 14, 1846 died at his home near Borodino, N. Y., on June 3, 1918, aged 72. He was born the son of a farmer and beekeeper, and from his very infancy he was himself a beekeeper. For almost half a century he unceasingly taught the principles and details of good beekeeping thru the apicultural journals to a great audience of both beekeeper learners and beekeeper experts. Among all the correspondents of bee journals no writer perhaps, has been more closely followed than Mr. Doolittle. The readers of Gleanings thru many years have expressed in thousands of letters their appreciation of him as a teacher. So universally was his opinion sought that Gleanings Editor, early in 1900, asked him to conduct a department in Gleanings entitled *Conversations with Doolittle*. In that capacity Mr. Doolittle has been a continuous instructor to the American beekeeping public for more than 18 years. He has been a regular contributor to this journal from the first year of its publication, 1873.

"From his earliest years Mr. Doolittle was a very close observer, and his statements as to the actual operations that take place in the hive (or what we now technically call bee behavior) can be regarded as authentic. He came to be accepted as an authority on all manner of domestic economy of the bees. Mr. Doolittle was a large man in every way, of magnificent physique and commanding presence, the possessor of a fine voice, a ready and witty speaker, a good storyteller, and an excellent writer. In the telling of witty stories that illustrated valuable points in beekeeping, he surpassed any beekeeper we have ever known. At the great Buffalo beekeepers' convention in 1897, we recall that he was frequently called on, and each time he brought down the house with roar upon roar of laughter and applause. His stories always had a good point. One of Mr. Doolittle's most emphatic teachings was that the beekeeper must follow nature—that no beekeeper could succeed if he did not follow nature's rules. One of his chief theorems was that good queen-cells must be reared in strong colonies built up to the swarming pitch, and, as a corollary of this, he often said that good cells could not be built unless boney or sealed stores were supplied daily. He rightly and stoutly held that no queen breeder could succeed unless he observed these two rules. He was first to prove that good cells could be built under only two impulses-the swarming impulse and the supersedure impulse. Good queen breeders now recognize these two propositions as fundamental.

"Altho Doolittle did not invent artificial queen-cups, he was the first man to develop the process. His method of making artificial cell cups started a new era in queen rearing. While cell cups are now made in a wholesale way by machinery, the basic principle is Doolittle's. He was also the first man to demonstrate that queens can be reared in an upper story with a laying queen below. All in all, Doolittle's method of rearing queens is essentially those of all modern methods now in vogue, and this one contribution to beekeeping has done more to make better queens and consequently better colonies than any other one thing in beekeeping practice. His book on *Scientific Queen Rearing* is acknowledged today as containing the best of modern methods of queen-rearing.

"Years ago Doolittle originated the slogan "rich in stores." He talked it first, last, and all the time. He insisted that unless a colony at the beginning of the season had a great abundance of stores it would not build up as will a colony that has plenty of stores. Here again he was absolutely right, and was ever preaching this fundamental doctrine of good beekeeping. He developed a unique system of swarm control for the production of comb honey. This system is fully outlined in his book published under the title of *The Management of Out-apiaries.*

"One outstanding feature of Mr. Doolittle's beekeeping was that he was not only a good instructor, but he put his teachings into successful practice. Some men, like Langstroth, the peer of all instructors, never could make money from their bees. Others, like Quinby, one of the best authorities in his day, have made money, even with box hives. Doolittle always profited from his bees, and always succeeded in getting crops. Mr. Doolittle was more than a successful beekeeper and natural-history student. He was a big-hearted friend, a good citizen, and a Christian gentleman. Long will the good live after him that he has done."

The
Alley Method of
Queen Rearing

from The Beekeeper's Handy Book

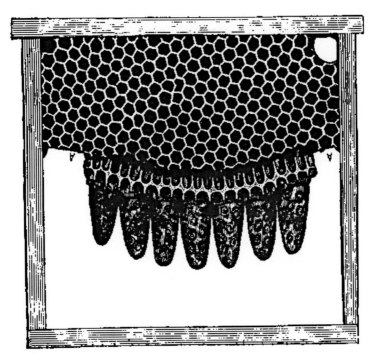

by Henry Alley

The Alley Method of Queen Rearing

From The Beekeeper's Handy Book, Published 1883 by
Henry Alley

Table of Contents

Table of Figures

THE IMPORTANCE OF QUEEN-REARING AS A BRANCH OF APICULTURE.

TWENTY-FIVE years ago I purchased my first colony of bees, and with that event began my interest in apiculture. My colony, being in a box hive, I transferred to frames, and commenced to rear queens and to experiment generally with bees. In the course of a few years, upon the introduction of the Italian bee to this country, there came a large and increasing demand for bees of this race.

Many bee-keepers began to rear queens and to offer bees and supplies for the apiary for sale, a large number of whom soon failed for want of patronage, or were compelled to abandon the business on account of the cheap queen traffic. Of all who as late as ten years ago were engaged in this branch of the business, I can call to mind but one beside myself. Few have made queen-rearing and the supply trade a success.

Shortly after the introduction of the Italian bee, the "American Bee Journal" sprang into existence, and simultaneous with its appearance began one of the most important industries of the age, viz.: practical apiculture. At this early stage of its history, queen-rearing was in its infancy, while but few bee-keepers had any practical knowledge of this interesting and vastly important branch of the business, and apparently very little advancement has been made up to the present time as compared with the other branches of apiculture.

After a thorough examination of the latest works on the subject, and a careful study of all the various Bee Journals, I find only the old methods taught which were in use many years ago. Hence the beekeeping public continue to rear queens in the old way, the result being that a vast number of inferior and even worthless queens are put upon the market every season.

The present and future interests of apiculture demand a more thorough and practical method of rearing queens, and I shall endeavor in this work to give my readers such information as shall tend to give a new impetus to this branch of bee-keeping, and also aid, if possible, in doing away with the cheap and worthless queens produced under the lamp nursery system; and to offer to the bee-keeping public, for their careful consideration and adoption, a thorough, practical and scientific method of queen-rearing, which is the result of many long years of practical experience, and much hard study.

In order to become a successful instructor one must first attain a complete knowledge of the subject to be taught, and unless it has been thoroughly and fully mastered in all its details, failures only can result.

In presenting this work to the bee-keeping fraternity, I do not wish to assume the position of teacher, but rather to place before its readers in as plain and practical a manner as possible my method of rearing queens, leaving to their judgment the careful study, and candid criticism of its contents, feeling assured of a favorable decision regarding its merits and value; knowing that if its instructions are carefully studied in all their details, and put to a practical test, the result will be successful. By careful attention to all the rules laid down herein, I hope better queens will be produced, a matter of great importance to the bee-keeper whether he keeps bees for pleasure or profit; and of vastly more importance to the bee-master who follows it as a vocation and depends upon the same for a living.

Chapter I

THE SELECTION AND CARE OF BREEDING QUEENS.

All bee-masters know the importance of having a strong, vigorous and healthy queen for the mother bee; hence it will be admitted that all queens used for this purpose should be carefully selected, perfect in every respect, and of undoubted purity and prolificness. The particular strains from which we wish to breed should be thoroughly tested, to determine their qualities regarding purity, gentleness, honey-gathering, and wintering. This is very important and essential as some do not combine all the very desirable requisites above mentioned, and I would lay it down as a rule to breed only from such.

THE ADVANTAGE OF USING SMALL HIVES AND COMBS FOR THE BREEDING QUEENS.

If a large number of queens are to be reared, the mother queens should not be kept in full colonies, as the risk of killing them in securing eggs for cell-building is too great, and many valuable queens are lost in this way. To guard against such loss my breeding queens are kept in miniature hives (Fig. 1). The queen being more easily and speedily found on small combs, there is much less risk of injuring her, and there will be eggs enough in one of them at any time to start fifty or more cells.

Alley Method

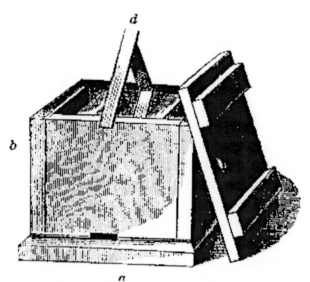

Fig. 1. Fertilizing or miniature hive.

COMBS TO USE IN OBTAINING EGGS.

In selecting the combs for the queen to lay in, to be used expressly for cell-building, take only such as are nearly new or that have been made use of for brood but once or twice. I do not use the combs in standard frames for this purpose, as in the course of the season a large number of nice brood-combs would be either badly mutilated or destroyed in so doing. Small pieces of comb the size of the nucleus frame, described in another place, are generally at hand and far preferable to larger combs. One standard Langstroth frame of comb will fill four or five of such small nucleus frames.

FOUNDATION FOR CELL-BUILDING.

It would be a good plan to fill frames with foundation, have it worked out in full colonies, and used for brood once or twice, then cut up and fastened into the small frames. I have found foundation very good for starting cells, even when it had not been used for brood at all; but care should be taken that the cells on both sides are drawn out to nearly the proper depth. These combs, when used to obtain eggs for cell-building, will be filled so that they can be removed as often as once every twenty-four hours. They should be properly numbered and dated for future use as wanted; the other four combs in the miniature hive being used for storing honey and brood to keep the colony prosperous in young bees. A good prolific queen will fill this small comb in. less than twenty-four hours, but it is better to let it remain that length of time. The advantage of using such small hives will be seen at a glance, as it will not be necessary to open a full colony every time eggs are needed from which to start cells. Again, the exact age of all eggs is easily and exactly determined, and the apiarist may tell at any time just when to prepare his bees for cell-building; the time when the queens will hatch from the cells may also be determined within a few hours. This is a matter of great importance; saving as it does, much time, labor and anxiety, as well as for other reasons which any intelligent bee-keeper will readily comprehend.

THE DISADVANTAGE OF KEEPING BREEDING QUEENS IN FULL COLONIES.

It may be convenient for those who wish to rear a few queens to open a full colony for eggs, but as they can seldom be found of the right age in sufficient quantities to start a large number of cells at any one time, the former plan is much to be preferred by large breeders.

By placing a comb, selected for the purpose, in the centre of the brood-chamber of a full colony, we can sometimes find eggs which are suitable for the purpose, on the fourth day after, which should invariably be used on that day for cell-building. Sometimes the bees will not allow the queen to use such combs in which to deposit her eggs for several days, as perhaps breeding is not going on vigorously, and the queen may not reach the empty comb for one or two days after it is placed in the brood-chamber; consequently the hive would have to be opened often to ascertain this fact. Under such circumstances, queen-rearing cannot be carried on as systematically and successfully as by the miniature hive system.

MINIATURE HIVES FOR BREEDING QUEENS.

A colony in such a hive can always be depended upon for eggs at any time for cell-building. I find that a comb put in at night will be filled with eggs the next day. Then it may be removed and another inserted in its place. Once in twenty-four hours is often enough to change them. By this plan one queen will furnish eggs sufficient to rear 10,000 queens in the course of one season. The above plain and simple rules are the first steps which should be taken and put into practice systematically, if one wishes to rear first-class queens, and make the business successful and remunerative.

ROOM FOR HANDLING BEES; WHAT COLONIES TO SELECT FOR CELL-BUILDING; HOW TO PREPARE THE BEES FOR THIS PURPOSE.

Every bee-keeper should have a room specially adapted for the purpose of handling bees, as many of the operations about the apiary cannot at all times be carried on in the open air, especially in wet and cool weather.

In fact, most of the work about preparing bees for cell-building can be performed only in a convenient and handy place. In view of this I have, in another place, given a description of such a room.

Alley Method

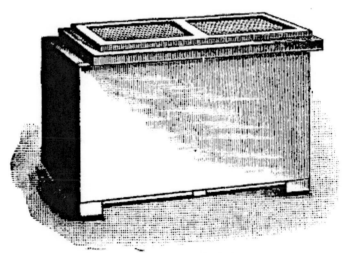

Fig. 2. Swarming box.

Always select the strongest colonies for cell-building, and never the weak or feeble ones, as such would not rear strong and hardy queens. You may perchance have some strong colonies in the apiary, having queens which you wish to supersede; colonies having old, uneven or crooked combs; odd-sized frames, or those in box hives. These may be used to good advantage; thus ridding the apiary of such undesirable stocks, and should be selected in preference to those in good hives and in fine working condition. The combs can be transferred to other frames, and the brood given to weak colonies. Having made your selection, take them into the bee room, give the bees a few puffs of smoke to cause them to fill with honey and remove the combs, examining them carefully to find the queen; after caging her, brush all the bees into a box, a Langstroth hive cap will answer every purpose. If the bees should attempt to crawl out blow a little tobacco smoke on them, and they will remain quiet. The bees should now be put into another box, say one that will hold three pecks, the top and bottom of which should be covered with wire cloth (Fig. 2, description farther on), in order to give the bees plenty of air. The top or cover of this box may be secured by Van Dusen clamps or some other simple arrangement. I use four screws for this purpose, but seldom fasten the top on unless the box is to be transported some distance.

THE LENGTH OF TIME TO KEEP THE BEES QUEENLESS.

The bees should be kept in this box at least ten hours. Soon after being put into it they will miss their queen, and keep up an uproar until released. This pre-pares them for cell-building. I find it a good plan to keep them in a cool, dark room, or cellar, until needed, as they will keep more quiet and there is less danger from suffo-cation. If the bees were properly drummed before being removed from the hive, they have filled themselves with honey sufficient to last during the time that they are kept confined in the box; but to guard against starvation, which might occur, as the bees do not in all such instances fill themselves readily, I give them a pint of syrup in the glass feeder (Fig. 8, described farther on). The bees must be kept queenless for from ten to twelve hours, else the eggs given them for cell-building will be destroyed. This I have learned from practical experience.

ROYAL JELLY; HOW PREPARED; COMPOSITION OF.

When eggs are given too soon after the bees are made queenless, they are very apt to remove or destroy them. This, however, is not the case when larvae are provided instead. I have had some experience in keeping pigeons as well as bees, and have noticed that there is a natural food or chyme (white, milky and very nourishing) secreted in the crop of the hen, with which to feed the young and tender birds during the early stage of their development. Reason and experience teach me that the same rule applies equally well to bees, and that when made queenless and confined in the swarming box they prepare or secrete the white, milky food, which we find in the bottoms of the cells around the eggs given them for rearing queens, and which is of the same nature as the royal jelly upon which the young queens feed while confined to the cell; also that it is necessary that they be kept queenless until instinct impels them to make this important preparation for cell-building.

In support of the above I would say, first, that this secretion is found in all animal bodies (under certain circumstances) for this purpose; second, that the hen pigeon is incapable of secreting this food until about the fourteenth day of incubation, showing that the secretion is not made until needed, and lastly, the fact that bees, after being kept queenless ten or twelve hours, and then given the prepared eggs, will place this food in the bottom of the cells within an hour, going to work contentedly, knowing that they have means to produce other queens, and showing no further symptoms of queenlessness.

Again, the amount of jelly-food furnished a larva from which a queen is to be reared is much greater than that from which a worker is produced, and the composition of each is entirely different. It may not be generally known that a colony deprived of its queen almost invariably selects a larva which is usually over twenty-four hours, and frequently from two to three days old, instead of an egg from which to rear another. Such queens must necessarily be "short-lived" as they are not reared in accordance with natural laws. Otherwise, so far as known, they follow nature. Every queen cell should be so abundantly supplied with royal jelly that after the queens have hatched there will be more or less left in the cells. This is the case with the best cells produced by the bees under the swarming impulse, and I claim that just as good cells can be produced by the method which I have instituted. Quite a distinguished writer made the statement some time ago in one of the bee journals that "artificial (or forced) queens left no jelly in the bottoms of the cells." He evidently jumped at this conclusion without thoroughly testing the matter. I admit that as artificial queens are generally reared, this will be the result, but when the reader has become thoroughly conversant with the directions given in this book, and has carefully put them into practice, he can produce those that show a goodly amount of jelly in the bottom of every cell from which a queen has hatched.

Chapter III

CELL-BUILDING.

Everything being now in readiness for cell-building, we go to the hive in which the breeding queen is kept and take from it the middle comb, placed there four days previous. We shall find in the bottom of the cells, if examined closely, a white, shiny substance. This is the just-hatched egg. Unless a powerful glass is used we shall be hardly able to see the small grub at this early stage of development.

Alley Method piece of comb

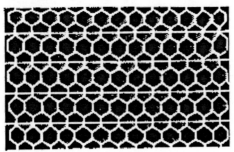

Fig. 3. Comb containing eggs.

Fig. 3 represents a piece of comb containing eggs, with lines running through each alternating row of cells, and showing the manner in which the comb should be stripped up for cell building. This piece of comb was also photographed, and is about one-half the natural size.

We now take this comb into the bee-room which has been warmed to prevent the eggs from being chilled. Cut it into strips with a thin, sharp knife (an old-fashioned table knife ground thin answers the purpose), running the knife through each alternate row of cells as seen in Fig.

3.1 After the comb has been cut up, lay the strips flat upon a board or table, and cut the cells on one side down to within one-fourth of an inch of the foundation or septum, as seen in Fig. 4 A very thin, sharp and warm knife must be used, or the cells will be badly jammed and mangled. While engaged in this work I keep a lighted lamp near at hand with which to heat the knife, as a delicate operation of this kind cannot be performed well with a cold tool, nor so quickly.

Alley Method strip of comb

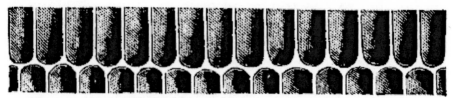

Fig. 4. Strip of comb on which cells are to be built.

Fig. 4 represents a sectional view of one of the prepared strips as cut from fig. 3, and ready to place in position for cell-building.

HOW TO HAVE CELLS BUILT IN ROWS; THE NEW METHOD.

Now we come to a process which I have for many years been trying to effect, and its discovery has proved invaluable to me. In truth, I would have paid one hundred dollars at any time during the first sixteen years of my experience for a knowledge of this fact.

I never was troubled in getting all the cells which I needed built in the old way, but to have them so evenly spaced that each could be cut out without injuring or destroying its neighboring cell, puzzled me for many years; by persistent thought, persevering labor and experiment, however, the matter was at last settled satisfactorily. This one fact alone is worth one hundred dollars to any queen-dealer, and ten times the cost of this book to any bee-keeper, even though they rear but few queens; and I feel assured that all my readers will admit this when they have tested it.

HOW TO PREPARE THE COMB FOR THE QUEEN CELLS.

The strips of comb being all ready, we simply destroy each alternate grub or egg, as seen in Fig. 5 In order to do this, take the strips carefully in the left hand and insert with the right the brimstone end of a common lucifer match into each alternate cell, pressing gently until it touches the bottom, and then twirl it rapidly between the thumb and finger; by this means the egg or grub will be destroyed. This gives plenty of room for large cells to be built, and the bees to work around them and also permits of their being cut out without injury to adjoining cells, Fig. 6 (a full description of which is given elsewhere).

Alley Method Strip of Comb

Fig. 5. Prepared strips with the egg removed from each alternate cell.

Alley Method

Fig. 6. This cut represents one of the frames used in my fertilizing hives, one-half of the comb being cut out to make room for the queen-cells. A A represents the strip of comb containing the eggs (on which the cells are built) fastened to the comb. The frame was photographed (as shown in cut) smaller than its natural size. The cells are exactly as the bees build them by my new method.

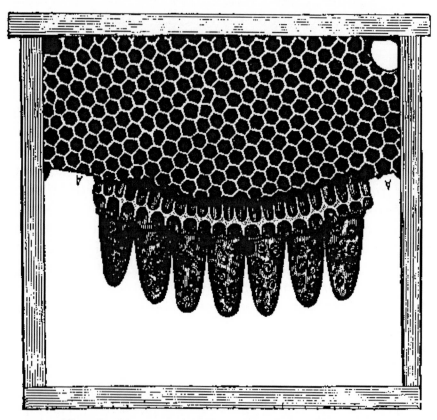

Fig. 6. The new way of having cells built.

All queen-dealers know that this cannot be done when the cells are built by the old method as shown in Fig. 7. I think I hear some "doubting Thomas" say, How will you place or secure this slender strip of comb in a frame so as to prevent its falling down? I would reply, have at hand a sheet iron pan about six inches long, three inches wide, and three inches deep, being rather larger at the top than bottom (or any other sort of iron vessel that will hold hot beeswax and rosin) and so arranged that you can place a lamp under it to heat it. Keep in this pan a mixture of two parts rosin and one part beeswax. Heat this enough to work well, being very careful not to over-heat it, as it will destroy the eggs in the cells if used too hot, and if too cold it will not adhere properly; the right degree of heat will soon be learned by practice. I keep on hand a number of frames of comb which are free from brood or eggs, into which the prepared strips A A are fastened, as shown in Fig. 6, page 15. You will notice that this comb is cut with a slightly convex curve. By putting the prepared strips in after this manner, still more room is given to each cell owing to the spreading caused thereby.

Alley Method old way of having cells built

Fig. 7. The old way of having cells built.

Fig. 7 represents a cluster of cells built by the old method; a piece of comb containing eggs was inserted and as none of the eggs were destroyed the bees built cells in clusters as shown, by this cut. The cluster shows five cells three of which may be saved by transferring.

HOW TO FASTEN THE PREPARED STRIPS IN POSITION.

To fasten these strips, dip the edge which has not been cut into the preparation previously mentioned, and immediately place it in position, so that the mouths of the cells point downward, pressing it with the fingers gently into place, taking care not to crush or injure the cells in so doing. A number of such combs, say eight or ten, and more if necessary, should be kept on hand for this purpose and used as often as cells are needed. This is a great convenience and saves destroying or mutilating other combs. When the cells are cut out, the strip on which they are built should be taken with them.

HOW TO RELEASE THE BEES FROM THE SWARMING BOX.

Everything is now ready for the queenless bees in the box, impatient to be released and anxious to commence cell-building. This box has a strip of tin nailed on the upper edge of one end flush with the outside; the cover has a similar strip nailed on the under side, which corresponds with the one on the box when in place, Fig. 2. (See description at end of volume).

The combs containing the brood prepared for cell-building are now placed in the nucleus hive with other combs sufficient to fill it. Place this nucleus on the stand from which the bees were taken, and at such an elevation that the bottom edge of the alighting board will come just even with the top of the box in which the bees are confined. Next let the bees out by drawing the cover back just enough to allow the worker bees to pass between the strips of tin into the hive.

HOW TO SEPARATE THE DRONES FROM THE BEES.

If there are drones with the bees either black or otherwise objectionable, they will be retained in the box and can be easily destroyed, after the bees have all left. In case the drones are needed (or if there are none with the bees which is often the case), all may be turned out in front of the nucleus, when they will soon run in; this being on the old stand they will accept it as their home and begin cell-building at once from the eggs given them. In no case should any worker brood be given to the bees, thus compelling them to concentrate their whole forces on cell-building. Some capped drone-brood should be given them, if at hand, as it would greatly encourage the bees, and is really an advantage.

WHEN TO PREPARE THE BEES FOR CELL-BUILDING.

I usually prepare the bees in the morning for queen-rearing, and give them the eggs at night. By the next morning they will usually become reconciled to the new state of things and from twenty to twenty-five queen cells will be started; this of course depends upon the number of eggs given them. Just here let me caution all queen-breeders against giving the bees too many eggs, or allow-ing them to build too many cells at one time. If not per-mitted to complete over twelve cells, the queens will be found as good as, if not superior to, those reared under the swarming impulse. If you wish to rear queens of which you and your customers will be proud, you will find the secret is in not allowing any colony to build more than this number. If this precaution is taken, good queens will be the rule and not the exception. Of course, to accomplish this your breeding queens must in all cases be as near perfect as possible, other things being equal. We seldom find more than twelve first-class cells, and often a less number, in a colony which has just cast a swarm, the Cyprians and Holy lands being exceptions. The queens of these races are very prolific and hardy, consequently they rear a much larger number.6

Now we have everything in good order and condi- tion, and cell-building is progressing satisfactorily, except perhaps too many cells are being built for the quantity of bees in the colony. If such be the case, and it does some- times so happen, all over twelve or fifteen should be destroyed. This may be done by means of a match, as before stated. Sometimes, in the hurry of preparation, an egg is passed by and not destroyed, but if the work in the first place is properly and thoroughly done no trouble of this kind need be apprehended.

Chapter IV

FEED WHEN FORAGE IS SCARCE.

Alley Method Fruit jar feeder

Fig. 8. Fruit jar feeder.

If the honey harvest is abundant during the period of cell-building little care is needed until the cells are about ready to be cut out, but if not the nuclei will need liberal feeding (say one pint of syrup each night and morning) to stimulate them properly. In such cases, feeding should certainly be resorted to or inferior queens will be the result. For this purpose I never have seen a feeder so convenient and so good as one constructed as follows: take either a quart or pint improved Mason's fruit jar, Fig. 8, remove the glass top and substitute a tin one in its place, cutting the tin the exact size of the glass top and punch about twenty small holes therein for the food to pass through. Bore an inch and a half hole in the cover of the nucleus, and place the jar bottom upward over it. There should be a space of one-half an inch between the jar and the top of the frames so that the bees can get at the food readily. The bees will take a pint of food in the course of two or three hours, if there are twenty or more holes for it to pass through. For slow feeding make about six holes. If honey is scarce, use granulated sugar and water, say five pints of water to six pounds of sugar, and mix either hot or cold; make a fairly thick syrup. When thoroughly dissolved, flavor with a little pure honey. Do not use glucose or grape sugar under any consideration.

WHAT TO DO WITH SURPLUS QUEEN-CELLS.

At this point we will consider that the bees have been at work four days on the cells, and that they are sealed over or nearly so. If desirable they may be left seven days longer where they are, and then cut out and either placed in nuclei or put in the queen nursery to hatch.

Where queens are reared on a large scale, the combs containing the cells just sealed may be taken from a number of nucleus colonies and given to one, as one colony can take care of one hundred cells as well as a smaller number. Care should be taken to have a colony on hand prepared to care for such surplus cells until it is safe to remove or cut them out.

It will be necessary to give such colony a frame of brood occasionally, to keep it well stocked with bees. These combs should be examined at least once a week in order that no queen-cells are built, as a queen might hatch out some day very unexpectedly and destroy all the cells in the hire. You will remember that each frame containing cells has the number of the breeding queen and the date of starting the cells marked on the top. If proper care is taken to keep a correct record of this in a day book kept for the purpose, you will know exactly when the cells are ready to hatch and the time they should be transferred to nuclei or to the nursery.

CUTTING OUT QUEEN-CELLS.

When the cells are sufficiently matured to be safely removed from the nuclei, cut them out, taking with them the strip of comb on which they were built. They should be immediately taken into a warm room and separated. A lighted lamp is kept at hand with which to warm the knife. Occasionally a small piece of one of the cells is shaved off. The fracture thus made may be easily repaired by placing a small piece of foundation over it, plastering it on, having the knife quite warm, being very careful to make sound work of it or the bees will reopen the cavity and remove the nearly matured queen. After this is done place the cells in the nursery or nuclei.

We will suppose that we have one hundred queen-cells on hand and one hundred nuclei ready to receive them. When the cells have been sealed seven days they may be safely cut out, but it is better to let them remain until the morning of the eighth day after they were sealed, and then give them to the nuclei. Of course, cells are being built in other hives and will soon hatch. The first lot of queens should be fertilized by this time, and disposed of, if the weather has been favorable, and room made for a second lot; but supposing the first lot not to have been fertilized on account of unfavorable weather, and there is not room for the second lot in the nuclei already in operation, what shall we do with them? No queen-dealer can afford to lose a fine lot of cells, especially if he has a large number of orders on hand; every cell and perfect queen must be preserved in some way. How can this be accomplished? I will give my method which is a simple and good one. I provide against such a contingency by having a queen nursery of my own invention, Fig. 9.

Alley Method Queen Nursery

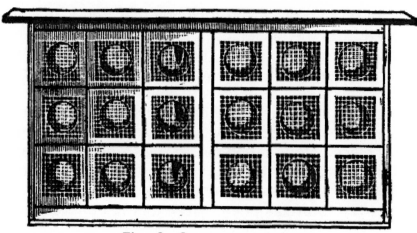

Fig. 9. Queen nursery.

Alley Method Nusery Cage

Fig. 10. Nursery cage.

QUEEN NURSERY AND HOW TO USE IT.

I use eighteen cages in one standard Langstroth frame; each cage, Fig. 10, has a place in it for a sponge to contain the food, and another for the cell. An inch and one-half hole is made in each cage, both sides of which are covered with wire-cloth. Each cage is provided with food sufficient to last a queen one week. The cells are cut out and placed in these cages which are then placed in the centre of a full colony. As eighteen of these just fill a frame they will stay in place without any fastening whatever. A bungling workman cannot make them so that they will work just right. The cages must be exactly alike, and then there will be no trouble in having them stay in place. They should be cut out with a circular saw run by steam power. I make it a point to do all such work in the best manner possible.

By the use of this nursery, my queens are all hatched in the brood-chamber and in nearly the natural way, and by the natural warmth of the bees; no lamp nursery nor other artificial devices being used, and none of which ever should be used in queen-rearing. When one digresses from the paths of nature, in this business, the more unsuccessful he will be. When the queens are in the nursery in the centre of the brood-chamber, they are perfectly happy and contented, and will live there safely for several weeks. In no other way have I been able to preserve them so long. Each cage must be supplied with food, as a colony with a laying queen will not feed virgin queens, and oftentimes even a queenless colony cannot be depended upon for doing so, as I have learned to my sorrow.

INTRODUCING VIRGIN QUEENS; HOW LONG TO KEEP A COLONY QUEENLESS BEFORE INTRODUCING A QUEEN.

This is another important point which properly comes under the head of queen-rearing. It has been said that virgin queens cannot be successfully introduced. Those who assume this certainly mistake. I am obliged to introduce hundreds of them every year, and have no trouble in so doing. I seldom have occasion to introduce them to full colonies, but that it can be safely performed I have no doubt.

In order to introduce such queens successfully the colony should remain queenless three days (seventy-two hours); then give them a pretty good fumigating with tobacco smoke. Remember, the bees must remain queenless three days at the least, and during the meantime no queen must be near them, otherwise the operation will prove a failure. Virgin queens can also be introduced successfully by daubing them with honey and using no tobacco smoke. Put a little honey into a tea-cup and roll the queen in it so as to daub her thoroughly, then drop her from a spoon into the hive among the bees. They will at once commence to remove the honey and when they have done so the queen is safely introduced. This is a much slower process than by fumigating them with tobacco smoke, but quite as successful. Do this just before sunset. When tobacco smoke is used to introduce them, throw some grass against the entrance to keep the smoke in and the bees from coming out. Blow in a liberal amount and then let the queen run in at the top through the hole used for the feeder.

ANOTHER WAY TO INTRODUCE VIRGIN QUEENS.

Make a cage wholly of wire-cloth or such an one as is used in mailing queens. Cut a mortise from the main apartment to the outer edge as seen in Fig. 11, cage the queen and fill the mortise with Good's food; by the time that the bees have removed it, they will have become acquainted with the queen. Bear in mind that the bees must be queenless three days before introducing virgin queens just a little tobacco smoke is used to scent the bees at the time the cage is put in, I think the undertaking will be rather more successful. Laying queens can be introduced by the same process. A colony made queenless for the purpose will always begin to rear a number of cells. When the new queen is introduced, they immediately stop cell-building; all are destroyed, and the bees commence to pay royalty to the new queen given them.

Alley Method introducing cage

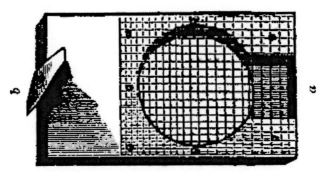

Fig.11. introducing cage.

Chapter V

THE OLD AND NEW METHODS OF CELL-BUILDING.

I presume the reader has followed the descriptions carefully and understands now how to have queen-cells uniformly built, so that none need be destroyed in cutting them out. By the old method a large number of fine cells must necessarily be destroyed in transferring them, as they are built so irregular and oftentimes so close together that three or four are rendered useless. It is also exceedingly difficult to determine when the cells will hatch, as the bees will use eggs or larvae in various stages from which to rear queens. Again the bees will continue to start cells three or four days after the brood is given them. My method does away with all this trouble, and we can determine to a certainty, within a few hours at least, the time when the cells will hatch. This is one of the great advantages to be derived from its use, and again, there is no guesswork about it, and no eight or ten day queens need be anticipated as none can be reared, simply because there are no old larvae for them from which to rear queens.

HOW TO REAR QUEENS IN FULL COLONIES AND STILL HAVE NONE QUEENLESS.

I have asserted that my queens are reared in full colonies and none are ever queenless. Well, none of my standard colonies are ever without queens.

I first determine as nearly as possible the number of queens I intend to rear during the season, and then select enough good strong stocks for the purpose. One half-dozen colonies will rear a large number of queens in a season, as will be seen by what follows. I deprive the first half-dozen colonies used for queen-rearing of all their brood and queens, giving the former to weak colonies to build them up, as by this process I can soon make them strong. I prepare as many colonies in this way as I wish to keep building cells at one time and as I think will meet my wants for one season. I would state here that no colony should be permitted to build two lots of cells; I mean by this that the first lot of cells should not be removed and eggs given them to build others. I consider it poor policy to do so, although such a colony will build a second lot of cells; but they would produce inferior and almost worthless queens, and a queen-dealer who would do so would soon lose his reputation.

HOW TO OBTAIN MORE CELLS.

When I want a fresh lot of cells, I take a strong colony and remove all the bees, by the process described on page 8. Replace the combs, and put the hive exactly where one of the colonies has been standing which has just finished a lot of cells. Then brush or shake the bees from the combs of the latter hive (care being taken not to shake the comb on which the queen-cells are built), in front of the hive from which the strong colony has just been taken, and give them the queen from the first colony.

Bees enough should be left with the cells to keep them from chilling, or the frame on which they are built can be placed in a hive where other cells are being built, or add one frame of honey and one of brood, and form a three-frame nucleus. When the cells are ready to hatch, transfer all but one to other nuclei, and thus gradually form the needed nuclei for the season.

HOW OFTEN THE SAME COLONY MAY BE USED FOR CELL-BUILDING.

Having taken care of the cells and brood in the full colony, we now have a fresh lot of bees on our hands, which will be ready in a few hours to build other cells, while the bees which have just completed the first lot will continue the work in the hive from which the fresh bees were taken. In three or four weeks the same process can be repeated, as the old hive will then be full of newly hatched young bees. After getting bees enough to start the first half dozen lots of cells, no more hives need be made queenless and every colony will be kept strong during the season, as they will have a laying queen all the time. It will be seen by any intelligent bee-keeper that it will not require half as many colonies to rear one thousand queens, as by the old process. This I also claim as original.

DESTROYING DRONE BROOD AND WORTHLESS DRONES; THE USE OF THE DRONE TRAP.

Where the extractor is used drone-brood not needed may be destroyed after the honey has been slung out, by uncapping it, being careful to shave off the heads of the drones; for this purpose I use a thin, flat knife such as Mr. Peabody sold with his extractor. Excessive drone-breeding can be kept down by such a process, when there are only a few hives kept. I cannot recommend this process in a large apiary and some other device must be resorted to.

Most any one, I think, has sufficient ingenuity to construct a drone trap for the purpose of destroying worthless or surplus drones, I find it rather difficult to describe one so that all may understand it, but shall have an engraving made and give a description of it at some future time. All that is needed is to place a gauge at the entrance so that the drones cannot get out, and make an outlet for them to pass through into a box from which the workers can escape and they cannot. The swarming box, only on a smaller scale (of which a description is given at end of volume), will answer every purpose. Care must be taken, if a trap is used, not to smother the colony; this will not be the case if the bees have an easy means of escape from the hive into the trap. The outlet for the drones to pass into it must be made large enough for them to pass through freely.

HOW MANY QUEENS OUGHT A COLONY TO REAR?

I have frequently cautioned queen-dealers and those who rear queens simply for their own use, against rearing too many queens in one hive at the same time. This is so very important that I must be excused for repeating it. By my plan one hundred queens can be raised by a colony as well as twenty-five; but the more queens reared the poorer they will be. The correct number, as my experience teaches me, is about twelve queens to a colony. I have found that worthless queens are reared under the swarming impulse as well as by artificial means. If a queen is removed from a full colony they will build from twelve to twenty queen-cells. Very few of these will prove to be valuable queens, as a colony thus made queenless will not start cells from the eggs. They will select larvae two or three days old for cells as their impatience leads them to diverge from nature's course every time. The queens reared from the latter cells always prove shortlived and almost worthless. I am aware that a large number of the queens reared are produced by the latter process. This statement is founded on the writings of many who rear queens.

HOW TO REAR VALUABLE QUEENS.

To rear valuable queens, I think the hive should not contain over twenty-five eggs to start with, and in about twenty-four hours after the cells are started, at least ten of them should be destroyed, so that not over fifteen remain to mature. Of course we cannot afford to sell such queens for seventy-five cents or one dollar each. Out of twelve cells that hatch, probably four of that number would never mature or they would be destroyed in some other way: lost in mating, killed in the hive, or by some other mishap to which queens are always liable and exposed.

Chapter VI

HIVE TO USE FOR YOUNG QUEENS.

For many years I have used the small hive described on page 4, for queen nuclei (for fertilization only), and find them as convenient and handy as any. They are made large enough to take four combs, but I seldom use over three to each hive.

Many large queen-dealers use the same kind for this purpose, and find them just as good as larger ones, much handier and less expensive. In the fall the bees are united with other colonies, and the combs packed away in barrels for use another year.

HOW TO FORM NUCLEI.

To prepare these hives for the cells, or young queens, I proceed as follows: if a box-hive is to be broken up, with which to fill them, it is taken into the bee-room. The bees are treated the same as though, they were in a frame-hive, viz.: induce the bees to fill themselves with honey by closing the entrance, smoking them, and rapping smartly on the hive for ten minutes or more. After driving all the bees out that can be induced to leave the hive, proceed the same as in transferring, placing the bees in the cap of a hive or box until ready to be placed in the nuclei.

Alley Method ventilation screen

Put the comb containing brood and that containing honey into the small frames. Give each hive at least one frame of brood, one of honey and one empty comb. Place two of the frames in the hive then put in about one pint of bees, putting the third comb in last, then place the cover on. Confine the bees in the hive forty-eight hours before permitting them to fly. If there are but few drones in the hives and we wish to destroy them, it can be quickly and easily done by the following plan: go to each of them early in the morning of the day on which they are to be liberated, take an empty hive with you, give the bees a small dose of tobacco smoke, let them remain quiet a few moments then examine the combs separately and very carefully, and pinch the head of every drone found. Put the combs with the adhering bees into the empty hive, placing it on the stand of the hive just examined. The bees must have plenty of air while confined in these small hives. For this nail on the front of each, Fig. 12. It should be made roomy enough so that the bees can come out from the body of the hive into it, and get all the air they need and return as often as they choose.

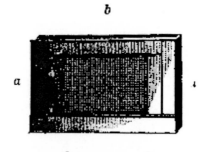

Fig. 12. Screen used for purpose I use a small screen to ventilation.

WHERE TO PLACE NUCLEUS HIVES.

Nuclei used for fertilizing queens should not be placed very near colonies building cells, as the queens, when returning from the marriage tour, are quite liable under such circumstances to enter the wrong hive, and young queens, even though not fertilized, are always welcome to a queenless colony. Nuclei for fertilizing queens should not remain queenless long at a time; if they do there is great danger of fertile workers gaining possession of the colony and they are the pests of the apiary. I have seen many valuable queens destroyed by them.

WHY BEES BALL AND DESTROY QUEENS.

When queens have returned from the flight in search of drones they are sometimes seized by the bees, hugged or smothered to death (called balling), stung or injured, wings torn, or one leg stung and rendered useless. In almost every instance where this hugging takes place the queens are nearly ruined, this being more than they can endure without injury to their fertility. When this takes place one may know that fertile workers infest the colony.

HOW TO INSERT QUEEN-CELLS.

Having everything properly prepared we are ready to give to each nucleus a queen-cell. This can be done without taking out the combs or cutting them, as is the general method practiced by most queen-dealers, and given by the bee journals as the best. I generally find plenty of room between the combs without disturbing any of them. If not, I slip one of the frames a little to one side, place the cell in position point down ward, of course, and gently press the comb back against it. It will thus be held securely in place and will hatch as well as though inserted in the centre of the comb. By doing this, the combs are not mutilated and the operation is quickly performed.

If young queens are to be given to the nuclei instead of cells, proceed according to directions given on page 25. Bear in mind that very young queens are more acceptable than those four or five days old. It is much more troublesome and more difficult to introduce older ones, and the latter will be destroyed unless scented by being fumigated with tobacco smoke or by some other means, the idea being to deceive the bees, which can be done by scenting them all alike.

A queen-cell can be given to a colony immediately after removing a queen from it, and should the young queen emerge from the cell within an hour she will generally be kindly received, and thus safely introduced.

Occasionally, a queen hatching so soon after the cell has been introduced will be killed, but this seldom happens.

Cells may be given to queenless colonies at any time, but queens should not be given to any colony until it has been queenless three or more days.

THE AGE AT WHICH QUEENS ARE FERTILIZED.

We read quite frequently in the bee journals of queens becoming fertilized when only three days old. This may be true, but in all my experience, I never knew one to take her wedding flight when less than five days old. In from thirty-six to forty-eight hours after this they usually commence to lay.

Early in the season they generally come out, between the hours of one and three P.M., and sometimes as late as four P.M. After the first of September they will fly as early as 11.30 A.M., and not much later than two P.M. unless the weather is very warm and pleasant.

HOW TO FORCE THE DOTING QUEENS TO FLY.

In localities where forage is scarce, some means must be adopted to stimulate the bees and cause the queens to fly when they are not disposed to do so. This can be accomplished by feeding the bees. The nucleus feeder Fig. 13, which I have used for twenty years, will hold one ounce of syrup and is admirably adapted for this purpose. Such colonies as have queens old enough to fly are fed during the forenoon and the queens will fly in the afternoon and generally be fertilized; whereas if they are not fed they will not leave the hives sometimes until a week later.

Alley Method Cone Feeder

Fig. 13 Cone Feeder

QUEENS FERTILIZED BUT ONCE; HOW FAR TO KEEP THE RACES APART TO INSURE PURITY.

I am satisfied that no queens are fertilized more than once. They may fly more than once, but if they make the second flight and bear all the indications of having met a drone, it is pretty certain that they were unsuccessful the first time.

There are some who entertain the idea that a race of bees cannot be kept pure unless they are kept isolated several miles from all other races. I have tested this matter carefully and pretty thoroughly during the last twenty years, and have found that one-half mile is as good as a much greater distance.

While drones will sometimes fly a mile or more, the queens will not. This fact can be ascertained by watching a queen when she takes her wedding flight. She seldom is gone over five minutes and nine out of every ten will return within five minutes. Of course if the young queens are only one-half mile from a large apiary of black bees, there must be an abundance of Italian drones in the same yard with them. This being the case not one queen in twenty-five will mismate. This rule applies equally well to all races.

HOW TO KEEP LAYING QUEENS.

Sometimes queen-dealers and often other apiarists have occasion to keep laying queens on hand several days before using them.

Surplus queens can be kept on hand a long time, in the queen nursery, provided the cages are well supplied with food; this being the only attention needed. The sponges in the cages must be refilled with honey as often as once each week at the least.

The nursery should be placed in the centre of the brood-chamber of a full colony. To make room for it either remove one of the outside combs, or, in case the colony is strong, one of hatching brood. The latter may be given to some weak stock. When the nursery is taken from the hive and no other queens are at hand, fill the space then left vacant either with an empty comb or a frame filled with foundation.

A few queens may be kept by placing the cages between the cushion and frames, in such a manner that the bees will have easy access to them. A colony having a fertile queen cannot be depended upon to feed other queens under such circumstances, hence the importance of supplying the cages with food. A person must use his judgment regarding keeping queens in this way, when the nights are cool. There need be no apprehension regarding this point between June 1 and Sept. 1.

FEEDING NUCLEI; WHY THEY SOMETIMES SWARM OUT.

Nucleus colonies (in hives, described at end of volume), must be fed as often as once each week, unless forage is abundant, or they will "swarm out," even when they are well supplied with brood and honey, as such colonies are easily discouraged.

They will not do so if fed a small amount of syrup occasionally. Use only the best sugar-syrup (not honey), giving it to the bees the same day on which it is prepared. Never use poor food as it soon sours, and runs out of the feeder besmearing the bees and combs. This will surely cause them to desert the hives.

The following incident well illustrates this point: one morning I fed fifty nuclei with some syrup which had been prepared but twenty-four hours. The weather was very warm; the syrup became sour and ran out of the feeder besmearing the bees and combs, every colony of which swarmed out and united in one cluster. This resulted in a loss of nearly fifty dollars to me.

Chapter VII

UNPROFITABLE BREEDING QUEENS.

I have found that many of the young queens from some mothers are lost on their " marital tour," although such queens when successfully mated prove very valuable. This is a singular and unnatural phenomenon for which I cannot account. Why they fail to return to the hive is more than I can comprehend. Queens having this imperfection should be discarded as breeding queens at once, notwithstanding they may in many other respects be very desirable as queen mothers. We cannot afford to use queens of this class from which to breed others, when so many of their young queens are lost in mating. It is rather discouraging when examining a lot of nucleus colonies, where there should be a number of fine laying queens, to find none, they having been lost while on their first flight.

MOST PROFITABLE BREEDING QUEENS.

I bred from a queen last season not one in fifty of whose royal daughters was lost in mating. These are the only profitable ones from which to breed. For breeding queens select carefully only those which are very prolific; whose royal progeny are fair in size and handsome, whose worker bees are uniformly marked, gentle, good workers, and perfect in all other respects. Purity of stock cannot be maintained unless great care is taken in selecting the queen mother. Never use one whose workers have from one to three bands. The young queens from such an one would show a variety of markings, black, striped, and a beautiful yellow, the yellow ones being sadly in the minority. If beauty, purity and business qualities are desired, such a queen would be worthless. Always select one having the markings which show her to be pure, prolific and hardy. This subject is more fully explained elsewhere.

THE SELECTION OF THE DRONE MOTHER.

I have long contended that success in queen-rearing depends largely upon the drones used for mating. The same care should be taken in the selection of the drone mother as with the queen mother. Her worker bees should be well-marked, gentle, good honey-gatherers, hardy, and absolutely pure; the drones large, handsome, and very active. I never permit drones from all my colonies to fly promiscuously, or have any haphazard mating of queens. Only selected drones having the above-mentioned qualities are tolerated in my fertilizing apiaries.

HOW TO REAR AND PRESERVE DRONES.

It is well known to most bee-keepers that colonies having fertile queens will neither rear nor permit drones to live in the hive late in the season, and seldom when forage is scarce. If queen-rearing is going on, drones must be procured at any cost, and some means must be adopted to rear and preserve them for use in the latter part of the season. To do this I pursue the following me-thod: have at hand several extra frames of drone comb; insert one in the centre of the colony from whose queen you wish to rear them. Feed this colony liberally if forage is scarce. Examine them in the course of a week; if the comb is well filled with eggs and larvae remove it to a queenless colony.

Instinct teaches queenless bees the necessity of rearing and caring for drones, hence they can always be depended upon for them provided the brood is given them.

Replace the comb just removed with an empty one; continue this as long as the queen can be induced to lay drone eggs. Remember that queenless bees never destroy drones, while a colony having a fertile queen will invariably do so, unless encouraged to preserve them by being fed. It is a very difficult matter here in the north to induce queens to lay drone eggs in September, even when forage is abundant; hence drones to be used in September and October should be secured in the early part of August, as most colonies seem disposed to rear them at this time.

HOW TO JUDGE OF A QUEEN BEFORE TESTING.

The practiced eye of an expert in any vocation can detect imperfection where the novice cannot. My experience enables me to judge of the qualities of a queen; whether she will prove prolific or otherwise, as soon as she has laid several hundred eggs. The laying queen, if a good one, will deposit all her eggs in the cells in exactly the same position. Every egg will point downward, and will be large and plump when compared with those of an unprolific one, and every cell not otherwise occupied will contain an egg. I can also determine whether a queen is worth preserving or not the moment she leaves the cell. An inferior queen in gnawing through leaves a little ragged hole instead of cutting a large clean cap, Fig. 14, and leaving an opening nearly large enough to admit the end of the little finger, as a strong and well developed queen always does. It is worse than useless to preserve such queens.

Alley Method cell from a strong queen hatched

Fig. 14. Cell from a strong queen hatched.

DESTROY WEAK AND FEEBLE QUEENS.

After the cells in the nucleus or those in the nursery are hatched, examine the outlet to each, and if small and ragged destroy the queens at once. A good prolific queen will lay nearly or quite four thousand eggs in twenty-four hours. One that will not do this is not worth preserving. I never saw a queen that I considered too prolific for my own use (the opinions of some others to the contrary notwithstanding). I want queens that will deposit ten thousand eggs or even more in a day if they can be reared.

Very few bee-keepers are troubled with queens being too prolific. I should as soon find fault with my hens for laying two eggs per day when eggs are worth forty cents per dozen, as with a too prolific queen when honey is worth twenty cents per pound. I am aware that hens do not eat the eggs after laying them as the bees do the honey after gathering it; but if the hens laid no eggs there would be no profit, and if a queen is not prolific there is no income from that source.

The main object in rearing queens is to get hardy and prolific ones, the more prolific the better. A queen that will lay five thousand eggs in a day is worth one hundred that will lay but one thousand in the same time. My opinion is founded on experience and the result of careful experimenting, and I believe that a large majority of bee-keepers and those of extended experience are of the same opinion.

LARGE VERSUS SMALL QUEENS.

I find customers occasionally who want large queens. A person engaged extensively in queen-rearing will have them of most every size.

I must confess that I like the appearance of large and handsome queens; but they do not as a rule prove to be the most prolific or profitable. Queens of medium size are generally the best. They have proven so with me. Good queens are those that keep their hives well filled with bees. The color or size has no effect on their fertility.

An experienced bee-master can judge of the quality of the queens which he is rearing even before they leave the cell. If the cells are short and blunt when just sealed they should be destroyed at once, rather than wait and destroy the queens after they have hatched, as such queens would prove worthless.

The cells containing good queens are long and pointed, and heavily waxed with a rough surface. The bees when constructing the cells seem to understand the condition of the embryo queen, and whether she will be strong and vigorous or otherwise.

Cells having the blunt point and thin walls, so thin in fact that the young queen can be seen through them, generally contain very poor ones. Queens that delay long before becoming fertile in favorable weather will not prove of first quality. A smart, active queen will invariably leave the hive on her wedding trip when she is five days old, and in all my experience I never knew one to become fertile at a younger age.

BEST BEES FOR QUEEN-REARING.

A person rearing queens extensively, and keeping several races of bees in his apiaries, should note those that build the largest and finest cells, and rear the best queens.

The Holyland bees will build a large number of cells if permitted to do so, and the queens reared by them are very large and prolific. All things considered, I believe them to be the best of the four races for cell-building. The Cyprians come next in value for this purpose; there is, however, little difference between these. My acquaintance with the Cyprian and Holylands, however, has not been as extensive as with the Italian and black bees, and I will give my experience regarding Italians, as it may differ from that of others.

THE ITALIANS NOT A DISTINCT RACE.

The fact that the Italians are not a distinct race is well established and generally admitted; hence it becomes necessary, in view of this, to propagate the other races in order to keep the former up to the standard and maintain their established reputation as a superior race.

I have tested the different races and find that the Italians are the least adapted for cell-building. I do not understand why this is so. Prior to the introduction of the Cyprians and Holylands, I always gave the black bees the preference as nurses, whenever I could procure them.

The cells built by the Italians are small, many being similar to those described on page 45. I would not hesitate for a moment to destroy them as soon as sealed. It would be folly to permit queens from such cells to become fertile, as they would prove worthless, and much valuable time would be lost in testing them. Distinctly remember that rearing queens artificially is quite a different process from that pursued by the bees when allowed to follow their natural instincts. The Creator has instituted perfect laws governing insects, and bees comply with the requirements of these laws when allowed to rear queens in the natural way; the result being perfect queens.

In view of this and on account of the increasing demand for queen-bees, it becomes necessary to adopt some artificial means by which equally as good queens can be produced. In all my experimenting with bees for this purpose, I have imitated nature as perfectly as possible.

REARING QUEENS FROM THE EGG.

Most intelligent and experienced bee-masters agree with me in the opinion that queens should be reared from the egg, as, other things being equal, they prove the best in all cases.

I formed this opinion during my first year in the business, have had no reason to change it, and claim that this is the only way to rear queens which will compare favorably with those produced in the natural way under the swarming impulse.

Many years ago, I frequently heard an old beekeeping friend make the following remark: " If you want large queens start them from the egg." Anyone can satisfy himself of this fact by thoroughly testing it for a few weeks. Full directions for procuring and keeping a continuous supply of eggs for this purpose may be found on page 4, which should be carefully followed in order that you may be prepared to start cells at any time.

THE PROPER TIME TO COMMENCE QUEEN-REARING.

I make it a point to have the hives well stocked with bees and in a proper condition to swarm, having sealed drone-brood before I start my first lot of cells. Here in New England, in favorable seasons, queen-rearing may be commenced the first week in May provided the bees are properly stimulated.

It is well understood that natural cell-building depends upon the following conditions, viz.: strong and populous colonies, a good supply of drone-brood and young drones, vigorous queens, warm and genial weather, and a plentiful flow of honey. Hence, you will see the necessity of stimulating the bees for early breeding by giving them a liberal supply of syrup, and in every other possible way.

I consider the above the best indications of their readiness to commence cell-building.

Chapter VIII

THE NEW WAY VERSUS THE OLD.

I wish to point out some of the advantages which the new method has over the old. The latter having been fully described in the various bee journals and standard works on apiculture, the reader must necessarily be familiar with them.

The first point is to make a colony queenless. A few hours later, give it a comb containing eggs and larvae (either with or without holes cut in it), permitting the bees to build the cells where they choose. They generally build them in clusters (see Fig. 7), and so closely joined that they cannot be separated without destroying many of them. The eggs given vary in age from one to three days, consequently when the queens commence to hatch, it will be from one to three days before they have all left the cells; and when a larva three days old is selected, the queen will hatch in nine or ten days. Unless carefully watched this early hatched queen will destroy the remaining cells by gnawing through them near the base, and, when she has made an opening of sufficient size, will deliberately sting the imprisoned queen, the bees finishing the work thus ruthlessly begun. I used quite frequently to find five or more queens in the hatching-box or in the nuclei at one time, and of course many of them would be stung.

HOW YOUNG QUEENS MANAGE, WHEN TWO OR MORE HATCH AT ONE TIME.

Perhaps my readers are not familiar with queen-rearing, and do not know how the young queens manage affairs when several emerge at the same time. Where several cells are clustered together, as shown in Fig. 7, several of these queens are likely to hatch at the same moment. When this occurs and they meet, a mortal combat ensues; the conqueror coming out of the struggle unharmed, and the other receiving a fatal wound is left unmolested to die, unless some merciless worker seizes her by the wing and attempts to drag her out of the hive. Please remember that worker-bees never destroy a cell that contains a healthy queen. In all my experience I never knew such a thing to occur. Every queen will be permitted to hatch unless attacked by a hostile queen. I am, however, fully aware that many writers state that the workers do destroy them. When breeding by the method just described, I was obliged to spend many sleepless nights watching such cells as they could not be separated, and they would almost invariably hatch at night. The lamp-nursery system has many advocates, but I have never tried this plan as I consider it contrary to natural laws. Cells need the natural warmth of the bees, and it is almost impossible, in the lamp-nursery, to keep the temperature even. Such means will produce poor, weak queens.

After long experimenting, I discovered my method as described in this book. With this came a wonderful change.

The eggs being of one age when the cells are started, I can accurately determine the time when they will hatch, and they all do so within a few hours of each other, The cells are uniformly built and can be easily separated. By the use of the queen nursery they are hatched by the natural warmth of the bees in the brood-chamber. The fact of knowing just when the queens will hatch, would have saved me hundreds of dollars had I known it when I first engaged in the business. Now I am not obliged to sit up nights to watch the cells and save the hatching queens, and I feel certain that my readers, without one exception, will admit that the above is entirely new, simple, and practicable, and see at once the advantage to be gained by its adoption. If the directions given are strictly followed, no queen need be lost when hatching, as the time can be calculated to within a few hours, as follows: the egg will hatch in three and one-half days after being laid; four and one-half days later the cell will be sealed, and in eight days more the young queens will hatch out, making sixteen days, the time required to rear a queen from the egg just laid.

HOW TO REAR A FEW QUEENS.

As many bee-keepers desire to rear a few queens for their own use, thus combining pleasure and profit, I will give special directions for so doing. The general rules given in a preceding chapter should be followed, but all operations will necessarily be on a smaller scale. If you wish to rear about a dozen queens, go to a strong colony at sunset, remove the queen, and on the following night take away all the unsealed brood (replacing it with empty combs) which maybe given to some weak colony; then examine all the remaining combs, carefully destroying all cells which have been started. Now give them eggs prepared for cell-building in the manner described on page 12. Mark the date of starting on the hive. Five or six days later take a three-frame nucleus hive, and place two combs in it; now take the comb on which the cells are started, together with adhering bees from the first hive and place it in the centre of the nucleus hive between the two others; then select several combs of brood from strong stocks, giving them to the colony from which the cells were taken. Next, give them a queen, letting her run in at the entrance and she will be kindly received.

Or, in case we wish to rear other queens, remove the bees from another strong colony, replacing them with those that have just completed the lot of cells, giving them the queen just removed and proceed as with the former. It will be necessary to have one of the swarming boxes for the last lot of bees. The above directions apply only to full colonies and standard frames.

HANDLING QUEEN-CELLS.

This also is an important matter to those who rear queens. Great care should be taken in handling combs on which cells are built, as when the cells are not sealed the slightest jar may detach the pupa or nymph from its position and separate it from the jelly-food; and, although the bees may elongate the cells and save the queens, they will be permanently injured, but in most cases of this nature the cells will be destroyed soon after being sealed. The combs should not be handled except when absolutely necessary, and never tip or turn them bottom upward. In no case attempt to shake the bees from the combs, allow them to stand in the sun, or exposed in the cool air until they become chilled. Any rough or careless handling will result in injury to the embryo queens.

The wings of the young queens are not perfectly formed until within twenty-four hours of the time they hatch; and if the cells are subjected to such rough treatment, many of the queens will be crippled by having defective wings or legs, or perhaps the abdomen deformed. These precautions apply equally well when cutting out cells.

Queens may be hatched from those that have been chilled, but they will be weak and sickly; hence worthless.

In changing combs with cells on them from one hive to another, do not brush the bees from them, but let them remain to protect the cells from the extreme changes in the atmosphere. The bees adhering to such combs are kindly received by others under like conditions and cir-cumstances.

Queen-cells should never be exposed to the burning rays of the sun as the cell in which the queen is encased is almost air-tight, and such exposure invariably produces suffocation and death. If the temperature in the room where the cells are being prepared for the nursery or nuclei is slightly lower than that of the hives, the cells will not be seriously affected by it, but do not keep them out of the hives longer than is absolutely necessary.

REASONS WHY QUEENS FAIL TO LAY, DIE SUDDENLY, OR ARE SUPERSEDED.

During the past ten years complaint has been made by some of my customers that queens sent them failed to lay after being introduced. Had these reports originated with unreliable parties, I should not have considered them worthy of notice, but on the contrary the complaints came from customers in whose honor and integrity I have the greatest confidence.

This led me to investigate the matter fully, as to whether queens taken from strong colonies while they were in a high state of prolificness and fertility were not more liable to injury in transit.

The matter was certainly of sufficient importance to demand thorough investigation. Occasionally, the pur-chaser would report: "My queen came to hand and was safely introduced, but has not laid an egg up to date." Of course I felt chagrined to hear a report like this from a customer to whom I had sent a tested queen.

When dollar queens are sent such a report will not surprise any dealer, as such queens are neither warranted nor tested; the only stipulation regarding them being that they are fertile, and little further is known of them as they are generally shipped as soon as they commence laying.

I was well satisfied that if these reports were correct, the injury must have been the result of rough treatment during transit.

In order to substantiate my opinion and conclusion, I was obliged to experiment considerably; consequently I removed, on different occasions, several queens from nuclei and full, vigorous colonies, keeping them in the nursery for a few days before shipping, also keeping a correct record of these queens and their destination in order to ascertain if they reported all right. No complaint came regarding them, hence I concluded that I had discovered one of the causes and also the proper remedy for it.

The above is not the only reason why queens fail to lay. Unless properly introduced, they will be rendered worthless before they have been in the hive an hour. Sometimes they will be slightly stung, but not sufficiently injured to cause immediate death, although rendered incapable of laying. When the hive is opened the queen is, apparently, kindly received by the bees and successfully introduced, as the marks made by the sting are not always easily recognized.

Occasionally they receive a sting in the leg rendering it useless, and such injury is easily recognized. Nevertheless, the queen will continue to lay, but not to the extent that she would had she received kind treatment from the bees when introduced. Sometimes, several weeks will elapse before they show any indications of failing or having been stung, and then are superseded, or as the term is "missing."

Parties purchasing queens should not hastily condemn the dealer, but should carefully study the causes of the loss. There are many reasons why queens die suddenly, fail to lay, or are superseded soon after being introduced, the principal of which have been described above with the remedy for the same, and regarding the others I can only say, use caution in introducing them. I would advise the reader (if a dealer) to keep queens "to be shipped," in the nursery for a few days before sending them out. I am satisfied that should this plan be generally adopted, less queens will be lost or injured in shipping.

Chapter IX

TESTED QUEENS THE STANDARD OF EXCELLENCE.

This is an important subject and one of great interest, especially to every honey-producer. Queen-rearing has become a specialty, and honey-producers who wish to rear queens for their own use, and the queen-breeders who desire to infuse new blood to prevent in-and-in breeding want good reliable stock, and in order to obtain this call for tested queens.

I think that the importance of the term is often forgotten; some consider that it simply applies to a queen whose worker progeny bear the markings which indicate purity. If so, they mistake, and I think it due to the dignity of queen-rearing and apiculture that this matter be more clearly explained and definitely established.

There must be some standard of excellence and I consider that this is implied in the term tested queens. It is not my intention to detract from the value and importance of the business by devising means for producing cheap queens, but to establish a method by which the best may be produced, thereby promoting its interests and worth.

If there is one thing more than another which will degrade any business or profession it is trying to produce a cheap article. This invariably leads towards fraud and deception, and results in general injury and loss. Where is the dignity of the mechanic to-day as compared with the past ? This principle of doing cheap work has ruined it, and it is almost impossible to get honest work done by contract.

Now with regard to queen-rearing, teach the mass of bee-keepers some way to rear queens cheaply in large quantities, and the country will be flooded with poor and worthless queens. First-class queens cannot be reared and sold for one dollar, and those who expect to get such queens for that price will be disappointed.

Tested queens are those bred from the best stock and kept in the apiary until the value of their progeny regarding honey-gathering and purity has been thoroughly determined; and no queen should be shipped as tested until the above mentioned requirements have been complied with. All tested queens should be graded, the best being selected tested. I have such an one in my apiary for which I should refuse one hundred dollars.

SELECTED QUEENS.

Selected queens are those which give promise of being perfect in all respects before being tested. When I open a nucleus hive and find a large, handsome and prolific queen, one that is putting the eggs in every cell in exactly the same position, I mark her either selected or for testing; and if she is shipped before she has been kept long enough to test her progeny she is called a selected queen. Such queens (all things considered) are the cheapest in the end for bee-keepers generally.

WARRANTED QUEENS.

My apiaries are so located and arranged that very few of the queens will mismate; hence most of the queens, sent out as warranted, prove to be purely mated; they are reared from the best stock and just as carefully as the tested ones. The chances are that the purchaser will be well satisfied with the warranted queens, especially if he be a honey-producer, although I consider the tested queens far preferable for those who wish to breed queens.

DOLLAR QUEENS.

As before stated, I regard the production and sale of dollar queens (so called) an injury to apiculture and its interests. I do not rear such, have none for sale, and advise the reader never to purchase them of any dealer.

ROBBING NUCLEI; HOW PREVENTED.

When nuclei are kept in the same yard with full colonies there will always be more or less robbing during scarcity of forage, especially when feeding is resorted to; and any apiarist who has once experienced wholesale robbing in his apiary will never forget it. To prevent this, feed only white sugar syrup as there is no "enticing scent" to such plain, simple food.

Clear honey cannot be safely fed, no matter how much precaution is taken, and should not be used even though the honey costs nothing and sugar fifteen cents per pound.

I know of nothing more discouraging to the queen-dealer (unless it be unfavorable weather) than upon examining his bees to find the strong colonies robbing their weaker neighbors; and as it is not easily controlled when once commenced, every means should be used to prevent it, as " an ounce of prevention is worth a pound of cure." There are, however, times when we shall find the bees robbing in spite of all our precaution; and there are several plans either of which may be resorted to in such cases, and unless this is done every nucleus in the apiary will be ruined in a short time. When a colony is being robbed, close the entrance at once to keep the robbers in the hive from coming out and others from entering. After giving the robbers confined sufficient time to fill their sacs, release them sprinkling them with flour as they leave the hive and trace them to their home. The hive being robbed should be closed and opened repeatedly until all the robbers have vacated. Then give the robbing colony a dose of tobacco smoke, which will soon stop their marauding (for a while at least), and when they have nearly all returned to the hive close it for a time with one of the screens, Fig. 12, page 34 (described at end of volume), thereby checking and preventing further robbing.

If the hive being robbed is queenless and reduced in numbers, it should be removed to the stand occupied by a stronger one (also queenless), thus equalizing them. I find this method of equalizing nuclei a good one, even when there is no robbing.

The entrances of the hives should be protected with a piece of glass (four or more inches long and one inch wide) placed against them in such a way that the bees can pass out at either end, and secured against the hive with two small tacks.

If robbery is being carried on to any great extent, one end may be closed with a piece of paper; and if the robbers are still persistent, throw some grass against the glass in such a way as completely to cover the entrance.

The bees belonging to this colony will find their way in or out of the hive, while the robbers hesitate before forcing their way through this barricade; and if they should attempt it, their chances of escape are few, for it is not an easy matter for them to find their way out again, and the colony thus assailed takes courage, while the sentinel bees, with renewed vigor, seize upon the intruders as they enter, stinging them before they can escape, often defeating them. The glass placed against the entrance in no way interferes with the queen or bees when they wish to pass out or in. When feeding nuclei, great care should be taken that the feeders do not leak, and that they are properly filled. If any syrup or honey is spilled upon the ground, cover it with earth so deeply that the bees will not discover it.

FERTILE WORKERS.

One of the greatest and most troublesome pests of the apiary (especially to queen-breeders) are the fertile workers. They are generally produced by allowing a colony to remain queenless for a long time, appearing sooner in nuclei than in full colonies. Their presence is known by drone-brood in worker cells. Drones reared from these eggs are invariably dwarfed, and, in my opinion, incapable of fertilizing queens. It is quite difficult to introduce virgin queens to such colonies, although a cell may be safely given them at any time. The remedy is this: place a frame of well matured worker-brood in the centre of the brood-nest, and in a few days give them a well matured queen-cell, and by the time that the brood is all hatched the fertile-workers will be gone.

When you think a colony is infested with them, examine the combs, and if the eggs are laid in a careless manner, two, three, and often more in a cell, with many cells passed by, and the brood when capped projects beyond the worker-cells, you may be sure that it is some of their work. It would be well to destroy all such brood before giving the colony a queen-cell. Do not destroy the combs but rather shave the heads off with a sharp knife.

About the Author

The following is adapted from an obituary in The British Bee Journal April 1908 edition:

Mr. Henry Alley, of Wenham, Mass., U.S.A. was the well-known breeder of queens, and his methods were widely adopted until they were superseded by improvements. In 1883 he brought out " The Bee-keeper's Handy Book ; or, Twenty Years' Experience in Queen-Rearing." In 1889 he compiled and published "The National Beekeepers' Directory," which contained a classified list of beekeepers in the United States and Canada, with "Practical Hints Regarding the Successful Management of the Apiary," occupying sixty-three pages of the work. In 1891 he produced a work entitled " Thirty Years Among the Bees," which contained his latest improvements in queen-rearing and the practical, everyday work of the apiary. For several years he edited the American Apiculturist. He claimed to have produced pure Golden Carniolan bees. He is remembered as one of the old-time beekeepers in America, who did his part towards improving queen-rearing.

The
Miller
Method of
Queen
Rearing

C.C. Miller

The Miller Method of Queen Rearing

Condensed from 50 years Among the Bees
by C.C. Miller
Originally published by
THE A. I. ROOT COMPANY
Medina, Ohio
1911

Table of Contents

Queen-Rearing—Breeding From Best

My sole business with bees being to produce honey, I am not particular to keep a popular breed of bees, only so far as their popularity comes from their profitableness as honey-gatherers. I am anxious to have those that are industrious, good winterers, gentle, and not given to much swarming. For some years I got an imported Italian queen every year or two. Then for a good many years I preferred to rear from queens of my own whose workers had distinguished themselves as being the most desirable. The chief thing considered was the amount of honey stored. Little or no attention was paid to color, and unfortunately no more to temper. So I had bees that were hybrids, hustlers to store, but anything but angels in temper. Then, beginning with 1906, I introduced quite a number of Italian queens, in the hope that among them I might find one as good as my hybrid stock, without so much ill temper. The hope was not realized, but continued effort may bring success.

Importance of Selection

The queen being the very soul of the colony, I hardly consider any pains too great that will give better queens. The first thing is to select the queen from which to rear, for generally all rearing will be from the same queen, whether for the home apiary or an outside apiary. The records are carefully scanned, and that queen chosen which, all things considered, appears to be the best. The first point to be weighed is the amount of honey that has been stored. Other things being equal, the queen whose workers have shown themselves the best stores will have the preference. The matter of wintering will pretty much take care of itself, for a colony that has wintered poorly is not likely to do very heavy work in the harvest. The more a colony has done in the way of making preparations for swarming, the lower will be its standing. Generally, however, a colony that gives the largest number of sections is one that never dreamed of swarming.

I am well aware that I will be told by some that I am choosing freak queens from which to rear; and that it would be much better to select a queen whose royal daughters showed uniform results only a little above the average. I don't know enough to know whether that is true or not, but I know that some excellent results have been obtained by breeders of other animals by breeding from sires or dams so exceptional in character that they might be called freaks. I know, too, that it is easier to decide which colony does best work than it is to decide which queen produces royal progeny the most nearly uniform in character. By the first way, too, a queen can be used a year sooner than by the second way, and a year in the life of a queen is a good deal. I may mention that a queen which has a fine record for two successive seasons is preferred to one with the same kind of a record for only one season. At any rate, the results obtained in the way of improvement of stock as a result of my practice have been such as to warrant me in its continuance, at least for a time.

The danger from inbreeding must not be lost sight of entirely. With two or three hundred colonies kept in three different apiaries it is perhaps not great. Should signs of degeneracy at any time appear, it will not be difficult to introduce fresh blood.

Conditions for Queen-Rearing

Having chosen the queen from which to rear, I have kept in mind that unless conditions are favorable the royal progeny of the best queen in the world may be very poor. Queen-cells must be started when the weather is sufficiently warm, when bees are gathering enough to make them feel that there is no need to stint the royal larvae in their rations, and until near the point of emergence it is much better that the cells shall be in the care of a strong colony. So I do not begin operations for queen-rearing until about the time that bees inclined to swarming would begin to make preparations therefore.

In Hive With Laying Queen

It would be too long a story to enumerate all the plans I have used in queen-rearing. I have reared excellent queens, and many of them, by the Alley plan, and by the Doolittle cell-cup plan, together with its modifications by Pridgen and others. I think I was the first one to report rearing a queen in a colony having a laying queen; and I have reared them in stories under as well as over the story having the laying queen. Neither is it absolutely necessary to have a queen-excluder between the stories. In lieu of an excluder I have used a cloth with room for passage at the corners. Neither excluder nor cloth is absolutely necessary; distance is enough. That first reported case was on this wise:

Upon a hive containing a colony had been piled four stories of empty combs for safe keeping. To make sure that the bees would not neglect the care of the most distant combs, I put a frame of brood in the upper story. A few weeks later I found a laying queen in the upper story with the old queen still below. The bees that had gone up to that frame of brood were so far from the queen that they had reared a queen of their own. A hole in the upper story had allowed the flight of the young queen without invading the domains of her mother. For those who produce extracted honey this plan might be used to advantage.

I have reared good queens by the old and simple plan of taking away the queen of a strong colony. Of course this must be a choice queen. Previous to the removal of the queen the colony is strengthened. Frames of well-advanced brood are from time to time given from other colonies until it has two-perhaps three-stories of brood. None of this brood, however, is given less than five or six days before the removal of the queen. The queen is taken with two frames of brood and adhering bees and put on a new stand in an empty hive, an empty comb and one with some honey being added.

Time to Start Nuclei

In nine or ten days from the removal of the queen it is time to break up the queenless colony into nuclei. It might generally be left till a day or two later before a young queen would come out to destroy her baby -sisters in their cradles, but it is best to take no chances. If it were true, as formerly believed, that queenless bees are in such haste to rear a queen that they will select a larva too old for the purpose, then it would hardly do to wait even nine days. A queen is matured in fifteen days from the time the egg is laid, and is fed throughout her larval lifetime on the same food that is given to a worker-larva during the first three days of its larval existence. So a worker-larva more than three days old, or more than six days from the laying of the egg would be too old for a good queen. If, now, the bees should select a larva more than three days old, the queen would emerge in less than nine days. I think no one has ever known this to occur.

Bees Do Not Prefer Too Old Larvae

As a matter of fact bees do not use such poor judgment as to select larvae too old when larvae sufficiently young are present, as I have proven by direct experiment and many observations. It will not do, however, to conclude from this that all queen-cells started by a queenless colony left to themselves will be equally good.

Bees have a fashion of starting cells for a number of days in succession, and will continue to start them when larvae sufficiently young for good queens are no longer present. So some means must be taken to make sure that no nucleus has for its sole dependence one of these latest cells. If several cells can be afforded for each nucleus, there is little danger they will all be bad. Neither is there great danger if a cell is chosen which is large and fine-looking. Perhaps the safer way is to give the queenless colony a frame with eggs and young brood three or four days after the removal of the queen, and then they will not be obliged to use the older larvae of the other combs.

Placing Queen-Cells

Two or three frames of brood with adhering bees are taken for each nucleus. If one of the frames has a cell or several cells in a good location, well and good. If not, the lack must be supplied. But the cells must be where they will be sure to be well cared for. They must not be on the outer edge of a comb, with the chance to be chilled, neither must they be on the outer side of the comb, but on the side of the comb that faces the other comb. Any cells that are not just where they are wanted must be cut out. For this purpose I like a tea-knife with a very thin and narrow blade of steel.

Stapling Cells on Comb

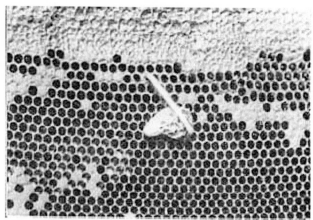

Fig. 85,-Queen-Cell Stapled on Comb.

A staple, such as is used to fasten a bottom-board to a hive, is used to fasten a cell in place. The cell is placed where it is wanted, then the staple is placed over it, one leg of the staple close to the cell, and the other leg is pushed deep into the comb (Fig. 85).

Making Bees Stay in Nuclei

Each nucleus is put upon a stand of its own, and the entrance is plugged up with leaves so that no bee can get out. One of the nuclei, however, is left without having its entrance closed, and this is put in the place of the hive which contains the queen, and the hive with the queen is put back on the old stand from which the queen was first taken. The entrances may be left closed until the shrinking of the leaves allows the bees to make their' way out, but I generally open them in about twenty-four hours, first pounding on the hive to make the bees mark their location upon emerging. Although queen-less bees are much better than others at staying wherever they are put, there will be still fewer bees return to the old place if the nucleus is fastened in twenty-four hours or longer.

Looking for Eggs

Twelve or fourteen days after forming the nuclei, I look to see if the queens are laying. I might find eggs in less time, but not always, and at any rate not in considerable number, and it saves time on the whole not to be in too much of a hurry. If no eggs are found a comb of young brood is given as an encouragement to start the young queen to laying, and a day or two later, if queen-cells are started on this young brood, a mature queen-cell is given.

Keeping Best Queen in Nucleus

Instead of having my best queen in a strong colony, as in the plan just given, she is usually kept in a two-frame nucleus throughout the summer, the nucleus being strengthened into a full colony in the fall for wintering. One object of this is to make the queen live longer. It is generally understood that a worker lives a longer time if it has little work to do, and probably the same is true of a queen. As laying eggs is her work, the less the number of eggs she lays the longer she ought to live, and in a nucleus she lays a smaller number of eggs than in a strong colony.

There is another reason for keeping her in a nucleus. Some who have tried to have comb built in the colony containing their best queen complain that they can get only drone-comb built. That may be divided by filling the frame with worker-foundation, but the better way is to keep the colony with the queen so weak that only worker-comb will be built. In a nucleus only worker-comb will be built.

Starting Brood for Cells

Having my breeding queen in a two-frame nucleus, I take away one of the combs, and in its place put a frame in which are two small starters four or five inches long and an inch or two wide. One of these starters is put about four inches from each end (Fig. 86). The nucleus must be strong enough in bees so that a week later this frame will have a comb built in it that will fill .most of the frame, the comb being fairly well filled with eggs and young brood (Fig. 88). It is taken away, and another frame with two small starters put in its place as before. Thus this nucleus will furnish once a week a frame of comb with brood of the best sort for queen-rearing. It will be a day or so after the frame is given before the queen lays in it, so that the brood will not be too old even if the bees were so foolish as to prefer it.

Fig. 86.-Starters in Breeding Frame.

Fig. 88.-Comb for Queen-Cells.

The comb being new and tender makes it probably an easier job for the bees to build queen-cells upon it; at any rate they always show a preference for such comb, and start on it a larger number of cells than they would on older comb.

Having now arranged for the right kind of brood and eggs to be ready on the same day of each week, the next thing is to find the right kind of bees to start the cells. Not only to start them, but to take the very best care of them. We can probably find no bees better fitted to produce good queen-cells than those that of their own accord have already engaged in the business. So a strong colony is chosen which has already started queen-cells in preparation for swarming. All queen-cells already started are destroyed, the queen is removed, and one of the frames is taken away, leaving a vacancy in the center of the hive. Most likely the colony has one or more supers, but these are not to be taken away.

Brood for Queen-Cells

We now go to the nucleus containing our best queen, take out the frame with the virgin comb, and replace it with an empty frame with its two starters, brushing back into the hive the bees from the comb taken out, and closing the hive. Looking at the comb taken out, you will see that instead of the oldest brood being in the center, it will be in the two places where the two starters were put. It was for this purpose the two starters at the sides were given rather than a central one. For by this means the waving contour will give opportunity for a larger number of queen-cells on the edge of the comb than would otherwise be the case.

Trimming the Breeding-Comb

For a little distance at the edge, the comb contains eggs only. This part is trimmed away, leaving the youngest of the brood at the edge of the comb (Fig. 89). One reason for this is that, other things being equal, the bees show a decided preference for building on the edge of a comb. Another reason is that I decidedly prefer to have cells on the edge, thus making them easier to cut out when wanted. The part cut away would only be in the way of both of us.

Bees Using Young Larvae Only

When a queen is taken away from a full colony, the bees start cells from young brood, and as I have already said, they continue to start fresh cells for several days, and until after there is no longer brood of the proper age, so that the last cells started will contain larvae too old to make good queens. But on these combs prepared as I have described, they do not do so. Rarely, if ever, will a cell be found elsewhere than on the edge of the comb, and I have never known the bees to start a cell after the larvae were too old. I do not know why there is this difference. I only know the fact. But it is a very convenient fact.

Age of Larvae for Queens

Scientists tell us that a worker-larva is fed for three days the same as a queen-larva, and then it is weaned. Theoretically, then, up to the time a larva in a worker-cell is three days old, it ought to be all right to rear a queen from. Practically, I do not believe a larva three days old is as good as a younger one. The only reason I have for so believing is the expressed preference of the bees themselves. Give them larvae of all ages from which to select, and they always choose that which is two days old, or younger. Indeed, it will be seen that in the comb from which I have trimmed the edge (Fig. 89) the larvae on the edge of the comb have been out of the egg but a short time, for I merely trimmed away the eggs, and possibly not all of them.

Placing the Breeding-Comb

The breeding-comb, thus properly trimmed, is taken to the queenless colony, and put in the vacancy that was left for it. On the top-bar of the frame is penciled the date on which the cells are to be cut out, allowing ten days from the time of putting in. Thus, if the frame be given June 27, the number 7 is put on the top-bar, July 7 being ten days later than June 27. No need to put the month on.

Beside giving the date, that figure marks the frame, so I can know at a glance which frame to take out. At the same time a memorandum of this date is put in the record book to remind me when to cut the cells.

Some one may ask, "But if you leave nearly all the old brood in the hive, will, the bees not start cells on them, with only the smaller part on your breeding-comb?" So I thought at first, and took some pains to have no very young brood of the old stock left. But I found upon trial that when I left all the young brood of the old stock, the bees ignored this, at the most starting upon it one, two, possibly three cells, confining their attention to the pre-pared frame I had given. Probably the hardness of the old combs and the lack of convenient places in which to build cells convince the bees that it is better to use the soft comb where room is abundant. Of course a cell or two on the old combs can do no great harm, for they will not be used.

More Than One Nucleus in Hive

The frames for nuclei are the regular full-sized frames, and a full hive may be used for each nucleus, but it is economy to have the hive -divided up into two or three compartments for as many nuclei. Three nuclei in one hive are mutually helpful in keeping up the heat, and thus it is possible to have the nuclei weaker than if each nucleus was by itself, while results are as good with the three weaker nuclei in the one hive as with three stronger nuclei in three separate hives.

Nucleus-Hive

For many years I have had hives divided into two or more compartments, and have had much trouble from the bees finding a passage from one compartment to another, but my latest nucleus hives have not troubled in that way. They are made from ordinary 8-frame hives together with the 2-inch-deep bottom-board. First, two pieces are nailed on the inside of the bottom-board, each piece 18 1/4 x 1 3/4 x 7/8. One piece nailed 4 1/2 inches from one side, the other 4 1/2 inches from the other side. These pieces do not lie flat in the bottom, but stand on edge, with 1 3/8 inches between them. Then the hive is fastened on the bottom-board with the four usual staples. Two division-boards, each 18 1/4 x 9 3/4 x 5-16, are now put in place and crowded down tight upon the two pieces in the bottom-board. These two division-boards are 4 5/8 inches from each side, leaving 2 1/4 inches between them. The four spaces at the top, at the ends of the division-boards, are closed by blocks 3/4 x 1/2 x 5-16, whittled enough to allow them to be wedged into place. Light 1 1/4-inch wire-nails are driven through from the outside to hold the division-boards in place. A block 10 x 2 x 7/8 is pushed into the entrance centrally, and held there by a nail lightly driven in front of it. That leaves an entrance at each end of the block for the two side compartments, but no entrance for the middle compartment. For this purpose an inch hole is bored in the back end of the hive midway between the two corners, its center being about three inches from the upper surface of the hive. Three boards of half-inch stuff cover the three compartments, and over this is an ordinary hive-cover.

At Fig. 90 will be seen a bottom-board for a nucleus hive. You will notice that the two pieces that run lengthwise through the center of the bottom-board are a quarter of an inch shallower than the rim of the bottom-board. If they were 2 inches deep instead of 1 3/4, the bottom-bars of the frames would rest directly on them. Of course the division-boards are deep enough to come clear down upon these two pieces.

Two nucleus-hives will be seen at Fig. 91. The one at the right faces us, showing the entrance at each side. The back of the left hive is toward us, showing the round hole near the top, which serves as an entrance to the middle compartment.

Large Space for Middle Frame

In one of these side compartments there is abundant room for two frames and a dummy, and three frames without the dummy can with care be crowded in. The central compartment will of course take only one frame. It seems as though 2 1/4 inches is quite too much space for one frame, but I use that space advisedly. Many years ago I made a nucleus hive with six compartments, and at that time not having had much experience I made each compartment 2 1/4 inches wide. Years afterward I made another nucleus hive, and smiling at my former ignorance and congratulating myself upon the superior knowledge I had gained with the passing years, I made the compartments more nearly in accord with the usual space occupied by each frame in a hive, making each compartment-I'm not sure whether it was 1 5/8 or 1 3/4. At any rate, the bees swarmed out of these limited quarters to such an extent that I could not use them, whereas they had not swarmed out of the 2 1/4 compartments. Neither have they swarmed out of these later ones. Having so much room in these central compartments, the bees sometimes build pieces of comb on the sides which I must clean away, but that is better than to have them swarm out.

Contents of Nucleus-Hive

A nucleus hive is tenanted by a two-frame nucleus on each side and a one-frame nucleus in the middle. Care is taken to choose one of the best frames of brood for the middle nucleus, and perhaps a few extra bees are brushed in. A third comb may be put in each of the side compartments, or a dummy, the same as the dummies used in the regular hives.

Making the Bees Stay

When populated, the entrances of the nuclei are plugged up with green leaves. These are generally taken away twenty-four hours later, after the hives are pounded to stir up the bees, but if they are neglected the leaves will dry and shrink so the bees can make their way out. It is better to form nuclei with queenless bees, for they are not so much inclined as others to go back to their old place.

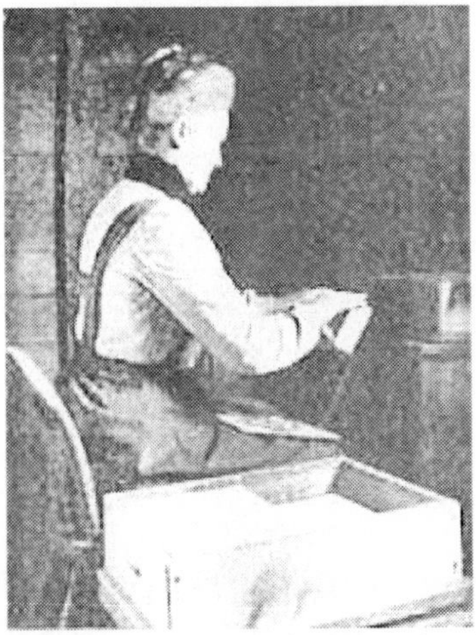

Fig. 87.-Putting Foundation in Sections.

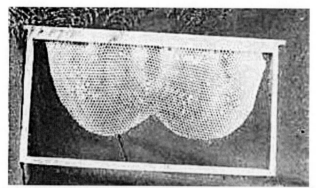

Fig. 89.-Comb for Queen-Cells, Trimmed.

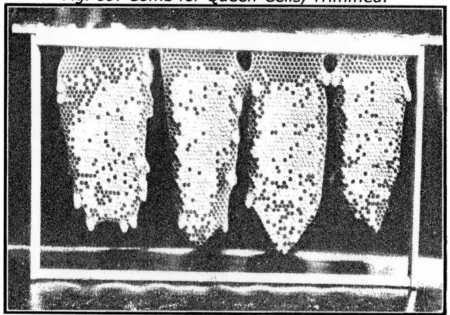

Queen-cells built naturally by the Miller plan (from Practical Queen Rearing)

Baby Nuclei

There has been much interest in the matter of having queens fertilized in small nuclei containing only 200 bees or so. About the year 1863 I had seen miniature nuclei in the apiaries of Adam Grimm, but they had not so few bees as the so-called baby nuclei of to-day. Of course, I had a number of queens fertilized in baby nuclei, but I did not go to the trouble of having hives specially built for them. I merely used an 8-frame dovetailed hive, putting in it sometimes a 1-pound section nearly filled with honey, and sometimes two such sections side by side. A frame of brood with its adhering bees was taken from some colony, the bees shaken or brushed into the nucleus-hive quickly, a virgin not more than a day or two old dropped into the hive among the bees and all hastily closed, the entrance having been closed in advance. Of course, the frame of beeless brood was returned to its old place. Three days later the entrance was opened, and in due time the queen was laying.

However it may be for the commercial queen-rearer, for the honey-producer there seems no great advantage in baby nuclei. He generally needs to make some increase, and it is more convenient for him to use 2 or 3-frame nuclei for queen-rearing, and then build them up into full colonies.

Regular Hives for Nuclei

One year I tried rearing queens on a commercial scale, producing them for Editor G. W. York of the American Bee Journal. I may say, parenthetically, that one season was enough to convince me that it was best to stick to honey-production, rearing queens only for my own use. But I had 50 three-compartment hives left on hand, and in spite of that, truth compels me to say that lately they generally lie idle, and I use a full hive for each nucleus, merely putting 3 or 4 frames in one side of the hive, with a dummy beside them. To be sure, it takes more bees than to have three nuclei in one hive, but it is a good bit more convenient to build up into a full colony a nucleus that has the whole hive to itself.

Queen-Cage

When we go to give queen-cells to the nuclei, we are provided with introducing queen-cages. The first introducing-cage I devised was the Miller introducing-cage, listed in the catalogs of supply-dealers. Then I got up one I liked better, three of which are shown in Fig. 92, the blocks containing the candy being separate from the cages. This may be called Miller cage No. 2. Two blocks 3 inches by 1/2 by 1/4 and a piece of wire-cloth 6 1/2 x 1 7/8, form the material for the cage. Lay the two blocks parallel on their edges, and nail on these one end of the wire-cloth, the end of the wire-cloth corresponding with the ends of the blocks. Fold the wire-cloth around the ends of the blocks and nail it on the other side, and you have a cage 3 x I 7/8 x 1/2, outside measure. The plug to close the cage is not so simple, for the cage is to be pro- visioned, and the plug holds the candy. Two blocks I 1/4 x 1/2 X 1 1/4, a piece of tin and a piece of section stuff each II/4 inches square form the material for the plug. Lay the two blocks parallel on their sides, with I/4 inch space between them. On these nail the piece of tin, turn over, and nail on the section stuff. Near one end drive a tack partly in to prevent the plug going too far into the cage. That makes all complete.

After using these for some years, I got up another that in some respects I like still better. This is shown in Fig. 87 1/2, and may be called Miller cage No. 3. Make a block 3 3/4 x 1 1/8 x 5-16. From one side of the block, at one end, cut out a piece 1 1/2 x 3/8, making the block as shown at No. 1 Fig. 87 1/2. Cut a piece of tin 1 x 2 inches. Stand the block on edge with the cut-out place uppermost, and in this cut-out place lay a lead pencil or similar object 11-32 in diameter. Over this bend the tin, letting it come out flush with the end of the block. Then laying the block on its side, still keeping the pencil in place, drive two 1/2-inch wire nails through tin and wood, clinching on the opposite side. When the pencil is withdrawn there is left a tube to be filled with candy. So much for the plug. The cage itself is made of a piece of wire-cloth 4 inches square, if one edge is a selvedge. If there is no selvedge, it must be 4x 4 1/2 and 1/2 inch folded over as a selvedge to prevent raveling. A block must be made, not to be part of the cage, but to be used to form the wire-cloth over. It must be a little larger than the first block, say 5 x 3-16 x 3/8. If the block were the same size as the first, there would be too tight a fit, and if the fit be loose it is easy to wedge in a thin slip, as a piece of wood separator. The wire-cloth is wrapped around the block and allowed to project at one end about 1/2 inch. A light wire is wound twice around, about 1/2 inch from the selvedge end (which is the part that does not project) and fastened. Another wire is similarly fastened about 1 3/4 inches from the first wire. Now the projecting part of the wire-cloth is bent down upon the end of the block, and hammered down with a hammer. That completes the cage, but for convenience in hanging it between brood-frames one end of a light wire 7 or 8 inches long is fastened into one side of the cage about 1/2 inch from the open end. To put it in a hive, I shove the frames apart, and holding the end of the wire lower the cage where I want it, and then shove the frames together. That leaves 3 inches or more of the wire above the top-bars, and when I want to take out the cage I take hold of the wire, draw the frames apart, and lift out the cage. The wire serves also to mark the spot where the cage is.

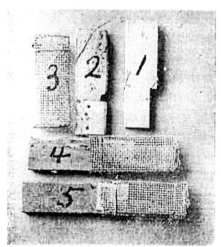

Fig. 87 1/2-Miller Cage No. 3.

When the tube is filled with candy, it may be pushed so far into the cage that the bees can not get at the candy. Then when it is desired that the bees shall get at the candy, the plug is drawn out until the candy is exposed. This is more reliable as to time than to have the usual cage with the candy covered with card-board. With the card-board there is no certainty as to whether the queen will be released in 24 hours or much longer. Sometimes it may be several days. With the No. 3 cage you know just how long the bees have the cage before they get to the candy, and after the candy is exposed you may count on the bees clearing out the candy in about 24 hours.

It may be objected that it is troublesome to open up the hive to change the position of the plug in the cage. That is true, and often, if not generally, the cage is not put between the combs, but thrust in the entrance, making sure that it is where it will be protected by the bees. After being there about two days, it is only the work of a minute to take out the cage, expose the candy, and put the cage back in the entrance.

Sometimes, if I want to have the work done automatically, I use a device that delays the work abort as much as the card-board, but is more uniform in the time it takes. I thrust into the center of the tube of candy its whole length a wooden splint about 1-16 of an inch square, and that delays the bees at gnawing out the candy.

When a queen-cell is to be caged, the No. 2 cage allows more room for the cell.

For making queen-cages, instead of the common painted wire-cloth that is used for screen doors, I like Better extra heavy bright wire-cloth. It is more substantial. But E. R. Root says queens have been poisoned in such cages, so have a care, although I have had hundreds of queens in them without noting any harm. Perhaps all tinned wire-cloth is not alike.

Distributing Queen-Cells

When the queen-cells are to be distributed, the first thing is to provision a number of queen-cages of the No. 2 style, with the usual queen-candy, tacking a piece of pasteboard on the end of the plug. Then we go to the nucleus where the cells are stored, cut out the cells, rejecting any that do not appear satisfactory, and put the cells in the cages. Some cells, however, are left uncaged. When we come to a nucleus that has had no queen for a day or more, there is no need of caging the cell. It is put against the comb in a good place, and fastened there with a hive-staple (Fig. 85). Coming to a nucleus with a queen which we wish to remove, we put the queen in a cage, and give the nucleus a caged cell, laying the cage against the comb and nailing it there with a l 1/2 or 1 3/4 wire-nail (Fig. 93). This nail is slender so as to push easily through the meshes of the wire-cloth. Then the young queens that we have removed are used wherever needed.

Brushing Bees Off Queen-Cells

Before cutting cells from the comb the bees must be removed, and it would mean the ruin of the cells to shake the bees off. Brushing with a Coggshall brush, although it might do with extreme care, would be likely to result in torn cells. Even something no stiffer than goldenrod or sweet clover needs much care. I like best a bunch of long and soft June grass-a very flimsy affair to use as a brush, but it is safe.

Advantage of Caging Cells

Of course the object of caging the cells is to prevent the bees from tearing them down. At the time of taking a queen out of a nucleus, if a cell were merely stapled on, the bees would be pretty sure to destroy it, for not yet realizing that their young laying queen has been taken from them, they feel no need of anything like a queen-cell. So the cage saves the time and trouble of waiting and making a second visit another day.

There is, however, another advantage in using the cage, making it somewhat desirable to use it in all cases. We often want to know what has been the fate of a cell, and can generally tell pretty well by its appearance. If it has the appearance of most of those in Fig. 94, we know that; a young queen has emerged and must be in the nucleus. If it is torn open in the side, like the one at the extreme right, we are sure that the young queen in it was destroyed by the bees.

If the cells have merely been stapled on, the bees are so prompt about removing them as soon as they are no longer of any use that scarce a vestige of them is left, so we have nothing to judge by. But when a cell is enclosed in a cage, the bees are very slow about removing it, so the cage gives us a better chance for judging.

Appearance of Vacated Cells

In Fig. 94 the first three cells at the left have the cap still adhering by a neck, showing that it has been only a short time since the queen emerged, providing the cell has not been caged; if it has been caged the queen may have been out some time. The fourth cell looks entire, as if it yet contained a young queen. But it is deceptive. The bees have a trick of fastening the cap back again as if it were a great joke, sometimes thus imprisoning one of their own number. A very close look will generally show a little crack, and a very little force will be needed to pick the cap loose. The next six cells show plainly that a young queen has emerged from each, and finding a cell of that kind is just as good evidence as a sight of the queen; only I would a little rather see the queen for the bare chance that she may not have perfect wings. As already mentioned, the cell at the extreme right shows by the hole in its side that no queen ever came out of it alive.

Miller Queen Nursery

Whatever the advantages of using queen-cells instead of virgin queens, there are also advantages in having the young queens hatch out in a queen nursery. So I have made considerable use of a nursery of my own devising, Fig. 88 1/2. It may take the place of a brood-frame in any hive, in the lower story or in an upper story, and it does not matter whether a laying queen is in the hive or not.

For this nursery I use a regular Miller frame, which lends itself to the purpose admirably, top-bar, bottom-bar and end-bar being all of the same width, 1 1/8 inches. If you haven't a Miller frame, you can easily make a frame having all parts the same width, 1 1/8 inches; only be sure the end-bars are at least 3/8 thick, and have the outer dimensions of the frame the same as the frames you have regularly in use. I'll give instructions for making a nursery with a frame of the Langstroth size, and if your frames are of different size you must act accordingly.

Make 7 pieces, each long enough to reach from top-bar to bottom-bar (with top-bar 7/8 and bottom-bar 1/4, that makes the length 8 inches), 1 1/8 wide, and 3/8 thick. Saw-kerfs must be made on each side of these 7 pieces. Beginning 1 1/4 inches from one end, on one side of the piece, with a very fine saw, make a saw-kerf by sawing about half-way through. Make a similar kerf 1 1/4 inches from the first, and then, each time measuring off 1 1/4 inches, make 3 more kerfs, making 5 in all. (Your last kerf will be more than 1 1/4 inches from the end, but that's all right.) Do the same thing on the opposite side, beginning at the opposite end. Make similar kerfs in each end-bar, measuring from the top-bar for one end, and from the bottom-bar for the other end. Of course these kerfs are to be made on the inside of the end-bar, and none on the outside. Now distribute these 7 pieces at equal distances from one end of the frame to the other, and if you are exact about it the distance between each two will be 1 25-32 inches. Fasten these 7 sticks in by driving one nail down through the top-bar into each, and two nails through the bottom-bar. Before nailing, make sure that each stick faces right, as mentioned further on. Nail upon one side of your frame a piece of wire-cloth to cover it (17 5/8 x 9 1/8). Have the nails not more than 2 inches apart all around and on each stick. I use bright wire-cloth, extra heavy, with meshes of the usual size in screen-doors.

Fig. 88 I/2-Miller Queen Nursery.

You now need 40 pieces of tin, 2 x 1 1/8 inches to go into all the saw-kerfs. Each piece of tin serves as a shelf, thus dividing up the whole into 48 compartments. You will now see the necessity of having the sticks face each other so as to have the kerfs correspond, as mentioned a minute ago. Look out for this before you nail the sticks in place.

To close these compartments, you need 8 pieces of tin, each 10 x 2 inches. That's 7/8 inch longer than the depth of the frame, allowing the 7/8 to be bent over at right angles on the top-bar. To hold these covers in place I use heavy pins bent over. Small screw-hooks of straight pattern might do better. Three are needed in each end-bar, and 6 in each upright. Of course these tin covers are put in at the top and slide down.

You will see that each of the compartments furnishes a large amount of room, 40 of them being 1 25-32 x 1 1/4 x 1 1/8 and the remaining 8 being larger. That gives abundance of room to put in the largest kind of a queen-cell. With each cell is given a ball of candy the size of a pea.

Pros and Cons of a Queen Nursery

If a ripe queen-cell is given to a nucleus or colony, there is no way to be sure that a queen that is all right will issue from it. She may be imperfect as to her legs, and, what is still worse, her wings may be so deficient that she never can fly. If she can not fly she can never be fertilized, and so is worthless. Indeed she is worse than worthless, for she is wasting the time of the nucleus. Sometimes, indeed, it happens that the occupant of the queen-cell is dead. All of this is avoided by having the virgins hatch out in a nursery. If a cell is cut into, and is given to a nucleus, the bees will at once destroy it, but in the nursery it will hatch out all right.

One may have a lot of queen-cells on hand with no immediate use for them. It will not do to leave them without cutting out beyond a certain time, for the hatching out of the first one means the death of all the rest. But if they are put in a nursery they are safe, and may be left stored in the nursery for some days after hatching out.

Over against these advantages stands the one disadvantage that in the nursery the bees are not allowed to come in immediate bodily contact with the cells, nor with the young queen after she issues from the cell. Some think this so serious a disadvantage as to overbalance all the advantages of the nursery. It is claimed that the clustering of the bees about the cells and the young queens does more than merely to keep up the temperature to a certain point, and that when this close contact is lacking something will be lacking in the resulting queens. Also that the young queens thus isolated and imprisoned are in a frightened condition, and that a young queen reared in such an atmosphere is not the same as one that has the feeling that she is all the while closely surrounded by friends.

So whether it be wise to use a nursery or not, it will certainly be wise not to put cells into it before it is necessary for their safety, nor to leave a virgin in a nursery any longer than necessity demands.

Quality of Queens

The question has been raised whether queens reared in the way I have described are as good as those reared by the latest methods. I think I can judge pretty well as to the character of a queen after watching her work for a year or two; I have kept closely in touch with what improvements have been made in the way of queen-rearing, and have reared queens by the hundred in the latest style; and I do not hesitate to say that the simple method I have given produces queens that can not be surpassed by any other method.

Beginner Improving Stock

I have been asked whether I would advise a beginner with only half a dozen colonies, one of them having a superior queen, to use the plans I have given to rear queens from his best queen. I certainly should, if he intends to give much attention to the business and increase the number of his colonies. The essential steps to be taken are simple enough; and even a beginner can easily follow them. But in a few words, here is what I would advise him:

Fig. 90. Nucleus Bottom Board

Take from the colony having your best queen one of its frames, and put in the center of the hive a frame half filled or entirely filled with foundation. If small starters are used in a full colony the bees are likely to fill out with drone-comb. A week later take out this comb, and trim away the edge that contains only eggs. Put this prepared frame in the center of any strong colony after taking away its queen and one of its frames. Ten days later cut out these cells, to be used wherever desired, giving the colony its queen or some other queen.

Now there's nothing very complicated about that, is there?

Fig. 91. Nucleus Hives

Fig. 92. Improved Miller Queen-Cages

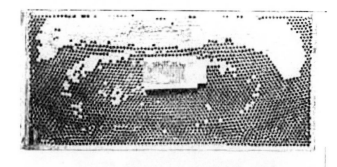

Fig. 93. Caged Queen-Cell

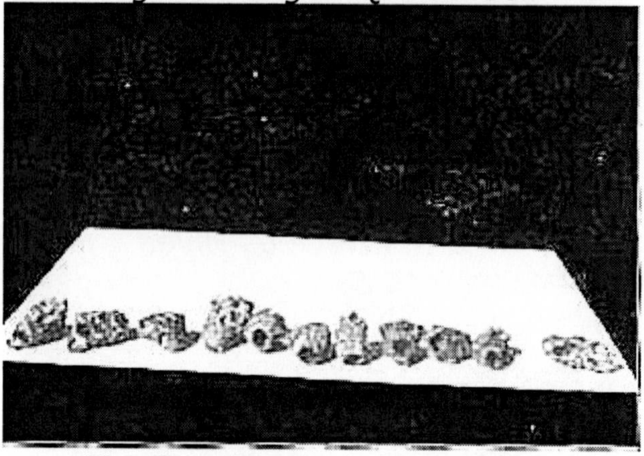

Fig. 94. Vacated Queen-Cells

About the Author

Cordially Yours,

C. C. Miller.

From Wikipedia:

Charles C. Miller (generally referred to as C. C.) was born in Ligonier, Pennsylvania on 10 June 1831. Miller's father, Johnson J. Miller, died when he was ten years old, leaving a family of six and little money. As a young man, Miller worked his way through grammar school (taking three years off to help support the family) and eventually moved from his native Pennsylvania to Schenectady, New York, where he worked his way through college.

Miller writes "This last undertaking was a bit reckless, for when I arrived at Schenectady I had only about thirty dollars, with nothing to rely on except what I might pick up by the way to help me in college. I had a horror of being in debt, and so was on the alert for any work, no matter what its nature, so it was honest, by which I could earn something to help carry me through.

"I had learned just enough of ornamental penmanship to be able to write German text [Miller's mother, Phoebe Miller, was from Germany, and he likely spoke German from childhood.], and so got $4.00 for filling in the names of 88 diplomas at two commencements. I taught singing school; I worked at Prof. Jackson's garden at seven and a half cents an hour; raised a crop of potatoes; clerked at a town election; peddled maps; I got $100.00 for teaching a term at an academy. Neither were my studies slighted during my course, which was shown by my taking the highest honor attainable, Phi Beta Kappa, which, however, was equally taken by a number of my class."

With such sacrifice, hard work, and dedication to become a physician, one assumes Miller would have easily made a career of medicine. Unfortunately, his disposition did not allow him to follow through with a practice. "It did not take more than a year for me to find out that I had not a sufficient stock of health myself to take care of that of others, especially as I was morbidly anxious lest some lack of judgment on my part should prove a serious matter with some one under my care. So with much regret I gave up my chosen profession."

Soon Miller was married, was teaching voice and instrumental music, and had become principal of a public school. He needed something extra to stimulate his vast intellect, to allow a bit of challenge, and to improve his health with "robust work and fresh air." A swarm stumbled onto his porch. He became a beekeeper. As a physician, Miller suffered greatly from stress. He wrote that he worried constantly that he would misdiagnose a patient and prescribe an incorrect medicine.

Beekeeping

Beginning as an amateur beekeeper as the result of a swarm that his wife hived into a sugar barrel when it landed on their porch in 1861, Miller expanded his business steadily. By 1878, Miller made his living from keeping bees. He eventually settled in Illinois. Miller expanded his honey farm to over 300 colonies of bees, and became North America's largest producer of comb honey.

Writings

Writing part memoir, part bee culture, Miller began his personal account of the honey business in 1886 with A Year Among the Bees, in which he announced he had "made the production of honey his exclusive business" for eight years. This was expanded into Forty Years Among the Bees (1903, 2nd ed. 1906), then into his classic, Fifty Years Among the Bees (1911), culminating with A Thousand Answers to Beekeeping Questions (1917). Miller also edited the American Bee Journal and produced a popular monthly advice column answering reader's questions. In honor of his fifty years of writing about apiculture and his gift of his personal beekeeping library, the University of Wisconsin named its apicultural library the Dr. Charles C. Miller Memorial Apicultural Library.

Isaac Hopkins on Queen Rearing

By ISAAC HOPKINS

Isaac Hopkins on Queen Rearing

By Isaac Hopkins, Auckland, New Zealand.
(Late Chief Apiarist to the New Zealand Government.)

From The Australasian Bee Manual, 1886 and 1911
versions

Reprinted 2011 by
X-Star Publishing Company
Founded 1961

Table of Contents

Table of figures

IMPORTANCE OF REARING QUEENS.

There is no branch of commercial bee-keeping deserving of more strict attention on the part of the bee-keeper as a means of improving his bees, than that of queen rearing. It is only in the judicious selection of their breeding stock, season after season, that bee-keepers can hope to make the continuous progress that is possible, and which their interests demand. The improvement of his bees should be the constant aim of each bee-keeper, he should never be satisfied with those he has, but be always striving after a better strain. That it is possible to improve the hive-bee by breeding out inferior characteristics, and breeding in more desirable ones, and so to produce a strain of a higher standard, has been proved over and over again, and no commercial bee-keeper can afford to neglect this part of his business.

DEVELOPMENT OF QUEENS

The cells in which queen, or perfect female eggs are laid and developed differ widely from those of the workers and drones; in the natural state, they are only built in the swarming season, or in cases where the colony has become queenless; in the former case the cells are laid out for the purpose on the under side, in a depression, or on the edges of the comb, as shown in Fig. 8, which exhibits, on an enlarged scale, the top view of a number of worker cells, with the egg and larva in the different stages of development up to the time of capping the cells (in the line marked a); a section of a queen cell (b) showing the larva and a supply of the royal jelly, and a similar one completed and closed (at c). They somewhat resemble a peanut in shape.

Fig. 8.—WORKER EGG LARVAE AND QUEEN CELLS.

The material of which these cells are composed is not pure wax; there is much pollen mixed with it. The outside surface is uneven and indented like the sides of a thimble. The number built at one time varies much, according to circumstances—sometimes only two or three, but ordinarily not less than five or more.

The transformations of the queen larva are completed in seven days from the closing of the cell, so that on the sixteenth day from the laying of the egg (six days shorter than the period for the worker, and nine days shorter than that for the drone) the fully developed queen emerges from the cell.

In the case of a colony becoming queenless in an abnormal manner, queen cells may be built over worker eggs or larvae in convenient places on the flat surface of a comb as shown in Fig. 9. The ordinary worker cells, with eggs in them, are shown at A; B is a queen cell partly built; and C one completed and closed. D shows a case, which sometimes occurs, of a queen cell built over drone brood.   Such cells—which may be known by the absence of indentations on their outer surfaces—are of course useless, as the nature of the drone egg is not altered by the form of the cell or the quality of the food given to the larva.

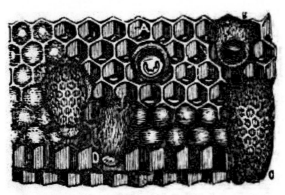

Fig. 9.—QUEEN CELLS BUILT OVER WORKER CELLS.

CHOICE OF BREEDING QUEENS.

The colonies chosen for breeding stock each season should be those that have given the most surplus honey, been the least inclined to swarm after the main honey flow has started, the gentlest bees, and the best defenders of their hives. Any of these qualities lacking at the start should be gradually bred into them. Remember that infinitely better work can be accomplished in the way of improving one's bees by the judicious selection of breeding stock in one's own apiary, than can be done by continually bringing in unknown breeding stock from outside. Even when but a few colonies are kept as a hobby, the rearing of a few queens will be found a most interesting study.

RAISING QUEEN CELLS.

The raising of queen cells is the starting point in queen rearing, and whether the bee-keeper assists the bees in this work by supplying artificial "cell cups," and transferring selected larvae to them or not, he is dependent upon the bees to bring the young queens to maturity. To ensure this he takes advantage of the natural instinct of the bee, which at once sets about raising another queen when deprived of the reigning one, and in this way he forces the colony by making it queenless to start queen cells.

By supplying it with selected eggs or larvae, and taking away all others, the bees are compelled to raise queens from these, so that the bee-keeper has almost complete control over their work, and by adopting certain methods he can encourage the bees to build more cells than would be built under natural conditions.

To describe in full the methods adopted by many commercial queen breeders for cell raising, and grafting of larvae into artificial cell cups, would require too much space, and special books giving full details are obtainable from those who cater for bee-keepers. "Doolittle on Queen Raising," "The Swarthmore Library," and the "ABC and XYZ of Bee-Culture," being the best. I shall therefore only briefly touch on these methods; and then explain the one most suitable for the average bee-keeper, by which he can be assured of raising the best queens obtainable under any plan.

THE DOOLITTLE PLAN.

Mr. G. M. Doolittle, if not the first to make artificial cell cups, was the first to perfect and make commercial use of them. He uses a small, round, smooth stick, pointed to the size and shape of the base of a queen cell. This he dips into melted wax three or four times, the first time about half an inch up the stick, and less each time, so that the base of the cup is thickest. The cell cups are then placed in wood bases, Fig. 52; A being a cross section of base, and B, the same with cell cup partly inserted.

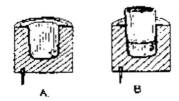

Fig. 52.—CROSS SECTIONS OF WOOD BASE WITH CELL-CUP.

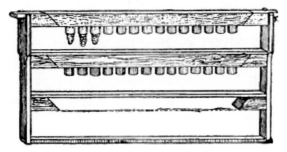

Fig. 53.—CELL-BASES SPIKED TO THE BARS OF A FRAME.

Fig. 54.—A CELL COMPLETED ON WOOD-BASE.

A portion of "Royal Jelly" from a newly-made natural queen cell is first inserted in each cell cup, and then the selected newly hatched larvae are transferred to them. The cell-bases with their cups are then spiked to the bars of a frame, Fig. 53, and the frame inserted in a hive containing a colony prepared to receive it, when the work of extending the cells, and caring for the larvae, is left to the bees. The final treatment of the cells till the emerging of the young queens is explained further on. The "Swarthmore" and other methods are exactly similar to the above, slightly modified in the appliances used.

THE ALLEY PLAN.

The late Mr. Henry Alley was one of the oldest and most respected commercial queen breeders in the world. His experience extended from about 1860 till his death a few years ago. He worked out a system of queen rearing which is at once simple, easy to follow, and closely in agreement with the natural method. On his plan, which I strongly advocate, I have raised thousands of far finer queens than I have ever seen bred or been able to raise under other methods.

SEASON FOR QUEEN RAISING.

Any time from the commencement of fine, settled weather in the Spring, when drones begin to fly, until the beginning of the Autumn, before the drones are killed off, queens can be reared. In the Auckland Province, and other parts of Australasia in the same latitude, from early in October to beginning of March. The greatest success is achieved just about the time when the bees are ready to swarm in Spring, and early Summer. The largest number of the finest cells will be built at that time.

In order to have the colonies chosen for queen rearing well forward in Spring, and the right drones flying in time, they should be stimulated by slow feeding, and if necessary strengthened by giving a frame of brood occasionally from other colonies. I am, of course, taking it for granted that Italian queens will be bred. As soon as the colonies selected for rearing drones are getting fairly strong, put in near the centre of each of the brood chambers a frame of drone-comb.

SECURING DRONE-COMB.

When the bees are in full swing storing honey in the surplus boxes, remove a couple of frames of comb from the centre of the surplus box and insert in their places frames containing a narrow strip of comb foundation. The bees will at once build the frames full of drone-comb, and may store honey in part, and breed in part. They can be removed till the brood dies, and then be given back to the bees to clean. Any quantity of good drone-comb can be secured in this way for future use.

SECURING SELECTED EGGS.

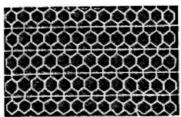

Fig 55.—SHOWING HOW TO CUT THE COMB.

To return to the drone-combs in the brood chamber. When the first drone brood is sealed over, insert a clean worker-comb in the centre of the brood chamber of the colony set apart for eggs. Combs of the previous season's building that have not been bred in are best for the purpose. On about the fourth or fifth day there will be eggs and probably some tiny larvae in the comb. If so, remove it to a warm room or workshop and insert another comb in its place. This work should be done on a fine day to avoid getting the eggs or larvae chilled.

CUTTING THE COMB.

Lay the frame of comb flat on a table or bench, and with a thin, sharp-bladed knife, cut as much of the comb containing eggs or larvae in the cells as desired into strips by running the knife (previously wetted with honey diluted with water to keep it from sticking) along every second row of cells, as shown by the white lines in Fig. 55, taking care to leave the intermediate row with the contents intact. The strips will be about one inch deep, but the cells on the side to be used for queen cells should be pared down one-half, and two out of every three eggs or larvae should be killed, so as to allow room between the queen cells, when built, to cut them out without injury. A tiny splinter of wood, with its end dipped in melted wax, is the best for killing the spare eggs in the cells.

The strips should now be fastened to the under-side of the bars of a frame prepared as in Fig. 56, cells pointing downwards, same as the cell bases in Fig. 53. Or, better still, they may be fastened to the lower edge of a comb, cut in the section of a circle from end bar to end bar, about one-third down from the top bar. One frame of eggs or larvae will usually afford strips for two or more frames. The wax used in fastening the strips must not be too hot or it will melt them and destroy the eggs.

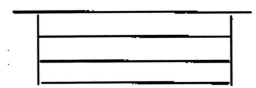

Fig. 56.—FRAME FOR FASTENING THE STRIPS OF COMB TO.

PREPARING A COLONY FOR QUEEN CELL BUILDING.

Select a strong colony—a strong two-story colony with plenty of nurse bees is best—and, first of all, make a nucleus colony with the queen and the frame she is on, and one additional frame of unsealed brood and another of food, with their adhering bees (see instructions for making nuclei).   Then remove all frames of unsealed brood without the adhering bees.   These, for the time being, may be placed in the top story of another hive containing a strong colony.   The frame of strips for queen cells may now be placed in the centre of the brood nest; or the hive may be prepared for them a few hours beforehand.   Being deprived of their queen, and having no eggs or larvae in the hive except those supplied, the bees must build the cells over them. The date and age of the eggs or larvae should always be marked on the frames, as it will then be known when the young queens will come to maturity.   If honey is not coming in freely at this time, the bees should be fed liberally with sugar syrup; there will be more and better cells built by so doing.

RETURNING THE QUEEN AND BROOD.

The Alley system of queen rearing has been objected to by some bee-keepers because they unreasonably supposed that a strong colony must be broken up for each batch of queen cells. This is wrong, for as soon as the queen cells have been sealed (or even before), the frame, or frames of cells, may be placed in the top box, the queen and brood be returned to the brood chamber, a queen excluder put on over it, and the upper box with the queen cells placed over the excluder. So long as the queen cannot get to the cells, they will be as safe, as a rule, as though she were not in the hive; so that at the most the colony need not be queenless more than four days.

NUCLEUS HIVES.

I have already in Chapter VI. on Hives, described the small nucleus hives to be used in queen-rearing which should take three or more of the regular Langstroth frames, three are the usual number. Good sized entrances should be made so as to afford ventilation to the interior when wire cloth is tacked over them.

Additional ventilation, which is advisable, may be provided by boring a two-inch hole through the bottom and covering it with wire cloth.

There is a very great advantage in using the same regular frame in the nucleus hives as is used throughout the apiary, and I am satisfied from experience that the economy of bees in caring for queen cells and the young queens until they mate, and are laying, as advocated by some writers at the present time, is entirely against the production of good queens. So far back as 1887 in conjunction with the late Mr. Obed Poole, the inventor of queen excluders, I gave small nucleus boxes, similar to those subsequently advocated by E. L. Pratt ("Swarthmore"), a good trial. Since then I have had experience with small boxes of larger dimensions, but I unhesitatingly advise the adoption of the larger nucleus hives mentioned above, well furnished with bees, if the object is to raise first-class queens.

FORMING NUCLEI.

Whatever number of queen cells are to be made use of, the same number of nucleus hives will be required—a piece of perforated zinc or wire cloth should be tacked over each entrance, and there should be some spare frames of empty combs or foundation at hand. A strong two-story colony will make five nucleus colonies, and leave sufficient bees with the old queen to make another. When the embryo queens are 13 days old from the egg the cells are ready to be given to nuclei.

With the cells and some cell protectors (Fig. 57) ready, select a colony to break up and find the queen, placing her with the frame she is in, in an empty hive or comb holder for the time being. Now, put one frame of brood (as much sealed brood as possible) with the adhering bees into a nucleus hive, and another also with adhering bees containing honey and some pollen if possible, and also an empty comb or frame of foundation.

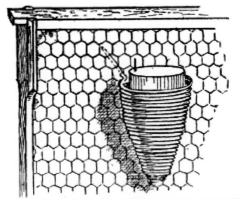

Fig. 57.—QUEEN CELL PROTECTOR

Place a queen cell in protector (Fig. 57) and fasten it on to the centre comb by pushing its projecting end through the comb. Sometimes the bees will tear down the cell, hence the need for protection. All being finished out on the cover and do the rest in same way.

The bees are now fastened in by the wire cloth over the entrance, and the hives should at once be placed in a cool, dark situation until sunset on the following day, when they may be put in their permanent position and the entrances be opened. Unless the bees are confined for a time the majority would return, and the nucleus hives be deserted. Nucleus hives are best set apart from the main apiary, and from each other.

THE EMERGING AND MATING OF YOUNG QUEENS.

The normal time for the young queen to emerge from her cell is on the sixteenth day from the laying of the egg, but the weather is often the cause for the time varying. If warm and favourable for several days, she may emerge late on the fifteenth day, or through cold weather, be delayed till late on the seventeenth day. When about five days old, if the weather be favourable she takes her "wedding flight" to meet the drone, usually about mid-day. If successful she commences laying in a few days, and is then ready for use in the apiary, but if the nucleus colony, which is now established, is required for other cells later, the queen before removal should be allowed time to stock the combs with eggs.

QUEEN NURSERIES.

In queen rearing there are frequently more queen cells coming to maturity than can be made use of at the moment; a nursery in which the spare ones can be placed for the time is very useful in such cases. The

Alley Nursery (Fig. 58) is again coming into use, and from a long experience with it I can speak very highly of its usefulness. Similar nursery cages (Fig. 59) may be used for the wood based cells, but the holding frame must be made differently. The cages (Fig. 59) are made out of a smooth batten seven-eighths of an inch thick; $2^7/_{16}$ in. wide, and each cage, being $2^{11}/_{16}$ in. long, they can be cut off the batten after all are bored. The large central hole is 1½ in. in diameter, and the two smaller ones on the edge are $1^{11}/_{16}$ in., and $^5/_8$ in. In diameter and bored through to the central hole; the latter is then covered with wire cloth on each side to make the cage complete.

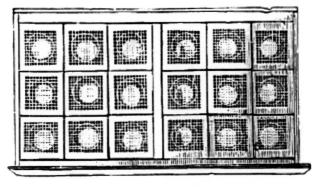

Fig. 58.—ALLEY QUEEN NURSERY.

The queen cell is placed in the larger hole on the edge, and candy food in the smaller one. The frame with cells should be suspended in the centre of an upper story of a hive till the cells are required or the queens emerge. The same cages can be used to introduce the young queens.

INTRODUCING QUEENS.

As a rule it is not difficult to introduce an alien queen to a colony, be she a virgin or laying, so long as certain rules are observed. The ordinary conditions to ensure safety are—that the colony must first be made queenless, that is, the old queen must be removed. In the next place the new one, when first placed in the hive, should be protected in such a way that while the bees can see and even feel her with their antennae, they are prevented from stinging her, as they might do before becoming used to her.   And lastly, the colony should be fed if there is no honey being gathered while the queen is being introduced. There are exceptions to the second clause. In the busy season, when honey is coming in rapidly, if the queens can be changed without much disturbance of the hive, the new one is likely to be accepted just as readily if she is turned loose on the frames as she would be were she protected for a day or two. I have often introduced them in this manner with success. On the other hand, I have had great difficulty with some colonies when trying to get them to accent a queen when introduced in the usual way.

Place the queen to be introduced with as little handling as possible and without any bees; and plug up the entrance to cage with candy. After the queen you are superseding, together with queen cells (if any) have been removed, hang the cage from the top bars between two of the centre frames (as in Fig. 62) and pressed against some honey the queen can feed herself. Close down the hive and don't disturb it again for three or four days, long before which she will probably have been released, when the cage can be removed.

Fig. 59—NURSERY CAGE.

INTRODUCING CAGES.

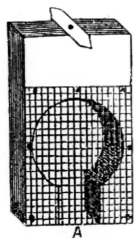

Fig. 60.—ALLEY'S INTRODUCING CAGE.

There are several kinds of introducing cages, but I think the two shown in Figs. 60 and 61 are about the handiest of any.

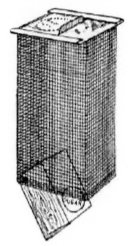

Fig. 61.—TITOFF INTRODUCING CAGE.

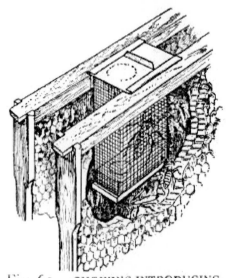

Fig. 62.—SHOWING INTRODUCING CAGE IN USE.

CANDY.

I have spoken of Candy for use in the queen nursery and introducing cages; the following is the best way of making it. Crush up some loaf sugar very fine; if a good deal of candy is needed, as when commercial queen rearing, a good coffee mill is best for this work. The sugar should be like flour; beware of buying ground sugar, as there is frequently starch or some foreign matter mixed with it that is death to bees. Warm some honey, but be sure it comes from a clean hive, and mix a little (it requires very little) with the ground sugar. Knead it well and add more sugar until the ball becomes firm but moist; when the ball of candy is left on a board for 24 hours it should only flatten out a little, it is then right.   It will do for nursery, introducing and shipping cages.

DRONE TRAPS.

When queen rearing, if there be any undesirable drones flying in the apiary they may be trapped and destroyed by placing drone traps (Fig. 63) in front of the hives containing them.   Some also use them for trapping queens at swarming time, I have never used them for this purpose, so cannot speak of their useful-ness or otherwise in this respect.   No apiary should be without a few of them.

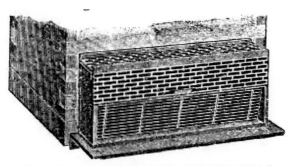

Fig. 63.—DRONE AND QUEEN TRAP.

ITALIANISING AN APIARY.

I have already advised beginners for the sake of economy to start with black bees, and as soon as the apiary is fairly established to Italianise all the colonies. If you have decided to try your hand at queen rearing, then purchase, say, three tested, or select tested, Italian queens from a reliable breeder as early as you can get them in the season, or at any time during the swarming season will do. Introduce them to strong colonies and follow the instructions herein given for rearing queens. If, on the other hand, you prefer to purchase all the queens at first to Italianise your stocks, then order, say, two tested and the rest untested queens, and after you have Italian drones flying in fairly large numbers start queen rearing with eggs from your tested queens, and so gradually change the untested queens that turn out to have been mismated to purely mated ones of your own raising, unless you are not particular about having a few hybrid colonies in your apiary.

CLIPPING QUEEN'S WINGS.

The chief object in clipping the queen's wings is to prevent swarms absconding. Much has been said for and against the practice, but it has been largely adopted of late.

HOW TO CLIP.

When clipping, the queen should be held as in Fig. 64, by the abdomen, while the two wings on one side only are clipped off near the stumps.

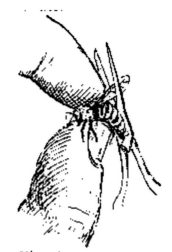

Fig. 64.—CLIPPING WINGS.

On the other hand, if the bee-keeper is not at hand when the swarm issues, the queen, not being able to fly, may get lost by falling on the ground, and so being unable to reach her hive again. A tin fence, 1½ in. wide, tacked round the alighting board, and having the upper half-bent inward, would prevent the clipped queen from falling to the ground, and so enable her to return to her hive.

In every case when a queen is purchased the wing should be clipped to prevent subsequent misunderstanding. It sometimes happens that a short time after a queen has been accepted the bees for some reason will supersede her, raising another queen from her eggs, the young one not being distinguishable from her mother. If the new queen should get cross-mated her bees will be hybrids, and the bee-keeper, not knowing the queen he introduced has been superseded, naturally accuses the bee-breeder of fraud in sending him a cross-mated instead of a pure queen. Now, by clipping the wing it can be seen at once if the original still reigns.

SUPERSEDING QUEENS.

The supersedure of queens after they are past their prime or in some other way have become defective, may be done by the bees, by installing another and younger one in her place. It is now the practice however, among the majority of experienced commercial bee-keepers, to carry out a system of superseding themselves, and not trust to the bees to do it, as they believe in the latter case that queens are frequently kept till long after they have passed their profitable age. The consensus of opinion is in favour of supersedure at or near the close of a queen's second season, and I feel certain that (with few exceptions), owing to the genial winter temperature of Australasia, and the prolonged breeding season, queens are at their best in this part of the world in their second season, and rapidly deteriorate after.   I therefore recommend the replacing of queens not later than February of the second season, excepting, of course, in very special cases, where a queen may have exceptionally good qualities, as reflected by her bees.

SWARMING CELLS.

Some bee-keepers object to making use of spare queen cells from a colony that has just swarmed, on the grounds that they are likely to produce queens whose bees would have an abnormal propensity to swarm. This, in my opinion is poor reasoning; they apparently overlook the fact that to swarm is natural in all strains of hive bees. Their apprehension could only correctly apply to strains which already are prone to swarm, and from which no sensible apiarist would breed in any case. There certainly can be no reasonable objection to making use of spare swarm cells from a good strain of bees, using the same discretion in choosing them that one should do in queen rearing. Such cells from a strong colony produce the very finest of queens.

WOOD BASES FOR QUEEN CELLS.

There is no doubt that the wood bases to cells as described, are a very great convenience, and it has occurred' to me that some such bases might be attached to the queen cells built on the Alley plan. If shallow holes were bored into similar pieces of wood as is used in the Doolittle plan, the Alley cells when built might be glued to them with melted wax; at all events, it is worth trying.

FEEDING IN QUEEN REARING.

Though I have already mentioned this matter, I wish to impress upon all who undertake queen rearing for the first time, the importance of feeding in all stages of queen rearing when no honey is being gathered, and the same applies especially when introducing queens. I shall give the formula for making sugar syrup later on.

Hopkins Method of Queen Rearing
_____

ANOTHER METHOD OF RAISING CELLS.

The method I am about to describe was, I believe, first tried and described by an Austrian bee-keeper, but so far as I am aware, I was the first to give it a trial in this part of the world at the Government Apiary, and with excellent results as Fig. 66 indicates.

A new bright (wired) comb of the previous season's construction was put into the hive of one of our breeding queens; when fairly full of eggs and newly hatched larvae it was removed and laid flat on a bench. A thin-bladed knife was run along each side of every fourth row of cells, cutting down to the mid-rib only. The three intermediate rows of cells were scooped out with the blade of a broad bradawl, as shown in Fig. 65, an easy matter, leaving every fourth row intact. Two out of every three eggs or larvae in the standing rows were killed, as in the Alley plan, and also all eggs and larvae between the rows. This is important. The cells on the opposite side of the comb were not touched.

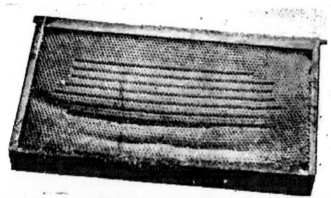

Fig. 65.—COMB PREPARED FOR QUEEN CELLS.

Fig. 66.—COMB OF 80 GOOD CELLS.

A strong two-story colony was in the meantime prepared for cell building in the manner already described, an empty half-story was placed immediately over the brood chamber, an empty frame being laid flat on the brood frames, and the prepared comb (prepared side downwards) laid flat on the empty frame. The latter was covered with a light mat, and the upper story replaced. In due course we obtained sixty good cells in our first experiment, and over eighty as shown in Fig. 66 in our second trial. The above illustrations were made from photos taken by myself, the cells being foreshortened in the view, look smaller than they really were. As soon as the cells are well started a queen excluder may be put on and the queen returned in the manner previously described.

CAUTION.

The comb lying flat over the brood chamber is subject to considerable heat, and we found in one case part of the comb had sagged down owning to the softening of the comb, and weight of the bees. We then wound wire around the frames between the standing rows of cells, which checked the sagging. Either wire or thin splints of wood will do. We obtained some very fine queens by this method, and as a wholesale way of raising cells, I consider it immensely superior to raising them on the swarm box plan with a small force of bees from artificial cell cups and transferred larvae. Plenty of ventilation should be provided when raising queen cells in this manner.

Big Batches of Natural Cells by the Hopkins or Case Method.

Many extensive honey producers who desire to make short work of requeening an entire apiary, and who do not care to bother with mating boxes or other extra paraphernalia, make use of the Case method, which has been somewhat modified from its original form. This method is advocated by such well known beekeepers as Oscar Dines of New York and Henry Brenner of Texas. The plan was first used in Europe.

To begin with, a strong colony is made queenless to serve as a cell building colony. Then a frame of brood is removed from the center of the brood nest of the colony containing the breeding queen from whose progeny it is desired to rear the queens. In its place is given a tender new comb not previously used for brood rearing. At the end of four days this should be well filled with eggs and just hatching larvae. If the queen does not make use of this new comb at once, it should not be removed until four days from the time when she begins to lay in its cells. At that time nearly all the cells should be filled with eggs and some newly hatched larvae.

This new comb freshly filled is ideal for cell building purposes. The best side of the comb is used for the queen cells and is prepared by destroying two rows of worker cells and leaving one, beginning at the top of the frame. This is continued clear across the comb. We will now have rows of cells running lengthwise of the comb, but if used without further preparation the queen cells will be built in bunches, that it will be impossible to separate without injury to many of them. Accordingly we begin at one end, and destroy two cells and leave one in each row, cutting them down to the midrib but being careful not to cut through and spoil the opposite side. Some practice destroying three or four rows of cells, and leaving one to give more room between the finished queen cells. (see Fig 24b.)

We now have a series of individual worker cells over the entire surface of the comb, with a half inch or more of space between them. The practice varies somewhat with different beekeepers beyond this point. However, this prepared surface is laid flatwise with cells facing down, over the brood nest of the queenless colony, first taking care to make sure that any queen cells they may have started are destroyed. In general, it is recommended that the colony be queenless about seven days before giving this comb. By this time there will be no larvae left in the hive young enough for rearing queens, and the bees will be very anxious to restore normal conditions. Some beekeepers simply take away all unsealed brood, rather than leave the bees queenless so long.

As generally used, this method requires a special box or frame to hold the prepared comb. This is closed on one side to prevent the escape of heat upward and to hold the comb securely in place. Figure 24a. Some kind of support is necessary to hold the comb far enough above the frames to leave plenty of room for drawing large queen cells. It is also advisable to cover the comb with a cloth which can be tucked snugly around it, to hold the heat of the cluster. By using an empty comb-honey super above the cluster, there is room enough for the prepared comb and also for plenty of cloth to make all snug and warm.

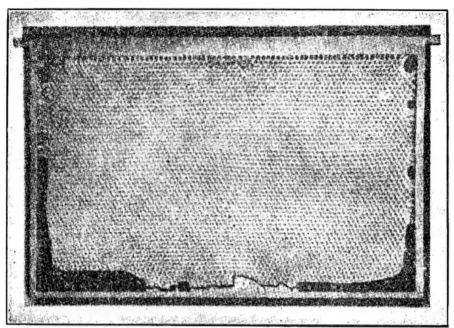

Fig. 24a. Frame for holding comb horizontally above brood-nest for getting queen-cells by the Case method.

Fig. 24b. Comb Prepared for Queen Cells

Fig. 24c. Comb of 80 good cells

Strong colonies only should be used for this, as for any other method of queen rearing. If all conditions are favorable, the beekeeper will secure a maximum number of cells. From seventy-five to one hundred fine cells are not unusual. (see Fig. 24c.) By killing the old queens a day or two before the ripe cells are given it is possible to requeen a whole apiary by this method with a minimum of labor. According to Miss Emma Wilson, it is possible to get very good results by this method, without mutilating the comb, although it is probable that a smaller number of queen cells will be secured. By laying the comb on its side as practiced in this connection, the cells can be removed with a very slight effort and with a minimum of danger.—*Frank Pellett, Practical Queen Rearing*

NECESSITY OF THE PRACTICE.

IN order to obtain good results from the apiary we must, as a matter of course, have good bees and plenty of them; and in order to have good bees we must first of all have good queens to breed from. That there is often a vast dissimilarity in the characteristic qualities of different colonies in the same apiary, no one who has had even a short experience will deny. How often do we find the bees of some colonies constantly irritable and disposed to sting, while those of others may be handled with impunity; or some giving a good return of honey, while others are doing little or nothing. Such cases may be seen in every apiary where a careful breeding of queens has not been systematically carried out. That it is possible on the other hand, by means of such a system, to develop to a greater extent the good qualities, and to breed out the bad ones in the honey-bee is no longer a matter of doubt. Such advanced apiarists as Alley, Heddon, Doolittle, and others in America, who have gone about the work in a conscientious as well as a scientific manner, have undoubtedly succeeded in developing a superior strain of bees. There is nothing to cause surprise in all this, when we consider the analogous case of the results obtained by select breeding of horses, cattle, and all our domestic animals. The breeder of bees has one advantage as compared with breeders of horses and cattle-he has not to wait so long for the results of his experiments; the bee-keeper can do as much with is bees in the way of crossing and improving races in from four to five years as a cattle breeder could probably accomplish with his stock in twenty or thirty years. Considering the many advantages to be gained by cultivating the best qualities in our bees, I am induced to look upon the rearing of select queens as one of the most important braches of modern apiculture, and I would therefore advice every bee-keeper to make it one of his special studies.

But Independent of all considerations about improving the breed the modern systems of apiculture cannot be carried out in its entirety if this important branch of it be neglected, therefore it becomes imperatively necessary for the bee-keepers of the present day to rear and keep a stock of queens on hand sufficient for their needs. It has been shown in the preceding chapter that in the successful practice of either the natural or artificial methods of increase a supply of queens is required, and some spare ones should also be kept ready to make up for losses that may, and in large apiary certainly will, occur during the season when surplus honey is being taken. The sudden loss of a queen at this time would cause a delay of about 24 days before the hive would be furnished with a laying queen again, that is, if the bees have to rear one for themselves from a newly-hatched larvae, and it is easy to understand what effect this would have upon the colony, and how necessary it is that we be prepared for such contingencies.

A WORD CONCERNING DRONES.

When endeavouring to improve our bees by cross-breeding we must of course be as particular about raising select drones for mating purposes as about the queens themselves. As the mating takes place in the air (see Chapter III.) and is not, at least as yet, under our control, our only security is to have our select young queens mated when only select drones are flying. The periods of the year when we are most likely to succeed in this way are the early spring and the late autumn; in the former by managing to breed our select queens and drones in advance of all others, in the latter by making the colony which produces the best drones queenless before the drones are killed off, and thus secure that these shall be flying when there are none alive in the other colonies. At other times throughout the season there will of course be drones from all the hives upon the wing.

ENTRANCE GUARDS.

It is however, claimed by some breeders, that with the aid of entrance guards, or "drone excluders," the drones which are not intended for mating purposes may be restrained from flying. These guards are made of perforated zinc, and fitted so as to cover the entrance to the hive. The perforations, being 5/32 of an inch wide, are large enough for the worker bees to pass through, but too small for the drones. Another kind of entrance guard is shown below. It is an invention of Mr. Alley's, and answers the double purpose of a trap as well as an excluder. This guard has an upper compartment, into which lead two wire-cloth cones, seen through the side openings in the figure. The drones, after failing to make their way out through the perforated zinc, finally force their way up through the cones; but not being able to return the same way, they become prisoners in the upper compartment. Should workers go up through the cones, they can still make their way out through the perforated zinc on top. From various conflicting reports which have come under my notice, I am inclined to doubt the efficacy of all these guards, as they are now made for the, for the objects intended.

Fig 97-Entrance guard..

HOW TO SECURE CHOICE QUEEN CELLS.

During the past seven years I have paid great attention to the rearing of queens, both for home use and for sale. I have tried several methods for raising queen cells, but none have given me so much satisfaction as the one I first saw described in Gleanings in Bee Culture for August, 1880 by Jos. M. Brooks and which I have since practiced. It is very similar to Mr. Alley's method, explained in his "Handy Book," a copy of which should be in every bee-keeper's library.

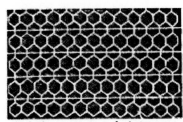

Fig 98.-Comb containing eggs. (picture of comb with eggs showing cuts every other row of cells)

To Secure good queen cells early in the season, we should select as soon as breeding has commenced in the early spring, two or more, as may be required, of our best colonies, and work them on in advance of the rest by slow feeding or, if need be, by giving them frames of emerging brood from other colonies, taking care to keep them covered up well. As soon as the one chosen for raising drones is sufficiently strong, insert a clean empty drone comb-to be obtained in the manner explained in Chapter VIII.-in the centre of the brood-chamber. Note the time when the drone-brood is capped, and in eight or nine days after, place a frame of clean new worker-comb in the centre of the brood-chamber of the hive containing your choicest queen. I would here point out that the cleaner the comb is the better; I find combs built the previous season that have only contained honey, give the best results. The colony now being pretty strong, with plenty of brood in the combs, the new one inserted will soon be in charge of the queen, and in three or four days will be full of eggs. As soon as the eggs commence to hatch, which will be in three days after they were laid, remove the comb to a warm room, and if more eggs are required, insert another in its place. Lay the frame of comb flat on the table or other convenient place, and with a sharp, thin-bladed knife, dipped in thin starch or diluted honey to prevent its sticking, cut the comb into strips, by running the knife along every second row of cells as shown by the white lines in fig. 98, taking care to leave one row of cells containing eggs intact in each strip. Some empty frames will next be required having two thin laths of wood nailed inside longitudinally so as to divide the depth into three compartments as shown below. Next take the strips of comb, and after destroying the eggs in every alternate cell on one side of the strip-which may easily be done by pressing it with the head of a wax match-fasten the strips under the top and two centre bar of the frame with a little melted wax, allow-ing the cells in which the alternate eggs have been destroyed to point directly downwards. The object of destroying each alternate egg is to prevent the cells being built too close together. A space intervening gives facilities for cutting them out subsequently without injury. Care must be taken, when fastening the strips, that the wax is not too hot or else it may melt the comb and kill the eggs. Having filled as many frames as may be required (I generally find one comb sufficient to afford strips for three frames), the next step to be taken is to remove the queen, every egg, and all ucapped brood from some one or more strong colonies, and place the frame of strips in the centre of the brood-chamber in each case. Mr. Alley recommends preparing the colony by removing the queen, etc., some twelve hours or so before giving them the selected eggs. Mark the hours or so before giving them the selected eggs. Mark the date and age of the eggs on the frame, and also upon the cover of the hive. A memorandum book is very useful in connection with this work for keeping records in. The queen and brood removed can be util-ized in forming a nucleus colony by caging the queen, removing a strong colony from its stand, placing the hive containing the brood and caged queens in its place, and shaking the bees from a couple of frames down near the entrance, to secure some young bees with the old ones that will return from the removed hive; the queen can be released in twenty-four hours.

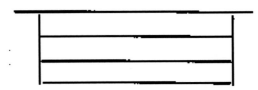

Fig 99.-Frame for raising queen cells on.

We have now, by removing the queen, forced the colony to turn its attention to raising others, and by depriving it of its own eggs and larvae, have compelled it to raise queens from those supplied to it. We have also, by taking away all its uncapped brood, lessened its labours, and thereby obliged it in a manner to give more heed to the matter in hand.

It is often stated that better queens, as a rule, are developed under the swarming impulse than can be raised by the forcing process. The reason given is that the larvae from which queens are to be reared, when the bees are preparing to swarm, receive the attention of the nurse bees, with this object in view, from the time of hatching, and are abundantly supplied with the "royal jelly"-so much so, indeed, as to apparently have more than they can consume, some usually being found in the bottoms of the cells after the queens have emerged. This surplus jelly being found in a cell is considered a good sign that a strong, healthy queen has developed from it. I have no doubt that this is all correct; but if these conditions can be brought about by the forcing process, there appears to be no good reason for supposing that the queens raised in that way will not be just as good; and by the method I am describing this can be effected, as I have proved time after time. The main considerations are to develop the queens in strong colonies, and to let the nurse bees have as little to do as possible, that their whole attention may be devoted to rearing the queens from the selected eggs or larvae we have supplied them with. The larva of a worker bee several days old can be transformed into a queen, but all breeders agree that such queens are of little use.

In less than twenty-four hours after the eggs have been given to the colony, several queen cells will be started over them. Some colonies will build more than others, but I think we may reckon the average at about fifteen with Italian bees, though I have had as many as thirty-five in a frame. There will be more built when honey is plentiful; and if little or none is being gathered, the bees should be fed while cell-building is going on.

Twelve cells are considered enough for one colony to care for, and this is near the number that is usually found in a hive from which a strong colony has just cast a swarm. As soon as the cells are forward enough to be plainly seen, destroy all except about ten or twelve of the largest and best looking ones.

Having now as far as possible fulfilled on our part every condition necessary to ensure the rearing of good queens, we must be content to leave the rest to the bees for a few days. The cells, when fully formed and capped will have something of the appearance of Fig. 100.

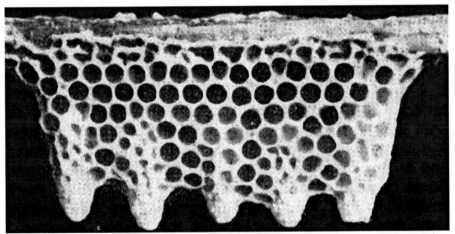

Fig 100.—Drawn queen cells (picture from Better Queens)

It will be remembered that the date and age of the eggs-three days-was marked on the frame so that we can calculate the day when the queen will be at maturity; that will be on the thirteenth day after inserting the eggs. Being able to know within a few hours when the queens will emerge is one of the great advantages of this system of queen-rearing. By the old methods and even when cells are build under the swarming impulse, it is impossible to say correctly how old the embryo queens may be.

As soon as the cells are capped, a frame or two of emerging brood may be given to the colony to strengthen it. It will have been noticed by those who have had any experience in queen-breeding, that there is often a marked difference between queen cells; some are long, pointed, and dense looking, while others are stunted and thin walled. The latter are always reckoned to contain poor queens, and it will be well to shun them, and make use of none but well-formed rough-looking, long pointed ones. One the morning of the twelfth day after the eggs were given the nuclei can be formed.

FORMING NUCLEI.

A nucleus colony in connection with queen-rearing is a small colony formed for the special purpose of caring for a young queen during her maidenhood, or until she may be required to do duty in another colony. A nucleus hive, is a small hive suitable for the colony, and is rarely used except for queen-rearing purposes. Some queen-breeders use a very small hive with much smaller frames than their common ones for keeping their queens in till mated, but for several reasons I consider it best to have but the one frame in both the queen-rearing and the ordinary hives. In the first place, a nucleus colony can be formed in a few minutes from any hive by simply transferring two or three frames and the adhering bees from it to the nucleus hive. Then again, a nucleus colony can be built up at any time or united with another where the frames are all alike, with very little trouble. And lastly, we have only the one sized frames to make. I have always used a nucleus hive such as I have described, and would not care to use any other.

The required number of nucleus hives being ready-their entrances covered with wire cloth to confine the bees-take the frame of cells and cut out carefully all but one; then return the frame to the hive until the queens shall have emerged, when it may be removed and a frame of combs or of foundation inserted in its place. Care must be taken that the queen cells are not injured or chilled; a small box, with some soft material to lay the cells upon, is handy to keep them in until they are inserted in the combs. Now go to a strong colony and hunt up the queen. This is sometimes a difficult task with a strong colony of black bees. If you have an empty hive alongside to place the frames in after you have examined them much trouble may be saved. Having found the queen, place her with the frame she is on in a hive by herself for the time being, and insert a queen cell in each of the other combs as you take them from the hive, remembering that you require some brood, a fair number of bees, and a fair share of honey in each nucleus. I usually put either one pretty full frame of brood, or two that are not so well filled, with the adhering bees, and a frame of honey, which may be taken from another hive, or else a frame of foundation in each nucleus. The frames of brood and bees should be taken as equally as possible to form the different nucleus colonies. A strong stock will generally furnish enough for five nuclei.

Something is required to support the combs while the cells are being inserted. To enable one to work easily and quickly the comb should be about on a level with the shoulders while stooping or kneeling beside the hive. A stand like that shown below is quickly made and is very serviceable.

The drawer near the ground will be found handy for keeping queen cells and small tools in.

HOW TO INSERT QUEEN CELLS.

When Cutting the cells from the frame, as much as possible of the base should be taken, clear up to the wood. With the frame which is to receive a cell placed on the support or stand in a convenient position, cut a small hole in the comb just large enough to put in the cell without pinching it in any way. In cutting the part for the base let it fit as nicely as practicable, as shown by the white line in the next illustration.

As soon as the cell is inserted, place the comb with the adhering bees in the centre of a nucleus hive; put in the other combs as already explained, and put on the mat and cover. Be sure that you have blocked up the entrance with wire cloth so that no bees can escape. Then proceed with the other nuclei in the same manner. When all are finished, take the nucleus hives to a cool shady place, or if they can be put in a dark, well ventilated room or shed, it will be better still. Keep them closed till the evening of the second day after that on which the cells were inserted, when they may be placed where they are to remain, and the bees liberated a little before dusk. By confining the bees in this way for a day or two they become reconciled to their new quarters, and very few fly back to their old hive. Before I adopted this plan I sometimes had a deal of trouble on account of so many bees deserting the nuclei.

The above method of forming nuclei and inserting queen cells is no doubt the best to adopt when queen rearing is only carried on, on a limited scale, and where the loss of a queen cell would be felt, but where, as in my own case, a saving of time is of greater consequence than the loss of a cell now and again, a knowledge of my plan may be of service.

MY METHOD OF FORMING NUCLEI AND INSERTING QUEEN CELLS.

With the nucleus hives, a few spare combs, and provided with some long pins, I go to a hive, and without troubling to look for the queen-except merely to glance over the combs as I take them out-I insert the cells as quickly as possible. Instead of taking the time to fit them nicely, I give a hasty look at the cell, cut a hole in the comb I think will suit, put in the cell and fasten it there by running two pins through the base of it into the comb, one each way-sometimes one is sufficient. Advantage may be taken of a depression in the comb and so save cutting a hole. In this way I can insert the cells and form the nuclei in a very short time. If the queen should be seen during the operation, she is placed with the frame she is on to one side until all is finished, when she is put back into the hive after contracting it with division boards, if necessary. Should she not be seen it only means the loss of one queen cell, which is more than made up for by the time saved in not waiting to find her. I have often spent a considerable time looking for the queen in a strong colony and then perhaps had to give it up. Professor Cook recommends inserting the queen cells twenty-four hours after the nuclei are formed, but says: "We may do it sooner but always at the risk of having the cell destroyed." I very rarely find one destroyed, and I think the risk likely to be greater when time is allowed for the bees to commence building cells before giving them one. Occasionally it happens that a nucleus colony will not accept a queen cell even when it has been queenless for some little time. When this occurs a cell should be protected in a cage when placed in the hive until the queen emerges, when there is likely to be no further trouble.

MATING YOUNG QUEENS.

To return to our nuclei. We left them just after liberating the bees. At that time the queens would be one day old; in four or five more they will take their wedding flight just when our select drones are about fourteen days old and flying. If our plans have been carefully matured there would be no other drones flying from our apiary at this time, so that there would be every likelihood of our queens mating as we desired.

Sometimes quite a number of young queens will be lost during their wedding trip, at other times very few. I have never been able to satisfactorily account for this difference. Whether it be that there are more bee-enemies about at one time than another I cannot say, but of this I am certain, that there are a less number lost when the nucleus hives are far apart and located some little distance away from the main part of the apiary. Mr. Alley says that the daughters of some queens are more liable to be lost than others but cannot account for it. In another place he says: "I bred from a queen last season, not one in fifty of whose daughters were lost in mating." Possibly some have a sense of locality better developed than others, and are therefore less likely to miss their proper home on their return from their first flight. At any rate it is a matter worth giving attention to.

When the young queens commence to lay, which they will do in a few days after mating, they are ready to be made use of unless we desire to test them, and when raising them for sale they should always be tested for purity and laying qualities for at least a month. By following up with cell building others may be ready to place in the nuclei when the young laying queens are removed, though there may not be the same chance to have the second lot of queens mated by selected drones unless it can be accomplished by the use of drone excluders, the utility of which, as I have before remarked, I am rather doubtful about. Even then no other bees should be near the apiary.

NECESSARY DISTANCE APART OF DIFFERENT RACES TO ENSURE PURE MATING.

This is a question upon which a considerable difference of opinion exists. Mr. Alley thinks that half a mile is far enough, while many other experienced apiarists considers that a mile or even two, is rather close. It is one of those still debatable questions connected with apiculture which may be argued on both sides for any length of time without being settled satisfactorily one way or the other by actual proof. I am inclined to think, however, that if we want to make sure of our queens being mated with drones of our own apiary there should be no other bees located nearer than one or two miles off.

Isaac Hopkins (1837-1925)

Isaac Hopkins's writing is eminently practical. Hopkins obviously had a lot of actual commercial scale experience and not just a lot of book knowledge.

Hopkins was first of all a beekeeper. In 1877 he was a big promoter of the use of the Langstroth hive in New Zealand. He was the Chief Apiarist to the New Zealand Government from 1905 until 1909 however he stayed working for the New Zealand government until 1913.

Through his position and especially through this book, he was instrumental in shaping apiculture in New Zealand and Australia; as a promoter of the use of the Langstroth hive, long before becoming Chief Apiarist and pushing for bee associations and laws to update beekeeping and control AFB. He remained active in the associations and the politics of beekeeping until his death in 1925.

Queen

Rearing

Simplified

Jay Smith

Queen Rearing Simplified by Jay Smith

Originally Published by
THE A.I. ROOT COMPANY Medina, Ohio 1923

Reprinted 2011 by

X-Star Publishing Company
Founded 1961

Transcriber's preface.

   Queen Rearing Simplified is one of the most popular queen rearing books of all time written by a man who raised a lot of good queens. It is no longer in print, so I am trying to keep Jay's wisdom alive here. There are many queen breeding books by scientists or small-scale breeders, but this is by a beekeeper who raised thousands of queens every year. I think that is much more applicable to practical queen rearing. This is a reprint so the old pictures are not the highest quality. Many of the opinions stated here are outdated in Jay's point of view by the time of Jay's second book, *Better Queens* written a quarter century after this one. You can buy *his other book,* from X-Star Publishing Company or you can find it at **http://bushfarms.com/beesbetterqueens.htm**

This book is dedicated to my full partner - my wife. –Jay Smith

Introduction

For several years past there has been a growing interest in Queen-rearing, as more beekeepers are coming to recognize the important part the queen plays in beekeeping. I have been receiving a large amount of correspondence on the subject of Queen-rearing from beekeepers wishing for detailed information on the subject. Their many questions have prompted me to attempt this book, and to explain such points as are not clear to those interested in Queen-rearing.

To assist the honey producer in rearing his own queens is my primary object; but I also describe methods adapted to the amateur as well as the commercial queen-breeder. To the beginner in beekeeping, however I would recommend a careful study of one or more of the following books on general beekeeping before taking up this work; "The A B C and X Y Z of Bee Culture" (Root), "Beekeeping" (Phillips), "Starting Right with Bees" (Rowe), "Langstroth on the Honey Bee" (Dadant), "Fifty, Years Among the Bees" (Miller). In addition to bee books one should read all articles in the bee journals by able writers and especially those written by Geo. S Demuth, who is now generally recognized as our highest authority on beekeeping. For a description of different methods of Queen-rearing read Pellet's "Practical Queen Rearing."

I presenting this volume to the beekeeping public, nothing radically new or revolutionary is offered. The system described has been taken from many sources, so it is impossible to give credit to all who have contributed through their books and their writings to our bee journals.

More is due to Mr. G.M. Doolittle than any other, for to him we owe the invention of artificial cell cups and the art of grafting.

I shall deal mainly with the successes I have had and not with the failures. I have two reasons for doing this. One is that almost any beekeeper has failures without having to refer to a text-book on the subject; and the second is, that I wish to keep this book within modest dimensions. If I should chronicle all of my failures, a book so voluminous would result that a Webster's Unabridged might look like a vest-pocket edition in comparison.

The object of this book then, is not to present many new methods but to place before the reader, with the aid of the camera, such methods with variations as I have used for twenty-one years, and to describe them in detail so that any one wishing to rear queens can succeed, and, if failure comes, he may refer to this book, and find the cause of it. Many have reported indifferent success with the grafting method of queen-rearing. Upon investigation, it was frequently found they had followed all of the rules laid down with *one* or *two* exceptions. These very exceptions brought the failure. I hope this book may be of help to such. Frankly, I do not know whether it will or not. The reader must be the judge. Again, if this little volume interests some overworked business or professional man or woman, and, through it, pleasure and recreation are gained, and he is thus better able to meet some of the harsher things of life, I shall consider my efforts have not been in vain.

Vincennes, Indiana, October 5, 1923.   JAY SMITH

Table of Contents

Chapter I. Importance of Good Queens.

In view of what has been said by the writers in the past, it would hardly seem necessary, if the best results are to be obtained in honey production, to call attention to the importance of having every colony headed by a good, prolific Italian queen.

You will note that I say a good *Italian* queen. Beekeepers are practically unanimous in the opinion that the Italian bees are much superior to Blacks in nearly all respects. They are better workers, swarm less, are more gentle and are much superior in cleaning out European foul brood. Unfortunately the black bee was introduced into the United States over two hundred years before the Italian, and therefore the Blacks have become pretty well established in all parts of our country. They are now found wild in trees and rocks in every state from coast to coast, and in many parts of Canada. Consequently, one very good reason why the honey producer should rear his own queens is to get rid of the black bees and hybrids.

Every beekeeper concedes the point that each colony must be headed by a good prolific queen, and all writers on the subject have emphasized it in the strongest terms, yet in truth very few of us fully realize the importance of good queens.

Put yourself to this test. When the season is over and you are taking off the honey, notice how much more honey some colonies produce than others. Then get out your pencil and paper, and figure how much money you would have made if *all* colonies had made as much honey as the best. The results are frequently startling. Then remember that there is positively no one element that contributes to the production of these big yields as much as *good young queens*. After you have these results tabulated, consider whether or not it would pay you to rear your own queens and become an expert at it, or have some members of the family or firm take up this most important branch of beekeeping.

Our best authorities are agreed that there is not so much difference in the inherent honey-getting ability of the different colonies as there is in the *condition* of these colonies; that is, they produce large honey crops because the conditions within the hives are ideal. There were plenty of young bees and brood at exactly the right time. These colonies seemed to devote all of their energy to honey-getting. They did not loaf. They did not swarm. They just *worked*, and these conditions were brought about by the fact that these colonies had good young queens, and not because they had inherited any exceptional traits or were constitutionally superior. That there is a difference in the honey-getting ability of different colonies is not denied; but it is difficult, indeed, to be able to prove that the reason a colony made the largest surplus was due to natural ability rather than to the condition within the hive.

Therefore, it is no easy matter for the honey producer to pick out the best queens, since it may be the opportunity that the queen had, rather than her natural ability. How, then, are you to select your breeding queen? First, be careful to see that *conditions* are the same in all colonies, and that the queens are of the same age. Then select the queen that has the most desirable qualities, such as prolificness and vigor, and whose bees are gentle, of pure blood, good honey-getters, showing little inclination to swarm.

Years ago I endeavored to breed up a honey strain by simply using as a breeder the queen whose bees produced the largest yield. I found that the honey-getting quality was not in the least improved; but that the bees were getting cross and dark in color. Then I adopted the rule of selecting the largest and most prolific queen whose bees were gentle of good color. I found that better results were at once obtained. Being more prolific, this queen was able to keep the hive full of brood and the bees at the beginning of the honey flow, which is the secret of successful honey production. If this rule is followed and in addition all colonies are requeened from the best, in order to have them as nearly alike as possible in every respect, then we may select as our breeder the one that has the above qualifications and also the one that produces the biggest crop.

Some have reported that a medium-sized queen is as good as a larger one. That has not proved true in my experience, which has been that the larger the queen, the better. A queen that is extremely prolific has to be very large in order to contain the necessary number of eggs in process of formation to enable her to lay the four or five thousand eggs per day, which is the performance of a really good queen.

When the virgin emerges from the queen-cells she should be large, long and pointed. In three or four days, she will be much smaller, but extremely active and nervous. After mating she rapidly becomes larger until she is twice her former size. The abdomen becomes long and broad near the thorax, gradually tapering to a point. Short, blunt queens are inferior.

We must always bear in mind that, no matter how good our equipment, how well we pack for winter, how generous are the winter stores, and how abundant the nectar in the blossoms, our efforts will bring only failure if we do not have a good queen in the hive.

Chapter II. Conditions Under Which the Bees Rear Queens.

In order to rear queens successfully we must study the conditions in the hive under which bees rear their own queens. There are three of them known to beekeepers as the Emergency Impulse, Supersedure and Swarming.

In nature it sometimes happens that a colony suddenly loses its laying queen. Perhaps, as on very rare occasions, the queen, while laying, dies before the bees have time to supersede her in the regular way. The inmates of the hive at once realize they must meet this emergency, and immediately go to work to rear another queen. Fortunately, nature has made it possible for them to produce one from the larvae of eggs already in the hive. They, therefore, choose a number of worker larvae and begin to feed them lavishly with predigested food known as royal jelly. They usually fill the cell with this until the tiny larva is floated out to the mouth; then the bees build a queen cell over it, pointing it downward. This new cell is frequently over an inch long, and is made larger inside than that of the worker. The bees feed the larva until about five days after it is hatched from an egg, and then the cell is sealed over by them. The larva within spins a small thin cocoon, changes from a larva into a pupa, and in about eight days from the time the cell is sealed the virgin queen gnaws off the cap of the cell and crawls out. For a few hours she is a weak, frail creature, downy and delicate. However, she develops rapidly, and in from two to four hours, realizing she is a queen, she, just as many monarchs in the human family, becomes very jealous of any who may have ambitions to possess her throne. It is interesting to note the events which take place in the hive for the next few hours.

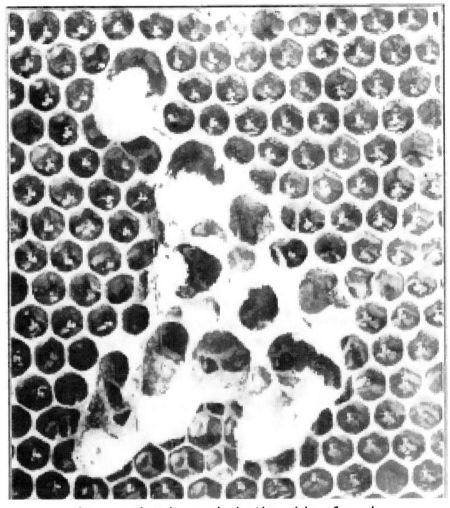

An opening is made in the side of each.

Having in mind the suppression of competition, the new queen roams over the combs. If there are any queen-cells from which the queens have not emerged, she supervises the destruction of them. The workers perform most of the labor under her directions, although she helps as best she can. She begins on the cells whose queens are most mature. She seems to reason these are the ones likely to give her the first trouble. An opening is made in the side of each, and, if the inmate is about ready to emerge, the queen backs down into the opening in the side of the cell and stings her helpless rival. The opening is then enlarged, and the dead queen is carried out by the bees. Other cells are visited and destroyed in turn. However, if there are queen-cells uncapped, these are left for a while, the newly emerged queen seeming to realize that she has plenty of time to handle their cases before they become any menace to her.

Now it frequently happens that, while this young queen is finding herself for the first two hours after emerging, other queens emerge, and several virgin queens will be in the hive at once. They seem to realize they are too young to do any satisfactory fighting, so by mutual consent they avoid each other's society and devote their time to supervising the destruction of queen-cells. However, as they grow from four to twelve hours old, they begin to seek out their rivals with the idea of doing battle. When they meet they clinch, and each tries to get a chance to sting the other. The fight does not last long, for soon one gets in the coveted position to give the fatal thrust of the sting in the thorax of her rival. The vanquished queen quivers a moment, and is dead.

Other "preliminary" fights are staged until only two queens are left. Then the "final" duel takes place, and the victorious queen reigns supreme.

In due time queen-cells in all stages of development are destroyed, and in six or seven or eight days the virgin queen flies out of the hive, meets the drone, and returns to become the mother of the colony, beginning her egg-laying within the next day or two.

A different order of events has been given by others, who state that the first thing a young queen does is to hunt up her rival and fight it out; but I have witnessed the occurrences many times as above described. Indeed, when occasionally grafted cells have been left too long in the hive, upon opening it I have found many queens safe and well, all busily engaged in tearing down cells. I have counted as many as fourteen superintending this work of destruction before any battle had begun. They have been given nuclei, thus saving them.

The most unsatisfactory manner in which bees rear queens is the Emergency method. The bees seem to feel their danger of extinction from having no queens. In their frenzy, a large number of cells are started. To make a bad matter worse, they take larvae that are too old, with an idea probably of rearing some sort of queen in the shortest possible time. (Transcriber's note, Jay changed his mind on this later and retracts this in Better Queens) We all know that in satisfactory queen-rearing, the younger the larva used, the better. By this method, the oldest larva chosen is the first to hatch, so the poorest queen in the batch is the one that heads the colony. However, as this is an emergency case, the bees seem to reason that, if this queen is not as good as she should be, they can take their time and rear a good one later on by the supersedure method.

Supersedure Method.

When a queen is beginning to fail from old age or some other infirmity, the bees seem to realize that she can not be with them much longer, so they take steps toward rearing for themselves a new mother. Queen-cells are started, sometimes only one, seldom more than four. In these shallow cups the queen lays eggs. As soon as hatched, the larvae are fed royal jelly, and as they receive the care and attention of the whole colony, good queens are, as a rule, the results. Sometimes the bees seem to wait until the old queen is so far gone that she lays several eggs in a queen-cell which results in the larvae not having sufficient food since they have to share it with their "cell mates." Owing to their being crowded in the cell, such queens are sometimes slightly misshapen. Usually, however, all but one of the larvae are removed before the cell is sealed. Under the supersedure method, however poor queens are rare and, as a rule, the best of the queens are reared. Usually the old queen disappears as soon as the virgin emerges from the cell; but sometimes, mother and daughter live peaceably together, both laying and usually found on the same comb.

Queens Reared Under the Swarming Impulse.

When a colony is preparing to swarm they start a large number of queen-cells in which the queen lays eggs. When the first cell is capped, if the weather is favorable, the swarm usually comes out. As swarming occurs when the colony is at its height of brood-rearing, the larvae are well supplied with royal jelly, so that the finest queens are reared. In rearing queens by any method, we can learn a great deal by carefully studying the conditions of the bees while building cells preparatory to swarming, for we wish to duplicate the performance.

Under the Emergency method, the bees build a large number of cells, but they do not give them the proper attention and skimp the larvae for food. Under the Supersedure method, they give the larvae plenty of food, but usually do not build more than three or four cells. Under the Swarming Impulse, they not only build large numbers of cells but supply the larvae in them lavishly with food. What is the reason for this? Is it because they have the "swarming fever" that induces them to do such good work? I believe not. My observations lead me to believe it is the *condition* of the colony and, in support of this theory, I have found that as many and as good cells may be built by a colony when not preparing to swarm as by one that is, provided the conditions are the same in all other respects.

What are these conditions? First a honey flow is on or just coming on, for bees seldom swarm at any other time. Second, they are strong in bees, especially young nurse bees. Third, the hive is crowded with brood in all stages; and fourth, the weather is reasonably warm. I believe these conditions enable the bees to rear not only a large number of queens but those of the highest quality. Understand, it is the *condition not the swarming fever*. As evidence to substantiate this statement, the following fact, which I have observed many times, is given. While having cells finished above an excluder, sometimes the bees take it into their heads to swarm, and as bars of cells are capped the swarm issues. Since the wings of the queen are clipped the bees return, and the queen is helped back into the hive. Removing the bar of cells frequently discourages swarming but sometimes they persist coming out every day or every other day for a week or more as the spirit moves them. I have never been able to see that, while they had this swarming fever, they gave the cells any better attention than before or after swarming. This fact satisfies me that it is the condition of the colony and the honey flow or the feeding that give good results in cell-building.

Under the Grafting method, we endeavor to get all colonies connected with queen-rearing in the condition above described. If we do, we can rear queens every bit as good as those reared under the swarming impulse; (Transcriber's note: Jay later changed his mind on this.) if we do not, inferior queens will result. By examining the cells one can easily tell which of the three methods the bees used in their construction. In the Emergency method, the queen is reared from a larva that has hatched in a worker-cell, so by looking into the bottom of the queen-cell, the worker cell may be seen. In the Supersedure method as well as the Swarming method, the cells are the same. The queen lays eggs in both; but during the swarming, many more cells are built than under the superseding impulse.

Queen-rearing apiary of the author.

There are several methods that may be employed where one wishes to rear but a few queens. Cells, saved from a colony that has just swarmed, may be placed in colonies to be requeened, whose queens have been removed. This is much better than to allow colonies to run along with inferior queens; but, by this method, little progress can be made in improving the stock since when you wish to requeen, your best colony may not be swarming. Consequently, you would have to use cells from an inferior colony. It has frequently been noted that the inferior strains of bees swarm the most. Blacks and hybrids are much more inclined to swarm than Italians.

In requeening by the swarming method, a piece of comb one inch in diameter should be cut out around the cell, using a good sharp knife, and being careful not to injure the cell. A hole of corresponding size should be cut in the comb of the colony to be requeened and the piece containing the cell fitted into it. Where but one cell is on the comb, the entire comb may be placed in the colony to be requeened. If this colony is of medium strength or strong, it makes no difference just where the cell is placed for there will be sufficient bees to give it proper incubation. The bees may be left on this comb or brushed off, but never should be shaken off since the undeveloped queen is almost sure to be injured. In giving a cell to a weak colony or a nucleus, it is important to place it near the center next to the brood. Frequently cells built on the bottom edge of a comb when given to a weak colony do not mature.

A second and very simple method of requeening is simply to remove the queen from a colony, and the bees will construct a number of cells by the Emergency method. Such queens are not, as a rule, as good as those reared under the Swarming or Supersedure Impulse. If care is taken to save only the largest and best cells, however, very good queens can be reared in this way. The principal point to commend in both of the above methods is their simplicity. If one has never reared queens, these will prove very interesting and are a step toward better ones.

The third system requires a little more skill, but will produce cells as good as the best if care is taken to have all conditions right. Go to the colony containing your breeding queen and insert an empty comb into the center of the brood-nest. Leave this there for two or three days or until the queen has laid a large number of eggs in the cells. Remove it, however before the eggs begin to hatch since our object is to get the bees to use very small larvae from which to rear queens.

Next, go to a strong colony and take the queen and all combs containing eggs or brood, but leave with the bees several combs of honey and pollen and give them the frame of eggs from your breeding queen. If it is desired to save this queen, she is given a frame of brood and adhering bees and put into a hive to start a new colony. Fill out the vacant space with combs containing some honey, if possible. If you have no extra drawn combs on hand it is better to take a few from other colonies and in their place give full sheets of foundation, for they will do better work at drawing foundation than would this new colony which is not strong enough for that purpose. The remainder of the brood is used to strengthen weaker colonies or to make strong colonies even stronger for the honey flow as occasion seems to demand.

Having now disposed of the queen and brood, let us go back to our queenless colony. Realizing their queenlessness, the bees will start cells as soon as the eggs begin to hatch. Very frequently by enlarging the worker-cell, they make it over into a queen cell even before the egg hatches. In this manner the newly hatched larvae receive abundance of royal jelly from the very start, which is necessary for the best results. This method has the advantage over the others just described since the bees can not use larvae that are too old for good results. However, it should only be used when there is a honey flow. In about six days after the cells are capped, they should be cut out with a sharp knife and given to colonies to be requeened which have been made queenless. When there are larvae of the proper age at the bottom of the comb, the bees prefer to build cells there, sometimes building a compact row of cells half way across the comb. In such cases some of the cells will have to be destroyed when being cut apart. In giving this comb of eggs to the colony, if there are no eggs at the bottom of the comb, it is well to cut away the comb so that the eggs will be at the edge. This is not necessary, however, for the bees will start plenty of cells if the comb is left intact. As the operation of forming nuclei to receive them, when that is desired, is the same as given under the Grafting method, it will not be described here.

A compact row of cells half way across the comb.

The Grafting Method.

If one keeps as many as fifty colonies or expects to do so in the future it will pay to learn the grafting method. This requires much more skill and practice than the ones above mentioned; but it has so many advantages over all the rest that it is used by nearly all queen-breeders and extensive honey producers who rear their own queens.

This method is more economical, for it is not necessary to have any colony queenless at any time. You have control over the situation and can rear queens in any quantity desired. It is exact, since you know within a very few hours when any cell will hatch. The artificial queen cups are much easier to handle, for with them it is not necessary to cut up and mutilate good worker combs. Larvae can be taken from your best breeding queen and the stock improved thereby. Last, but not least, the very best queens can be reared, if conditions are kept right. To rear a few queens during a honey flow is a simple matter; but to keep up a steady production throughout the season under variable weather and honey flow *is not* a simple matter. However, with experience and patience it can be done.

Root's Basswood Apiary.

Chapter IV. Rearing Queens on a Large Scale

I thoroughly believe that many beekeepers who have a thousand colonies or more and who do not rear their own queens could increase their honey yield fifty per cent by having a good queen-rearing outfit and being able to use it properly. Moreover, in localities where European foul brood is rampant the honey crop might be doubled or trebled, since there is nothing that eliminates this disease like strong colonies of Italian bees headed by young, vigorous queens.

Headwork.

Some of the most important work that can possibly be done in the winter months is reading bee books and journals, of which we have a goodly supply of the highest standard. Secure all the books you can and take all of the journals. If you do this and carefully study them, it will be the best investment you can make. Read, study and plan in the winter.

We should remember that successful business men work with their *heads.* They can hire hand work at a low figure; but headwork is always at a premium. A great deal of headwork is required of the successful beekeeper, and much of this work can be done in winter. During the honey flow we are too busy working with our hands to do much headwork. J.S. Knox, the efficiency expert, says that a man is worth $2.50 per day from his chin down-ward. If he earns more than this, it must come from above the chin. Consequently, he divides men into two classes, "Chin Uppers" and "Chin Downers." If we are successful we must be "chin uppers." For the beekeeper the best time to do his "chin upper" work is in the winter sitting before a comfortable fire, reading, thinking, study-ing, planning.

Moreover, as there is a great deal of work to be done with the bees during the queen-rearing season, one should plan to do all the work possible in the winter. Nailing up hives and nuclei, painting them, putting in foundation, dipping cells and similar work should be all gotten out of the way before spring comes.

Chapter V. Dipping Cells.

Since I know more about the way *I* rear queens than I do about the way any one else does it. I wish to take the reader with me through the season, while I attempt to show in detail how I rear queens. Possibly, you have methods of your own that you prefer. I do not claim to have a monopoly on all the good things in queen-rearing, but will be content if you find some little feature which I use that you consider worthy of adoptions, and which may be of help to you.

Let us start by dipping cells as this can be done in the winter. Wax is saved from the year previous. For this a solar wax extractor is an important item. During the summer months, many small pieces of comb are found that can be thrown into it. This makes the finest cell building wax. In the nuclei, bits of comb are built and when introducing queens, where a frame is taken out, the bees will construct more or less comb. All these can go into the wax extractor. From the wax extractor, the wax is placed in small molds, for use in dipping queen-cells. I have enough cell bars to last the season, so we always dip sufficient each winter to supply us through the entire summer.

Our cell-dipping outfit contains twenty cell-forming sticks, which work through holes made in two pieces of heavy tin. Metal is much better than wood since the latter swells when wet and the forming sticks do not work freely through the holes. These pieces of tin are fourteen or fifteen inches long, fastened one and one-quarter inches apart to small blocks of wood, which are to serve as handles when dipping the bars into the trays. Each piece of metal is pierced with twenty holes, one-fourth inch apart, and seven-sixteenths inch in diameter. The holes are exactly opposite each other on the two bars, in order that the cell forming sticks may slip up and down through them easily.

A solar wax extractor is an important item.

Two trays are used, one five by sixteen inches, the other two and one-half by fifteen inches. Water is placed in the larger forming a double boiler; while wax is placed in the inner tray and the whole set over the heat. The wax should be kept at the lowest temperature at which it will remain liquid. If it becomes too cool the cells will be lumpy; if too hot, they do not slip from the sticks. If one is not experienced, it is well, when the wax apparently reaches the proper temperature for successful dipping, to try dipping one stick, and, if the wax proves of satisfactory temperature, proceed to work.

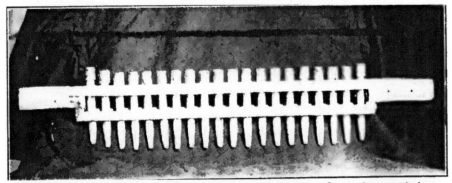

Our cell-dipping outfit contains twenty forming sticks.

Cells of the proper size and shape.

First, dip the ends of the forming sticks in cold water, then dip into the melted wax; again dip in the water and back into the wax for about four dippings, care being taken to have a firm thick base, with a thin even edge. By dipping the sticks in the wax and holding the bar up until a drop forms on the base of the cell, a thick base is procured. A thick base is necessary, for in trimming off the cells with a knife the cells would be injured if too short.

When completed, the cells should be about five-sixteenths of an inch across the mouth and one-half inch deep inside measurements.

Many beekeepers make a mistake in believing that the most important feature for successful cell acceptance is the grafting of the larvae into the cells cups; but a far more important feature is that of making cells of the proper shape and size. The ideal cell would be as the bees build them, large inside, with a small mouth; but it is not possible, or at least practical for the beekeeper to make cells of this shape. Upon several occasions, I have given cells that had been accepted and slightly built out in the swarm box to a colony for finishing, when by accident it contained a virgin queen. Of course, the larvae and jelly were both quickly cleaned out. I have given one bar of such cells to a swarm box and two bars of our dipped cells. The bees seemed to concentrate all their efforts on the cells already worked on by the bees and neglected my dipped cells. The bees prefer to make the mouth of the cell just large enough for a worker bee to crawl into, and it is frequently noticed that sometimes in the workers haste to back out of a queen-cell when smoke is blown into the hive, it is caught and has to do considerable scrambling and kicking before it can get out. I find the best cell for practical purposes is one whose size is between that of the inside of a natural queen-cell at its largest place and the mouth of the cell, this being five-sixteenths of an inch as given above. In our early experience, many of us, enthusiastic in rearing larger queens, sought to accomplish this by making larger cells; but being large at the mouth, the bees were loath to accept them, and it took considerable work on their part to build them over to the size they should be. When the bees get to work on the cells they mold them into the shape they want, regardless of the size and shape the beekeeper has made them. The smaller cells will give better acceptance than the larger ones; but do not for a moment imagine this cramps the larva and produces an inferior queen, for the bees enlarge the cell to suit their own fancy. For experimental purposes I have dipped queen-cells the size of a worker-cell, and excellent results were obtained. Cells larger than five-sixteenths of an inch are not accepted so readily as those of this size or smaller.

And the cell cups painted at the base.

Nothing but pure beeswax of good quality should be used. Upon one occasion, when everything was going finely, cells accepted and built out nicely, the bees in the swarm boxes began to balk until accepted less than twenty-five per cent of those given. I had all conditions right, as I supposed, the same as before-plenty of young bees, well fed. At length I noticed the wax of which we made the cells was not so white as some we had been using. I made up a new batch of cells from clear white wax, and as if by magic, all cells were again accepted and everything went on splendidly as before. Instead of heating the wax in a double boiler as we do now, this wax had been set directly over the flame and had become slightly scorched and darkened, so the bees would have none of it.

After the cells have remained in water long enough to become slightly hardened, they are loosened by giving each a slight twist, but allowed to remain on the sticks. They are then placed on the cell bar, the frame being supported on blocks. A small round paint brush is dipped in hot wax, and the cell cups painted at the base where they come in contact with the cell bar. A kettle should be kept at hand for melting additional wax to add to that in the inner tray, in order that sufficient wax may be had to make the cells the necessary one-half inch in depth. If the wax in ether becomes dark-colored or impure it should be discarded, and an entire batch of new clear wax placed in the tray. However, the darker wax may be used to paint the bases of the cells to cause them to adhere to the bar.

When the wax has become thoroughly cool, the frame is lifted off and all of the forming sticks come out of the cells easily. If properly done, the cells will remain on the bars even if subjected to considerable rough usage. When the cell bars are all finished they should be wrapped carefully in paper to be kept free from dust, since the bees will not accept dirty or dusty cells. If you have on hand the cardboard cartons in which foundation is shipped they make ideal containers for the cell bars.

Suggestions in Making Cell Cups.

Of course it is not advisable for the beginner to have a dipping outfit made as previously described. After mastering the grafting method, he may enlarge upon his equipment as he wishes.

The beginner can either dip his cells one at a time and mount them or he can purchase ready-pressed cells from dealers in bee supplies. Either one will give perfect results. These cells may be mounted on bars as needed, thus eliminating the necessity of purchasing a large number of bars. The base of these cells may be dipped in hot wax and stuck on to the bar when needed. To avoid the necessity of getting the swarm box, he can also use the queenless and broodless method described in Chapter XIII. However, I believe it pays to use the swarm box, for one can, as a rule, get better results. In this way it is possible to experiment until one gets his hand in without putting much money into equipment, and as he progresses can add to the equipment to fit his requirements.

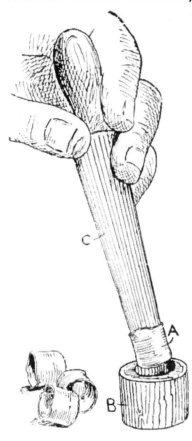

Pressed cell cup.

If one has difficulty in making his cells, one at a time or collectively, he can use to advantage the ready-made pressed cells sold by all dealers. Where only a small number are required the beginner will probably do better to buy what few he uses. The making of dipped cells is a nice art, and unless they are made just right, the bees will reject them.

Every thing in readiness, we await the coming of spring with a great deal of enthusiasm and no little impatience. Sometimes it seems spring weather will never come; but it does no good to worry and if you look backwards, you remember that spring has never yet failed to come.

Is there anything more interesting than to watch the bees bring in their first loads of pollen? If the beekeeper has done his duty toward them the season before, there will be no need of tinkering with them until later in the spring. The soft maples blossom and go; then come the pear and apple blossoms, and soon a few heads of white clover can be seen. It is now time to get busy at queen-rearing. Some seasons the weather permits grafting soon after the first blossom; but it does not pay to be in too great hurry to rear queens before the real queen-rearing season arrives. I know of no definite rule concerning the time for it. Each person will have to find out by experimenting until he knows his location well enough to be reasonably sure when to begin.

Many times in the Mid-West, the bees are strong and the weather conditions ideal for cell-building during apple blossoms; but later the weather turns cold, so that virgin queens can not get out to mate. As nearly as I can come to it, when the hives are getting nicely filled with brood, when plenty of pollen is coming in and the bees are gathering a little nectar, then it is time to begin grafting.

Before grafting, a supply of royal jelly is necessary. Some very successful queen-breeders report they get satisfactory results without its use; but I have never been able to procure as large acceptance or as good strong queens without it. *(Transcriber's note: Jay Smith changed his mind on this in* Better Queens.*)* J.W. George of El Centro,

California, gave to the beekeeping fraternity a valuable little kink when he explained that royal jelly can be bottled and kept in perfect condition from one season to another. I have practiced this to advantage, and find one of the great difficulties of queen-rearing is thereby removed.

If you have no royal jelly on hand, a colony may be made queenless until they build queen-cells, when you can get the jelly from them. After the first grafting, some of the jelly in a few cells you have produced may be used; but, in this way you continually destroy good queen cells.

As a container for royal jelly, I use a small porcelain jar with a screw cap. A piece of waxed cardboard in the cover makes it air-tight. Let me offer a suggestion as to where you can get one of these jars. Make a raid on your wife's manicuring outfit, and, if luck is with you, you will find one of these jars. To be sure that luck will be with you, better do it when she is out. This jar usually has some pink dope in it. Take this out, put it into a tin can, present it to your wife with your compliments and make off with the jar. Thoroughly sterilize this jar by boiling, for the bees seem to object to the funny smell that comes with it. If your wife does not have this, or if you do not have a wife, you can go to the drug store and find just the size and style that suit you. The dope looks as though it might be of use if you put it into the grease cups of your flivver, but I do not want to suggest too many dangerous experiments for you to try all at once. For a jelly spoon, I prefer to make one out of the bone handle of a toothbrush, which also may be found in the manicuring outfit. Break off the brush and whittle down the small end until it fits nicely into a worker-cell. This jelly spoon and the jelly jar are to be carried in the pocket of your trousers or dress, whichever you wear. While working with your bees during the season you will be running across colonies that have royal jelly to spare. Whenever a swarm issues, just take out the jar and spoon and get the royal jelly. I have found that I come across enough in my regular work so that I never have to make any special hunt for jelly. It is well to have two of these jars; keep one in your pocket and the other in the grafting room.

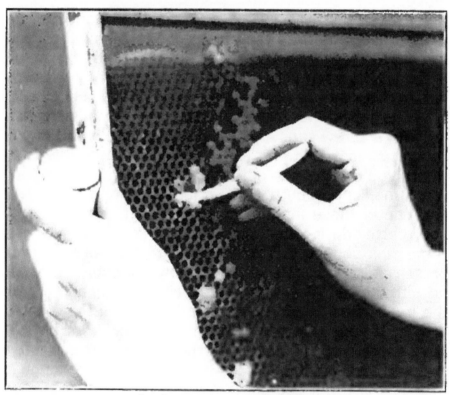

And get the royal jelly.

There are several methods used for getting cells accepted and started by the bees; but when all things are considered, I believe the swarm box has more desirable features than any other. For best results the swarm box must be kept in even temperature. It gives perfect results during cool weather, for, by placing it in the cellar, cave or basement, the outside temperature does not affect the bees. During hot weather it gives equally good results for the same reason. By using the swarm box it is not necessary to have any colony queenless at any time.

The swarm box.

The box is made eight inches wide, inside measurement, and should be the right length and depth to take the sized brood-frame to be used, allowing about an inch space below the frames. The bottom is covered with wire-screen cloth. Four legs one inch long are fastened to the bottom in order to provide plenty of ventilation. At each end of the box two strips of wood are nailed to support the cell bars. These should be placed so that the cell bars will be a little lower than the top of the brood frames. A cover is made with cleats running entirely around, forming a telescope cover one inch deep. In the center of this cover is made an oblong opening large enough so that three cell bars may be passed through it with ease. The object of this cover, which is little more than a rim around the edge, is to prevent the bees from boiling out over the sides when the bars are being put into the swarm box. A second cover slides on top of the telescope cover, thus closing the opening in it. The end-cleats on the telescope cover extend a little above to keep the upper cover in place. Two Heavy wire handles swing up over the ends so that when the swarm box is being carried, these covers cannot fall off. A round opening is made in the top of the upper cover by the use of an expansive bit just the size to take the cap of a mason jar. When feeding the bees, a Mason jar with a perforated cap is used in this opening which is also utilized for putting the bees into the swarm box by inserting a tin funnel and shaking the bees from combs into the box. The box is given two coats of paint to prevent it's warping so it will thus remain bee-tight. When made in such a way as to have the cell bars inside, the bees may cluster all around them and thus keep the temperature uniform, which gives much better results than the old style where the bars were placed in slots in the cover.

The old style swarm box is somewhat easier to make and for experimental purposes might be preferred by some. It is similar to the one just described except the cover, which is merely a thin board with an opening cut in the top large enough to hold the three cell bars. In using this it is well to place a cushion over the bars to keep out light and retain the heat. In this style it is necessary to make the filling hole at the end of the cover to avoid the opening made for the top bars

These covers cannot fall off.

Chapter VIII. Getting the Bees in Condition for Cell Building.

Let us remember that for the best results in cell-building we must have plenty of young bees which are being lavishly fed either from a honey flow or from receiving sugar syrup. This condition is necessary where even a few cells are being built by the colony. Now, as we wish them to build a large number of cells, the colony must be *exceedingly strong.* As we are starting early in the spring while only a little nectar is coming in, it will be necessary to feed sugar syrup in order to get the best results.

The method of feeding that I have found very satisfactory is to take a two quart mason jar and punch in the cap eight nail-holes 1-16 inch. Fill it with syrup, equal parts of granulated sugar and water. Many recommend a weaker syrup, but with me the heavier syrup gives the best results. There is no loss as the bees store in the combs all that they do not need. A bee-escape board is used for a hive cover, and the Mason jar is inverted over this hole. An empty hive body is set on, and a regular hive cover is placed on top of all. By this method of feeding there is no robbing, and the bees take the syrup night and day even if the weather is quite cold, which they will not do when this style of feeder is used at the entrance. However, our regular bottom-board feeder, as described later, has so many advantages over any other that I am now using it for all purposes.

As the swarm-box colonies as well as the finishing colonies are very strong, it is an advantage to slide the hive forward on the bottom-board, thereby affording better ventilation. This also provides an opening into which the syrup is poured. Many advocate giving them a very little thin syrup from an entrance feeder, using about a pint a day. This will answer very will if some nectar is coming in; but, when this is not the case, better results can be obtained by giving syrup in abundance. Two quarts of syrup, equal parts sugar and water, per day will give excellent results. Enough must be given to cause the bees to build white comb, and this enables them to draw out the cells to perfection as well as to secrete an abundance of royal jelly. My experience has been that the bees do every bit as well when the feed is given them all at once every night as they do when they take it through three or four holes from the Mason jar. It must be borne in mind that there is no waste in giving them more than they use at the time, for all surplus is stored in the combs. The colony that is to furnish bees for the swarm box must be very strong in bees. If it is not in this condition, it must be built up by giving it frames of emerging brood from other colonies. If a hive smaller than the ten-frame jumbo is used it should have a double brood-chamber, and both stories should be full of bees and brood. The hive should contain ten or twelve pounds of bees. Rearing good, vigorous queens without strong colonies and plenty of feed is an impossibility.

By this method of feeding there will be no robbing.

The colony that is to furnish bees for the swarm box must be fed at least three days before the swarm box is filled. It will do little good to feed them just before they go into the box. I do not know why this is true, but it seems to take a few days for the bees to assimilate the food and make it over into royal jelly. So I use the term, "fat bees." You must fatten the bees before they can do good work at feeding larvae. Poor, hungry bees will not accept cells. If there is a pretty good honey flow on, no feeding will be required.

This also provides an opening into which the syrup is poured.

Since we have the swarm-box colony in fine condition with abundance of bees and supplied with food, we will prepare the finishing colonies. As we expect to start three bars of cells in the swarm box, it will be necessary to prepare three colonies to finish them, for one bar of twenty cells is enough for even the strongest colony. However, if the finishing colonies are sufficiently populous, they will do exactly as good work at finishing twenty as can be done by a colony preparing to swarm, in building cells in their own natural way. The method of preparing the finishing colonies is similar to preparing the swarm box colony. They must also be kept running over with bees. At the beginning of the season when the colonies have not had time to build up to maximum strength, a large amount of brood is required to put them in condition to do the best work. All empty combs in the brood-nest should be removed and replaced with brood from other colonies.

They must make a two-story colony. The second story must be added with a queen-excluder between the two hive bodies. If the colony is of sufficient strength to care for nine frames of brood, the frames are placed in the upper story after shaking off all the bees. If the finishing colony is not strong enough to take care of the extra frames of brood, it is best to give it the nine frames of brood with adhering bees. There is some danger of these strange bees killing the queen below, and in order to prevent this, place a newspaper between the two bodies on top of the queen-excluder and let them unite the same as when two colonies are united. In this case you have the advantage of both brood and bees. In two weeks most of the brood will have emerged, and the combs will be filled with sugar syrup or honey. Remove these and put in some more brood. Do not wait until you notice that the cells are not being finished as they should be, for, if you do, a lot of inferior queens will result.

Running over with bees.

These combs of honey with a little capped brood are excellent for giving to colonies that are short of stores, or they can be given to nuclei. Always keep unsealed brood on each side of the frames containing cells, in order to draw nurse bees to them. No matter how strong the finishing colony may be, it will do poor work at cell-finishing unless there is unsealed brood in the upper story.

We must bear in mind that, when feeding is necessary, it must be done several days before the bees are to build cells-two days at the shortest and three days are better, the same as with the colony that is to furnish bees for the swarm box. If fed three days before going into the swarm box, they will be in splendid condition to feed the larvae and to draw out the cells in the proper shape.

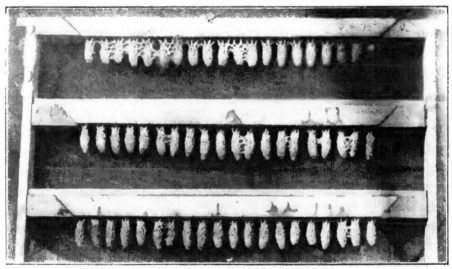

Showing September cells.

The very best of cells may be built any time of the year after brood rearing is well under way in the spring until it stops in the fall. In order to have good cells built out of season, it is necessary to put the colony that is to supply the bees for the swarm box and the finishing colony in the proper condition. We should remember it is the *condition* of these colonies that brings results, and not the time of year, the honey flow or swarming fever. Therefore, if we build up the colonies with brood taken from other colonies and feed these built-up colonies, we have as good conditions for cell-building as we have with a strong colony during a honey flow. It is evident that to build up colonies to cell-building conditions in early spring or late fall is expensive, as it will rob a number of colonies of their brood; but, if queens are needed at such times, it can be done with profit. The illustration shows some bars of cells built in late September after the flowers have all gone. But few of the virgins that emerged from these cells ever became laying queens for the weather turned cold and they could not venture out on their honeymoon.

One should bear in mind that it is the *best* queens that make the records at honey getting, so it pays big dividends to be extravagant with brood and feed. If all cells have dried-down royal jelly in them after the queen emerges, you have done all that can be done in the way of providing bees and feed to the cell-building colony. If some cells have no jelly in them, you are not rearing the best of queens. True, some may be first class for they may have had enough after consuming it all, but there are sure to be some that do not have enough and dwarf queens will result.

How many cells can a colony finish? During the summer of 1923 some experiments were made at our yard to determine the number of cells a colony should finish. In stating the number in this book, we wish to stand on firm ground and not advocate anything that might bring poor results to the beginner. If our experiments prove conclusive, we shall give them to the public later; but at this writing I believe that, when the colony that is to start or finish the cells is in perfect condition, it will build a large number of cells, and when not in good condition, it will not do good work on one cell. This is somewhat similar to the perplexing problems of "over stocking." When the honey plants are in good conditions, overstocking is almost impossible; but, when the plants are not in proper condition, a single colony can not make a surplus.

Root's grapevine apiary.

Chapter X. Filling the Swarm Box.

Two combs containing some honey and pollen are placed in the swarm box. These should be old combs and not too heavy, for, in the handling they are to receive, they will be liable to break down if new or if they contain much honey. These combs are placed one at each side of the box and are held in position by the two blocks that are to support the cell bars. If one has never used a swarm box, it is well to place it on scales for a few times until able to judge accurately the weight of the bees the box contains. A funnel such as is employed in the filling of pound packages, is used for putting bees into the box.

While good results can be had with no pollen in the swarm box, better results are obtained by having plenty of it in the two combs that are used. It is surprising to note the amount that the bees will consume while confined in the swarm box. If the two combs have an abundance, it will usually be eaten after the combs have been used three times. Before filling the box with bees, examine the combs, and, if they do not contain sufficient pollen, they should be removed and other frames containing plenty put in their place.

Set the swarm box in front of the colony from which the bees are to be taken, put the funnel into the hole and all is ready for the bees. It is quite desirable, upon all occasions when removing frames from the hive, to see that they are put back in the same position as found. If not, queen-cells are apt to be started, and when the virgin emerges, she will kill the laying queen. This subject will be discussed more fully under "Introducing Queens." A good method is to take out the frame nearest you and set it several feed away from the entrance. Then examine the next frame to find the queen. If she is not on that frame, set it back, lift out the next and then the next until the queen is found. Pick her up by the wings and put her on the frame you first took out. Set all frames back in their regular place except the one that has the queen.

The reason we set the first frame with the queen some distance away is to prevent the bees and the queen from crawling back into the hive, thus getting the queen into the swarm box.

Give it a quick shake downward and then upward.

Nothing I can call to mind creates such a strong desire to kick one's self as to get the queen into the swarm box. I know from experience. While you are working with the bees, they begin to fan and the first thing you know all the bees, including the queen, begin a grand march for the entrance or go over the top and into the hive. You do not know that the queen is there, and your grafting comes to naught. Instead of accepted cells you find them mostly torn out by the roots and made over into some fine worker comb containing eggs. In order to avoid this calamity, set the frame with the queen so far away that the bees will not heed the call of their companions.

We are now ready to put in the bees. Take out the first frame covered with bees, put one end of it down into the funnel, take hold of the other with both hands and give it a quick shake downward, then upward. Two little shakes, in which the comb is not moved more than two or three inches, will dislodge all of the bees except those that have their heads down in the cells. Set this frame back into the hive and do the same with the others until the desired number of bees are in the box. From five to seven pounds is the right amount. Try to get six pounds as nearly as possible. With a little practice you will not vary much more than a pound either way. There should still remain in the hive a sufficient number of bees to care for the brood. When the box has the required amount of bees in it, remove the funnel, place the cap of a mason jar in the hole, replace the comb containing the queen, close the hive and carry the box to the basement.

Caution. When a heavy honey flow is on, take care that the bees are not daubed with honey when they are shaken into the swarm box, for if they are they will suffocate and both bees and cells will be lost. True, if a little honey is smeared on them it does no harm; but too much is disastrous. When a heavy honey flow is on, shake the comb lightly so no nectar is displaced, and, if sufficient bees are not obtained in this manner, the bees not shaken off may be brushed off with a bee-brush. When more than five pounds of bees are put in the swarm box, it is advisable to set the box on two by four scantlings to afford more abundant ventilation.

Are contented and satisfied as though in their own hive.

The Dungeon.

In one corner of the basement I have what I call the "bee dungeon." This is a room made by stacking up extracted supers and hive bodies to the ceiling to make it dark. The opening that serves as a door is made in break-joint style so that no light can get in. It is wide enough so that a person can walk in carrying a swarm box in each hand, which is another advantage over a swinging door for plenty of fresh air can enter. Back in the dungeon the bees remain quiet as though it were night, away from noise, light and strong air currents, and are as contented and satisfied as though they were in their own hive. If no basement is available, any room in the honey-house where it is not too hot or cold will do; but it will pay to make a basement. It is the ideal place. I usually fill the box at one o'clock in the afternoon and leave the bees confined there in the basement until three o'clock. I find that two hours of confinement is all that is necessary, for as the bees are queenless, broodless and on strange combs, they realize their queenlessness to the fullest extent in that length of time.

Why the Bees Accept the Cells.

Now, while the bees are contentedly clustering to the lid of the swarm box, licking the honey off any luck-less individual that was daubed up when they were sha-ken from the combs, let us consider the condition brought about with the bees that causes them to do good work at cell-accepting. For some days previous, the young nurse bees have been feeding great hoards of larval food which is the same as the food we call royal jelly. We have sud-denly taken them away from these larvae, so they contin-ue to secrete the royal jelly but have no larvae to feed. They also realize their queenlessness. They are crying for a queen; they have the food with which to raise many queens, but they have no larvae with which to do it. It is our privilege to accommodate them in this respect, so now we will proceed.

Chapter XI. Grafting the Cell Cups.

The best place to do the grafting is in the honey-house or the room of a dwelling where there is plenty of light coming through a south window. A room is better than out of doors for several reasons. It is cool, and the larvae may be kept away from strong light, heat and drying winds. It is more comfortable for the operator, and he is away from robber bees. The grafting outfit is quite simple-a grafting needle that can be bought from dealers in bee supplies, a jelly spoon made out of a toothbrush handle, a little jar of royal jelly and a small individual salt dish in which to mix the jelly. With the jelly spoon, place some of the royal jelly in the salt dish and dilute it with pure water. It should be as nearly as possible like the thin larval food seen in the bottom of the worker-cell soon after the egg has hatched. When this is done, go to the hive containing your best breeding queen and take out a frame with as many young larvae of proper age as possible.

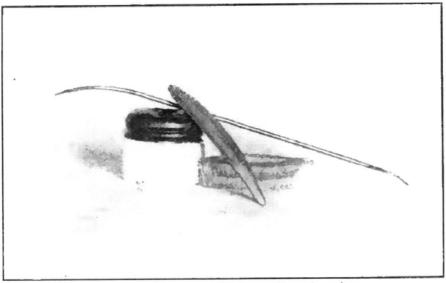

The grafting outfit is quite simple.

When no nectar is coming in, this colony should have been fed the same as the cell-finishing colonies already described; but, if even a very little nectar is coming in, no feeding will be necessary. The best results cannot be obtained by grafting hungry larvae. If they lie in the bottom of the worker-cells dry with no larval food around them, they are not fit to be used for grafting. They will not be accepted by the bees so readily nor can so good, strong queens be reared. Stunting the larva at the beginning of its development can not be overcome at a later period, no matter how ideal the conditions may be. If the larvae are floating in royal jelly, they are in perfect condition for grafting. If they are not, it indicates that the colony needs feeding. Should the colony not have a sufficient number of larvae of the right size, it is a good practice to insert occasionally an empty comb in the center of the brood-nest in which the queen may lay. If possible a black comb should be selected since the larva can be seen much better than in a new white one. Use a brush to remove the bees from the comb for, if the comb is shaken, the nectar will be scattered over the larvae, in which case they are not accepted so readily.

Carry the comb into the grafting room. Now take three bars of cells that were dipped during the winter. Be sure that the cells are perfectly clean. If they contain any dust or dirt they must be thoroughly washed and dried before being used, as the bees will not accept dirty or dusty cells. With the large end of the grafting needle place a little diluted royal jelly into each of the sixty cells. A drop about twice the size of a pinhead is sufficient. Endeavor to get this in the neat round ball right in the bottom of the queen-cell, for the bees accept them much better when it is placed in this manner. Keep the cells and the comb out of the bright sunlight as much as possible. When the weather was hot and dry, I formerly sprinkled water on the floor to keep the air moist that the jelly and larvae might not dry out and die. One day as I was doing this the Office Force was looking on in that inquisitive manner common to the gentler sex, and she said, "what is the use of dampening up the whole room when you merely want to keep those cells moist? Why don't you dip a bath towel into some water, wring it out and spread it over the cells?" "Yes, why didn't I? For the very good reason that I never thought of it. Thanks for the bright idea." I tried it. It works to perfection, so I have used it ever since. A moistened towel keeps the cells from drying out and protects the larvae from light and dust.

Now sit down in a chair with your back to the window so the light will come over the right shoulder. Place one of the cell bars on the side of the top bar of the brood-frame and parallel with it, holding it there with the thumb of the left hand. With the grafting needle in the right hand carefully slide the point under the larvae, choosing one that is about twelve hours old. Larvae that are twelve hours old are extremely small, and unless the one doing the grafting has very good eyes, he will be unable to see the clearly enough to do satisfactory work. A fine rule is to use larvae as small as can be seen; but, if the operator has exceptionally good vision, there is danger of getting them too small, for larvae less than twelve hours old are not accepted so readily as those older. I have never been able to determine whether it is due to the fact that these small larvae can not stand the handling or whether for some reason the bees do not like them so well. On the other hand, larvae much more than twelve hours old should not be used, for while they will be accepted, they often do not make such good queens as the younger ones. True, they have not yet received any food except royal jelly; but from experiments I have made, I am sure that the best queens can not generally be produced if older larvae are used. I believe it is due to the fact that, as the larva is grafted at an advanced age, the nurse bees do not have the same length of time to store royal jelly in the queen-cell as in the case of younger larvae, therefore the larva does not have sufficient food for it's fullest development.

Holding it there with the thumb of the left hand.

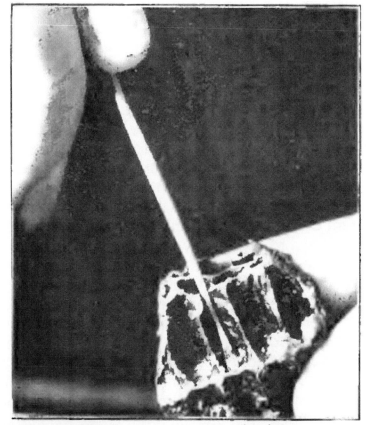

Slide the point under the larva.

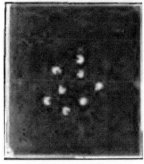

Larvae 12 hours old   *One-half of the larva over the edge.*

There is quite a knack in getting the larva on the grafting needle in just the right position. If it is entirely on the grafting needle, difficulty may be experienced in getting it off. The needle point should be placed under it in such a way as to leave about one-half of the larva projecting over the edge. When this is done, it is easy to remove the larva from the needle. Put the needle into the cell cup until the larva touches the drop of royal jelly, carefully draw the needle sidewise and the larva will remain in the jelly. At first this may be a slow operation, but in a short time you will be able to graft three bars in ten minutes or less. As soon as one bar is finished, place it back under the dampened towel. When all three are finished, they are ready to be placed in the swarm box. Take them to the basement and bring the swarm box out of the dungeon to the front where the light is better. Put the three cell-bars tightly together so they may be grasped with the right hand all at once. Lay them down with the cells upward. Now pick up the swarm box by taking hold of each end. Raise it about six inches from the concrete floor and bring it down with a jar. All of the bees that were hanging to the cover will fall to the bottom in a mass. Now remove the upper cover, take up the cell bars and place them in the box, allowing them to rest on the two cleats in each end, and slide the cover back in place. This can be done without a single bee's getting out. Have ready a quart Mason jar, with a perforated lid, filled with sugar syrup or honey diluted with about one-fourth water, which is much better feed for the swarm box. Sometimes bees will not take the syrup, but they will take honey. I do not put the feeder on when I first set them in the base-ment, for they will not take syrup or honey until they have been in there several hours, long enough to get the spilled honey all cleaned up, but feed them immediately after grafted cells are given them. Set the swarm box back into the dungeon, and the bees will do the rest.

Digression.

We have now passed over the most difficult part of queen-rearing, that of getting the cells properly accepted. Many have failed at queen-rearing because they could not get cells accepted with any degree of certainty. The question is frequently asked, "Why use the swarm box? Why not give the cells immediately to a colony?" The answer is, "Because the bees will not accept the cells." The condition brought about, as has just been described, enables the nurse bees to feed the larvae and draw out the cells in the best manner possible. Bees have many peculiar traits. One of them is that they will mechanically go ahead with a job that has been started. It is not difficult to get cells finished when they are once started. The difficulty is in the starting. So now, as we have brought about a condition by which the bees will start cells, it is a simple matter to get other colonies to go ahead with the job and rush it to completion; while, if we had given those same colonies these cells to start, they would have had none of it.

Reasons of Success.

Let me enumerate again the reasons why bees accept cells.

1.   Liberal feeding of the colony that furnishes bees to stock the swarm box. Remember, unless a honey flow is on, they must be fed generously at least three days before being placed in the swarm box.

2.   A sufficient number of bees in the swarm box. There should be at least five pounds.

3.   A large number of nurse bees, young bees. Re-member that old bees are poor nurses and will fail in accepting cells. (Transcriber's note: Jay believed this because the scientists told him this. After his own careful observation he changed his mind on this.)

4.   Well-fed, moist larvae in the colony which has the breeding queen.

5.   Clean cells, made of wax that has not been scorched in melting, are most essential.

6.   Cells must be the right size. Those that are too large will not be accepted.

7.   Be careful to keep the royal jelly at the right consistency. Royal jelly too thick or too thin will cause failure in whole or part.

8.   Grafting Larvae that are the right size and age.

9.   Be careful that the larvae have not been over-heated or dried by the sun's rays. If they are, the bees will always reject them.

10.   Keep the larvae from chilling.

11.   Careful handling while grafting so as not to injure or kill the larvae.

12.   Be sure the swarm box is kept in a place that is neither too hot nor too cold.

And lift out the bars.

The bees should remain in the swarm box until three or four o'clock of the day following. If taken out earlier the cells are not sufficiently advanced to insure their completion by the finishing colony. If left in too long, the nurse bees seem to exhaust their supply of royal jelly and the larvae are not sufficiently fed. They can be taken out at any time before the night of the following day; but as a general thing, the cells suffer if left in over night of the second day. From twenty-four to thirty hours is the proper length of time for the bees to be confined. Carry the box out to the hive from which the bees were taken, remove the upper cover and lift out the bars. If the work has been properly done nearly all of the cells should be accepted. One should average an acceptance of eighteen cells out of twenty and frequently all should be accepted.

When the bars are taken out the larvae should have an abundance of royal jelly literally swimming in it, and the cells be drawn out into proper shape. If conditions are right all sixty are accepted.

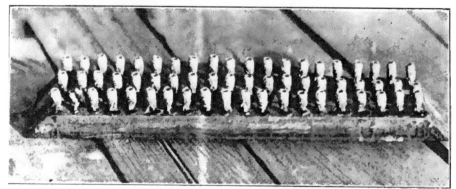

All Sixty are accepted.

Give the bar a very light shake to remove most of the bees that are clustered on it and then carefully brush off the remainder. Do not shake the bar too severely or some of the larvae will be displaced, in which case they will be removed from the cells by the finishing colonies.

For suspending the cell bars in the finishing colonies there is nothing better than a regular cell-bar holder made to hold three bars. Place one bar in the bottom section, spread the brood-frames apart, put in the holder frame, and give it to one of the finishing colonies. The best results are secured by placing frames of unsealed brood on both sides of the frame that holds the cells, for this draws the nurse bees right to the cells and they immediately take hold of the cells and carry the work on to successful completion. Replace the cover and see that the feeder is kept liberally supplied with feed so that the bees will receive an ample supply. Take the other two bars and give them to the other two finishing colonies. Go back to the swarm box, take off the lower cover, remove the combs and shake off all the bees possible and brush off the remainder. Pick up the swarm box, invert it and give it a rap on the ground to dislodge all the bees. Replace the combs, put on the covers and take it back to the basement to remain till needed again.

Chapter XIII. The Pritchard Forced Cell Starting Colony.

While I myself much prefer the swarm box for securing cells, there are others, like my friend Pritchard of
the A.I. Root Company, who prefer to use queenless and
broodless colonies for cell-starting. Instead of going to a
colony and shaking bees into the swarm box, Mr. Pritchard prefers to reverse the process by leaving the bees
in the hive, removing all the brood with the queen, and
placing them temporarily on another hive. He proceeds as
follows:

All the combs including the bees, brood and queen
of a medium colony (not a strong one) are removed from
the hive. Two combs of pollen and honey are selected and
set backing the hive one on each side, taking care not to
get the queen. Two frames for holding cell bars (without
the cell bars) are then put in the center of the hive. The
two combs of pollen and honey are shoved over next to
them. On either side is put a thin division board feeder
containing thick syrup. The remaining space on each side
is then filled out with dummies of division boards. The two
cell bar frames are now removed and all the other combs
of brood and bees are shaken into the space vacated
taking care not to get the queen. The brood and the
queen are now put in the upper story of a strong colony,
over a queen excluder. It is necessary to cage the queen.

In from half an hour to one hour's time, or as soon
as the bees in the made-up colony set up a roar of distress when they have discovered their loss of a queen and
brood, the two cell bar frames are supplied with prepared
cells, and are then put in the open space left, where the
bees are crying for a queen. The queenless and broodless
bees, supplied with an enormous amount of pap which
they expected to use in feeding their young larvae but
which has been all removed, immediately accept and
supply the prepared cells with pap. The two feeders con-

taining thick syrup and two combs containing honey and pollen will give the bees all that is necessary to supply them with material for making more pap.

The prepared cells will usually be accepted and lavishly supplied with pap in about 24 hours. It is not advisable to remove them before. When nicely started the cells are put in cell-finishing colonies as previously explained. The removed brood and the queen are restored to the colony.

While it might and could make a second batch of cells, Mr. Prichard does not advise it.

Mr. Pritchard says he prefers this method of getting cells started, because it saves the extra equipment of swarm boxes, toting them back and forth from the cellar, and because the bees during the time that they are starting cells are not confined. He thinks this is very important. The unconfined bees, he thinks, will do more and better work. By the plan described, he says he can get 200 cells started, each cell literally gorged with a big supply of pap. These cells, when given to cell-finishing colonies, will be completed in the regulation time, and every cell will be perfect.

The fundamental difference between the forced cell-starting colony and the portable swarm box is that Mr. Pritchard moves the brood and the queen, while I move the bees. He thinks that the unconfined bees do better work. I am not so sure of that. I succeed better with the bees that are confined in a cool place. Mr. Prichard will doubtless continue his way, and doubtless I will go on with the way that has given me the results I have secured.

It will be noted that the Prichard plan of making up a cell-starting colony amounts virtually to a swarm box left on the old location. Fundamentally the *principle* is exactly the same, but the procedure is different.

Of course it is necessary to feed these prepared colonies the same as the bees in the swarm boxes. This is important.

A Modification of the Doolittle Plan.

I have used a modification of the Doolittle method with the best of success and it may be preferred by many to the swarm box. With further experimenting and practice, possibly we ourselves may prefer it. This method is as follows: The colony for starting these cells should be one of extraordinary strength, being a two-story colony. Any standard hive will do, but we use the Jumbo hive, and in preparing this we see that it has twenty Jumbo frames of brood. The queen-excluder is kept between the upper and lower story. After all of the brood above has been sealed, we remove the lower story containing the queen and the brood to a location about ten feet distant. The upper story, containing only frames of honey and capped brood, is set on the bottom-board of the stand from which the hive containing the queen and brood was removed. Three frames are taken out to make room for three frames of cells. The hive containing the queen is now opened and the frame containing the queen is set outside. The frames are then lifted out and the bees from eight or nine of them shaken into the hive on the old stand. The queen with her frame of brood, is then set back into the hive on the new stand and the cover replaced. In a short time, half an hour or an hour, as soon as the bees have cleaned up the honey or syrup that has been daubed on them, they are ready to receive the cells. Three bars are grafted, placed in a frame to hold them and put into the starting colony. A second and third are prepared in the same way, which fills up the space. In this manner nine bars are started instead of three as with the swarm box; though it is not advisable to start so many unless the colony is of tremendous strength. After twenty-four or thirty-six hours these frames are removed and given to the finishing colonies, the same as when the swarm box is used. The other brood-nest containing the queen and brood is now brought back, set underneath and all is well. In two days' time the colony can be used to start another batch of cells. We use the same colony over and over for starting cells twice a week.

By this time the objection to the original queenless, broodless method is largely overcome. The queen can continue laying, as enough bees are left with her to keep up brood-rearing, and the work is much less than with the method whereby the queen has to be caged and bees brushed off from the brood. In using this, if there is no honey flow they must be fed the same as when the swarm box is used. Some will prefer this to the swarm box, and some will not. It has one advantage over the swarm box, in that no harm is done; while some have reported bees suffocated in the swarm box. Another time-saving feature is that if this colony is made tremendously strong as described, a larger number of cells may be started.

To say that a colony must be strong does not mean much; but if both stories are kept full of capped brood before being used, it will build the colony up to greater strength; and if it should run down, the upper story may again be filled with brood. But this should not be left in this story when giving cells until the brood is sealed, for the bees will not do so good work at cell starting where they have other brood to feed. Recently we had a field meeting in our apiary in Vincennes. Always willing to do their part, my bees decided it would be a good exhibition of they would swarm; so one of these cell-building colonies did so just before the crowd arrived. I put a piece of burlap on a saw-horse and placed the queen on top, and the swarm settled there where it remained for two hours or more, greatly admired by those present. Several disinterested parties weighted this swarm and found it weighted exactly twenty-five pounds. This would be about one hundred and twenty-five thousand bees, while seventy-five thousand must have still remained in the hive, thus making the enormous horde of two hundred thousand bees! I had estimated our starting colonies and finishing colonies at one hundred and fifty thousand each, but from that experience, I believe my estimate was too low. Of course, it is understood these bees were not the product of one queen as it would be impossible for one queen, no matter how prolific, to produce a swarm of that size; but from this we get some idea of a really strong colony, and for best results in raising the finest cells in large numbers, a strong colony as just described is very essential.

In the above modification of the queenless, broodless method it will be seen that the brood and queen remain together, and the queen continues to lay very similarly to the method in which the swarm box is used. The cost of equipment of the two methods is about the same. One requires the swarm boxes and funnel and the other requires an extra hive. Personal preference must determine which is used; both are excellent.

Extensive experiments are being continually conducted at our yard, and possibly in the future the present plan of cell-finishing above the excluder will be abandoned for something better. We have left one hundred and eighty cells with the queenless, broodless colony that started them until they were completed and some most excellent results have been obtained. Two colonies were shaken into one, and both colonies moved to a new location. Further experiments will determine whether or not this method will supersede the former with us. In any of the above methods, however the one big feature is proper feeding. Heavy feeding three or four days before giving the cells is a most important feature.

Many will find it more suitable to their systems of management to graft three or more times a week. Some graft every day. I graft Wednesdays and Saturdays. As I am to give my own system first, I shall follow it through. Let us consider that we did our first grafting on Saturday. Before time to dispose of the cells we shall again graft on Wednesday and Saturday. To make it clear, let us suppose the Saturday we first grafted was the 5th of the month. We must graft again on Wednesday, the 9th, again on Saturday, the 12th. When we put in the cells of the second grafting, some of the cells of the first will be capped over. The bar containing these should be moved up to the middle and the new one placed on the bottom shelf. The bees do better work when the newly accepted cells are placed on the bottom shelf. By this method, the bar of ripe cells is always on top and it will not be necessary to remove the frame to get the bar. Cells should be removed and introduced to colonies or nuclei on the 10th day after grafting. So the cells that were grafted on the 5th must be introduced on the 15th. This makes Saturday the busiest day of the week as we must both introduce cells and graft, but the light day's work comes on Sunday so we can enjoy a day of rest. While we try to arrange our work so there is as little work as possible on Sunday, there are a few chores that cannot well be avoided. Laying queens are removed from the nuclei on Mondays and Fridays so that by following this program everything fits nicely. All that there is to remember is the following program:

Program for the Week.

Monday: Remove laying queens from the nuclei.
Tuesday: Introduce ripe cells to nuclei.

Wednesday: Graft.
Thursday: Empty the swarm boxes.
Friday: Remove laying queens from nuclei.
Saturday: Introduce ripe cells to nuclei and graft.
Sunday: Empty the swarm boxes.

This program prevents mistakes, and, while no record or memoranda are used, it is almost automatic in its working.

In explaining this program at this time I am getting a little ahead of my story, but it seems necessary and details will be explained in due time. One feature must be watched carefully. In the finishing colonies queen-cells will sometimes be started on the frames of brood. If any queens are allowed to emerge, havoc will be wrought with all three bars of cells. It is well each time a bar is placed in the finishing colony to look the brood-frames over and cut out any cells that may have been started.

Chapter XV. Nucleus Hives

There are many styles of nucleus hives in use, and some have desirable features not found in the others. The small Baby Nucleus hive had a run for a while but is now generally considered a mere passing fad. It is so small that the bees are put into an unnatural condition, and they therefore perform in an unnatural manner. They seem to delight in pulling off all sorts of crazy stunts, such as absconding with a laying queen or absconding with a virgin; absconding when they run out of food or absconding when they have plenty. Another of their favorite sports is balling their queen when she returns from her mating flight. I have seen queens fly out from their baby nucleus and, unlike Lot's wife, they never looked behind them. The queens reasoned, "Why take a look at that little hive? I'm not coming back!" And frequently they did not, but would hunt around trying to find a real colony that would accept them. I used to keep a number of bars in a single hive for incubation. These colonies seemed to be particularly inviting to these truant queens, which were usually accepted. There was henceforth a great tearing down of cells, and it made me very dejected to see a double handful of dead queens lying in front of these hives. They never worked this game, however, where cells were finished above the excluder. These baby nuclei are easily robbed out, do not gather enough to live on and do not stand either hot or cold weather as well as the larger ones.

I have a vision of one day during a dearth of pasture in hot July when a baby nucleus absconded and went up into a tall tree and clustered. Was it worth going after? Maybe they had a laying queen, so I would try. The whole swarm was not much bigger than a walnut, shucks and all. At last, after climbing till I was completely tired out and had almost reached the, they took wing. While I was watching them disappear into the blue sky and was in a state of mind unnecessary to describe, along came a cheerful idiot who asked, "Say, Mister, how much honey did they make up there?"

Another objection to the baby nuclei is the fact that it is hard to tell a good queen from a poor one, for a good queen lays several eggs in one cell for want of room, exactly like a poor queen. Baby nuclei do not contain sufficient bees to incubate the queen-cells properly, thereby resulting in inferior queens. Yes, I have in use a hundred baby nuclei-as playthings for the children and for use as bird boxes. A woodpecker has appropriated one and, after peeking the entrance a little larger, crowded it full of acorns. For once the baby nucleus has secured a surplus.

The Root twin mating nucleus hive is midway between the Baby and the ones having standard frame. In it each compartment holds three frames, just the size for three to fit inside a regular Hoffman frame. These may be placed in a regular hive, and when the frames are filled with brood and honey they are taken out and placed in the nucleus hive. J.E. Wing, the well-known queen-breeder of San Jose California, prefers this hive to any other. He has special hives made to hold large numbers of these frames, for stocking those for the nucleus hives. In this way he overcomes one objectionable feature, that of fitting them into a regular Hoffman frame. Mr. Wing's system of management is favorable to these small hives for he practices migratory queen-rearing, moving to localities where there is a honey flow. He moves to one district where there is a heavy flow from honeydew. It should be remembered that the small nucleus hives give much better results when there is a honey flow than when there is a dearth of pasture so that feeding becomes necessary. The large "babies" give some better results than the smaller ones; but the ones taking a regular brood frame have so many advantages that they are being used by nearly all who rear many queens.

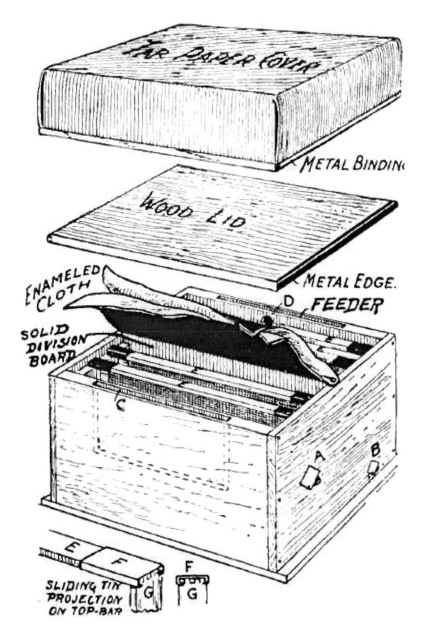

Root twin mating nucleus.

Each compartment large enough to hold two Jumbo Frames.

I strongly advise a nucleus hive that will take the regular brood-frame that is used in your hives. The one that I use is a twin hive, each compartment large enough to hold two jumbo frames and a division-board. The entrances are at opposite corners. A telescope cap is used with cleats that support it and give the air a chance to circulate, thus keeping it cool in hot weather. Usually only one frame is used with a thin division-board feeder to keep the bees from building comb in the vacant space. This gives the bees so much room that there is no absconding. It is comfortable in the hottest weather and has given perfect results. During a honey flow these nuclei are strong enough to fill up with honey. In fact in many cases, I have to give them sheets of foundation to keep the bees from going over the feeder and building comb.

Another nucleus hive among the best is that used by M.H. Mendleson, the veteran honey producer and queen-breeder of Ventura, California. This is a standard hive body, divided into three compartments. There is an entrance at each end, and one on one side. While working with this hive Mr. Mendleson sits on the blind side. In their mild climate he is able to winter these nuclei over with perfect success. There is nothing better than this style of nucleus hive.

Entrance Blocks for Nuclei.

The entrance block may seem like a trifling item; but, after experimenting with several, I feel that a description of one that has given me splendid satisfaction may be worth while. If the entrance is not very plain and easy to enter, the queen on her return to the hive will have difficulty in locating it readily. I have witnessed queens returning many times, and when I used an inferior type of block, the queen, after trying in vain to find the entrance, would go to another nucleus and try there.

The one that has given me perfect satisfaction is made to slope towards the entrance so that, if a bee alights within a few inches of it, the block guides her directly in. When the young bees fly out for the first time, they have no trouble in finding the entrance at once. Three nails are driven through the block from the upper side so that the points are barely exposed. When it is desired to confine the bees, as is necessary when forming nuclei, the block is placed over the entrance and pressed down. The nail-points catch in the wood so that the block cannot be pushed away by the bees. When they are first released, this may be moved enough to give an entrance just large enough for one bee to pass, and later it may be moved to allow the full entrance. In this small opening a single bee will stand guard and is able to keep off all oncomers in a manner similar to that of the Spartans at the Pass of Thermopylae.

Shade.

Shade is a very important item with nuclei. This is true in a greater or less degree, depending upon the style of cover in use. We use the deep telescope cover which shades nearly the entire side of the nucleus. In addition, the cover has an inner lining of thin boards with cleats on both sides so that there is a double air space, one above this inner cover and one below. Very little difference is noted in the behavior of the bees in such nuclei whether they are placed in the sunshine or in the shade. On the hottest days, however, the bees cluster out less where in the shade.

In case a single cover is used, shade is a necessity. Years ago when we used the baby nuclei, some were in the sunshine and many cells did not hatch; and when they did, many of the virgins were small, dark and inferior. A grove is the best shade, and if the trees are far apart to admit the sun in spots, it is all the better, for on dark days one can step out into these lighter spots to examine the combs for eggs, etc. In case the thermometer does not go above ninety degrees, the telescope cover with two dead air spaces gives perfect results.

Achord queen rearing apiary.

A number of methods are used to get the virgin queen safely into the nucleus in which she is to lay after mating. One is by allowing the queen in the cell which is kept in a strong colony for incubation to emerge in a nursery cage. I used that method for several years, but have discarded it since I could not get so good, vigorous queens in that manner. I found there are two reasons for this. One is because the bees are unable to cluster closely around the cells in order to keep the temperature right, and the result is faulty incubation of the tender pupa. This defect manifests itself in two ways, by smaller queens and darker queens. If the cells are kept too cold, it makes the queens dark. Sometimes in early spring the cells were a little below the proper temperature, in which case no harmful effect was noted but the queens were darker in color. When they emerged in the nursery cages and the weather was cold they were both dark and small, and a number of the cells did not hatch. Now, in theory, if the cells are placed in an upper story over a strong colony, the temperature will be the same as though no cage is used. But if the bees are ventilating the hive or fanning to evaporate the nectar or syrup, a current of air is blowing through the hive. If the bare cell is in the hive the bees cluster tightly around it, thus protecting it from air currents and keeping it at just the right temperature; but if the cell is in a cage, the bees pay no attention to it whatever, so that the cool current of air blows right though the cage and chills the cell.

The second reason why too many inferior queens are reared when the nursery cage is used, is because the new virgin does not receive the proper feeding when she first emerges, and this at the time that she most needs abundance of food of the right kind. When she emerges in the cage she has to depend upon the candy that is placed there. Sometimes she eats enough of this to keep her alive, and sometimes she perishes from an enforced hunger strike. Sometimes the bees feed the queens through the wires, and sometimes they do not. I have noticed some peculiar traits of the bees in this connection. They occasionally cluster around one cage and give that virgin all the attention in their power while they are balling another cage, and probably the rest are ignored altogether. I presume they had agreed to accept that one queen as their own and let all the rest go hang.

That the virgin does not receive the proper food and care while in the nursery cage and that her development is retarded, I have proved many times. As we know, seven to eight days usually pass from the time a virgin emerges from the cell until her mating flight. I noticed in many cases that a longer period elapsed before the virgin mated. Usually, in using the nursery cage, the virgin mated eight days from the time she was released from the cage, that is, the virgin remained in the cage three days; then it was eleven days from the time she emerged from the cell till she mated. The time spent in the nursery cage seems time lost as far as development is concerned. In many cases these queens never turned out to be first-class queens. When they are released from the nursery cage as soon as they emerge, not so much damage is done, but even then they frequently suffer on account of improper incubation at the time that Nature is putting the finishing touch on the pupa. Some had legs that were paralyzed or withered and wings underdeveloped.

Moreover, I noticed that the young virgin seemed to like the dried-up royal jelly that remained in the cell from which she emerged. She would eat it all, notwithstanding the fact that she had nice candy made out of powdered sugar and honey. Sometimes the queen would burrow into the bottom of the queen-cell and die there. From these observations I reasoned that the virgin needs royal jelly, and the only satisfactory way for her to get it is to have it fed to her by the nurse bees. I then conducted some experiments along this line. A number of cells were caged, and a number from the same grafting were given direct to nuclei. Those emerging in the nursery cages were introduced to nuclei in the same cage from which they emerged, taking about three days for the introduction. All from the cell-introduced lot with one exception were laying before the first from the cage-introduced lot. A number of experiments of similar nature were carried on, and all showed conclusively that the queens were injured by remaining in the cages at this time. So we can lay it down as a safe rule that in order to get the best results the virgin must emerge among nurse bees in order that they may give her the proper food and care. In an article in the American Bee Journal, I gave my experience with the nursery cage, and in a footnote Editor C.P. Dadant stated, "The editor applauds with both hands at these conclusions, for he had also tried the queen nursery years ago and did not like it." With the backing of such an acknowledged authority as Mr. Dadant, I fee sure of my ground in this connection.

The Cell Protector a Hindrance.

The cell protector was discarded for the same reason. The bees cannot properly care for the cell when it is in the protector. However, this is not so noticeable as in the case where the cell is caged; but there is little, if anything, to be gained by using the protector. It is a known fact that bees will accept a cell much more readily than they will accept even a newly emerged virgin. That being true, if they would not accept an unprotected cell, they would not accept the virgin when she emerged. In fact, from several experiments I am convinced that the cell protector is a hindrance rather than a help. At one time we introduced on hundred bare cells and one hundred with protectors. When we came to look for the virgins, we found about 30 per cent more in the nuclei where the bare cells had been given. Of course, they did not tear down the protected cells, but they killed the virgins as soon as they emerged.

The unemerged queens at this age are very tender and should be handled with the greatest care and should not be away from the bees longer than is absolutely necessary. A number of years ago I used to lay the cells on their sides in a box containing cotton batting. I found that, if they were left in this box for any length of time, many queens emerged from them would be crippled. Their legs and wings in particular would suffer. In conversing with Mr. Snodgrass of the Bureau of Entomology, Washington, D.C., he informed me that, if the pupa lay on its side for any length of time, the circulation stopped, which results in injury to the parts affected. However if the virgin will emerge within twenty-four hours, no harm will come if a cell is laid on its side for a short time. Keep covered from cool air or hot sun and by careful handling and maintaining as nearly as possible the temperature of the bee cluster it's perfect development is quite certain.

Fig. 1.-Larvae just as they came from the swarm box.

Fig. 2.-Cells are flooded with royal jelly.

Fig. 3.-Just as the bees were capping the queen cells.

Fig. 4.-Day after cell was capped.

Fig. 5.-Changing into a pupa.

Fig. 6.-Royal jelly is dry and brown.

This series of photographs shows the development of the queen from the grafting up to the time when the queen emerges. In Figure 1 you will note that the cells are remodeled to suit the bees. Wax was added by the bees, showing they were secreting it profusely. The bottom of each cell is covered with royal jelly. Thus in the twenty-four hours of confinement in the swarm box, the bees gave the larvae the proper send-off. In order to give a better view of the larva one cell was torn open.

Figure 2 pictures the cells one day later, after they had been in the finishing colony for twenty-four hours. The marvelous growth of the larvae in this length of time will be noted; but when we notice how the cells are flooded with royal jelly, the larvae really have no excuse for not growing. Figure 3 was taken just as the bees were capping the queen-cells. The larvae are getting too fat to curl up in them any more, so are beginning to lengthen out a trifle. The two end cells are already capped. Figure 4 shows the larva the day after the cell was capped. The royal jelly is still white and soft and would be in good condition to use in grafting, if thinned slightly with water. In figure 5 we see the larva changing into a pupa. The royal jelly is drying up and getting darker in color.

The next stop in the formation of the queen is shown in Figure 6. The change from the larva into the pupa is so very rapid as to seem marvelous. This takes place in twenty-four hours and in that short time, head, legs and short wings are formed so it appears a fully developed queen with the exception of wings and color. At first this pupa is exceedingly soft. While handling one, I accidentally dropped it on my foot. It splattered out much like a drip of clabbered milk, and no form of the pupa could be found. The pupa remains in this form with very little change, as far as appearances are concerned, for about seven days; but it becomes firmer and harder continually during that period. This picture was taken the tenth day after grafting, therefore one can see the condition of the pupa at the time the cells are to be handled. The royal jelly is dry and brown as shown in the top of the cell where it dried, leaving a space in the top of the cell. Where it is found necessary to handle the cell containing the tender pupa any time before the tenth day after grafting, great care must be exercised in handling it, as mentioned elsewhere, or crippled queens will result.

When the handling of cells before the tenth day after grafting is necessary, they should be placed in holes in a block to keep them right side up. As we handle them only on the tenth day, such care is not necessary. We place a cushion in the right end of the hive-seat, on which the cells are laid. A cover is tacked on it in such a way as to keep off the sun's rays and yet be easily raised when getting the cells Crippled queens are practically unknown to us since using this method.

Can be easily raised when getting the cells.

Chapter XVII. Cell Introduction

To establish nuclei and introduce cells to them is our next step in queen-rearing. Let us consider the subject a little. Two features in queen rearing have always been difficult; first getting the newly grafted cells accepted by the bees; and second, some safe and satisfactory method of installing the queen-cell or virgin in the nucleus.

This first difficulty, the manner of getting cells accepted, has already been described, and by carefully following the directions given one should have no trouble in having excellent cells built in abundance, so that the largest and best only are kept; but when all goes as it should, every cell will be so abundantly supplied with royal jelly that there is little choice. Having mastered this point in queen-rearing, let us now pass to the feature that has in the past been extremely difficult, and at certain times of the year has seemed insurmountable.

However, few things on earth are impossible if we know the underlying principles, find the cause of failure and apply the remedy. In cases of difficulty in any walk of life, even queen-rearing, a splendid motto to adopt is, "There is a way." When things do not go as we would have them, think of this motto, and proceed to find the way. There must surely be a way in this case, I figured.

In the past we have been led to understand that the reason why bees tear down cells given them is due to the fact that the cells are strange, with an odor different from that of the cells reared in the colony. Recognizing this, the bees proceed to rear them down. Now many things pointed to this belief. For instance when a cell was given to a colony, they would tear it down while at the same time they were building cells of their own; and many times when they had cells of their own, they would leave them unharmed and immediately tear down any strange cell given them. The natural deduction, therefore, was that they destroyed the cell given them because it was not their own. If such were the case, we could expect little relief, and our motto, "there is a way," would not apply.

Every spring when establishing nuclei there was a regular epidemic of cell-destroying. They would tear down cells as fast as given to them until they ran into laying workers. If virgins were introduced to them a little better progress was made, where we used our Push-in Cage. However, I was thoroughly convinced if the very best queens were to be reared, the virgin must not only emerge among the bees but these bees must be *anxious to receive* that queen in order that they may not injure her, but on the contrary, receive her gladly with out-stretched tongues and feed her abundantly that she may develop into a large prolific queen, the kind we all should strive to produce if we are to get big yields of honey.

So the question of how to get the bees to do our will in this respect was a puzzler. It was not practical to intro-duce the cell in a cage as we would a queen. If what we had been taught was true, that it was because it was a strange cell, there seemed little hope of overcoming this trait of bee nature; and to set about to overcome this rule seemed very difficult since there apparently was nothing to vie us a clue. There seemed to be no starting point.

In the spring of 1922, although there was a light stimulative honey flow, the bees were the worst at tearing down cells in all my beekeeping experience. I had planned to put out six hundred nuclei in two weeks, forming one hundred and fifty at a time to conform to our program. When the first one hundred and fifty were put out, nearly all the cells given to them at the time of forming the nuclei were destroyed. The second batch of cells suffered a similar fate, and there was no use in forming more nuclei as the first would not accept cells. The mystery as to why they tore down the cells was very impressive, to say the least.

I sat down under a tree to see if I could think out "the way." I carefully went over the experiences of the past fifteen years and finally the truth began slowly to unravel before me. I remembered that upon numerous occasions I had taken out cells from one of the finishing colonies, and having had no immediate use for them, had returned to another finishing colony, paying no attention about putting them back in the one from which they were taken. I recalled that, during those years, hundreds of bars of cells had been placed back into finishing colonies strange to them, and yet in all that time, to my knowledge, *not one single cell had ever been destroyed.* Evidently, then the reason the bees destroyed cells was not because the cells were strange. What, then, *was* the reason?

What was there about those strong colonies that had a prolific laying queen confined to the lower story that caused them to accept strange cells without question, while the nuclei would tear them down as fast as given them? Why did a strong queenless colony that had been used to finish cells accept a dozen bars of cells and never tear down a single one? Plainly, the fact that it was a strange cell had little or nothing to do with it. *Plainly, it was the condition of the colony.*

If this was true, what were those conditions and could they be duplicated in the nuclei? I became quite excited over the proposition. The only difference in the condition of the finishing colony and that of the nuclei was that the colony was being fed liberally while the nuclei were not; for even when artificial feeding was not resorted to, the strong colony was gathering enough from the fields to bring abundance to the colony, while the nuclei, being weaker in field bees, were not getting enough for the fields to supply them and were consequently drawing on the stores present in the combs. Could it be that the secret of successful cell acceptance was merely a matter of feed? I would soon find out.

I went to a number of colonies from which I expected to take brood for forming nuclei and gave them ten pounds of thick sugar syrup late in the evening, repeating it the second night, and you can imagine with what eagerness I awaited the results. I then formed nuclei in exactly the same manner as before. Presto! *Practically every cell was accepted!*

In my previous attempts at introducing cells to nuclei, out of one hundred, eight-nine cells were destroyed; while after feeding, only four were destroyed out of one hundred. Yes, "There is a way." To be sure that the conditions as regards the honey flow had not changed, I then formed some nuclei without feeding and some with feeding the results obtained were in every way similar to those of my first experiment.

My next step was to see if this would be as entirely successful with the established nuclei in getting them to accept cells. To our great delight we found this also worked to perfection. When ready to give a cell, we took a two-quart Mason jar with a perforated lid and shook about half a hint of syrup into the combs. We found there was practically no loss of cells; and that it was not necessary to have any honey in the frames when forming nuclei. We found that another admirable feature resulting from this feeding of the nuclei is that the bees are put in a condition to give the newly emerged virgin the very best of care. As a consequence they feed her as soon as she emerges so that she develops as rapidly as though reared in a populous colony during swarming time. Virgins thus reared mature, mate and thus begin laying one or two days sooner than in nuclei where no feeding is practiced.

This method of feeding to prevent cell-destruction is practiced at any time during the year when trouble is experienced in cell-acceptance. After a nucleus is estab-

lished and a good honey flow is on, feeding is discontin-
ued. Even during a honey flow, when a nucleus is first
established, feeding is necessary. In this case the fielders
return to their old house, leaving a lot of emerging brood.
The bees left in the nucleus can get no honey from the
field until they are older; therefore the nucleus bees are
hungry, and will destroy cells in their desperation. Feeding
prevents this. Our nuclei are now equipped with division-
board feeders, having a capacity of about one quart.
These are filled with syrup the day before giving them the
ripe queen-cell. Seldom indeed is a cell torn down.

Chapter XVIII. Why Nuclei Tear Down Cells.

Now that we have worked out this system whereby the cells are accepted in a satisfactory manner, let us ask the bees some questions to find out why they do things just as they do.

In the first place, when do they have cells in their colonies if left to their own devices? The greatest number of cells is found when the colony is preparing to swarm. And what are the conditions within the hive at this time? While there are a number of conditions that we cannot well duplicate in the nucleus, such as lots of bees and brood in all stages, yet there is one important feature—*plenty of feed.* Some nectar is coming in from the fields, the nurse bees are feeding the larvae, and all bees have an abundance of food. When in this condition they not only build cells of their own but will tolerate other cells if given to them. When preparing to swarm, the bees with ripe cells of their own will never tear down strange cells given them; but, let a rain come up and the weather turn cold so that no nectar comes in from the field, they begin tearing down cells whether their own or strange ones.

Now notice one point very carefully. They tear down *the more mature cells first.* Their instinct seems to lead them to realize that, if they tear down *all* the cells, they would have to start all over again and build from the bottom up, in case the weather turned warm, and nectar again came in so that they wish to swarm. If only the advanced ones were torn down, they would be retarded merely a few days. This occurs only occasionally during swarming season, but is almost invariably the rule when bees are superseding their queen. I have noticed many times during supersedure that the advanced cells are destroyed while new ones are still being started. The weather or nectar secretion seems to influence them in this respect.

This instinct, therefore, explains why, when we gave cells to nuclei, they tore ours down while at the same time they were building cells. They tore ours down because ours were advanced, and not because they were strange cells. If you wait until a nucleus has ripe cells of it's own, you can give it a strange ripe cell, and the bees will accept it without question. Many times when conditions are unfavorable, finishing colonies will tear down the cells they themselves have built. A heavy feeding will stop this destruction.

Therefore, we can accept as a rule of the bees that they will not tolerate cells when they are hungry; but, if they are lavishly fed either from a natural honey flow or by receiving sugar syrup, they will tolerate and accept cells. Moreover, there are some less important conditions that render cell acceptance certain. There should be capped brood in the nuclei and, if possible, brood in all stages; but, if these conditions are not present, a heavy feed will offset the lack of the former to a high degree. When there is no brood, cell acceptance is more uncertain.

You ask, "How about laying workers?" Well, we were talking about the laws that govern bees, and these do not apply to laying workers as the laying worker is a Bee Bolshevik and knows no law. However, a Bolshevik is more amiable on a full stomach; so some laying workers can be made to get into line by feeding, but it is better to give them a frame of emerging brood, after which a hearty feed gets them ready to accept the cell.

If this idea of feeding is applied to the different phases of beekeeping, the same results may be obtained during a dearth of pasture as during a honey flow. Many have observed how much more successful queen introduction is during a honey flow than at other times. Since discovering that feeding prevents cell-destruction, I have recommended to those having trouble introducing queens that they feed the colony heavily while the cage is in the hive. Several have reported that this gives as good results as they can get during a honey flow.

Some time ago the honey method of queen introduction was advocated. The plan was to daub the queen with honey and run her in. Some reported success; others, failure. Those who poured a pint or more of honey on the bees and the queen had better results. Now, it was not the daubing of the queen that helped; it was *the feeding*. In other words, for successful introduction, duplicate a honey flow by heavy feeding. The feeding should be done at night to prevent robbing.

In feeding with the division-board feeder it is necessary to keep the nuclei in strong condition, for robbers are always hanging around ready to pounce upon the nuclei if opportunity offers. By providing ventilation in the bottom of the nucleus hive and contracting the entrance to one beespace, and keeping it strong in bees, no serious trouble is caused by robbers. Sometimes, when they are exceptionally bad, we close the entrance entirely. In this way all the robbers that get on the combs are shut in the nuclei and can not go back home to spread the news that free plunder is to be had. Consequently, other robbers are not sent out to hunt the source of supply and marauding expeditions are restricted. After fifty or twenty minutes, the entrances are opened, when the robbers that were trapped rush for home. The nuclei are by this time reorganized and able to stand off all oncomers. In feeding either nuclei or cell-building colonies, it is necessary that they have some empty comb in which to deposit the food, for if all available space is occupied by stores they will fill up on this and become lazy and are easily robbed out.

Let us now take up our program where we left off. Before forming nuclei, we should have the nucleus hives in their places on the stands where they are to remain through the season. Ours are placed in rows running between the colonies, which are set four in a group for wintering in the quadruple cases. There is a big advantage to having the nuclei among the colonies, for much time is saved in drawing brood from the colonies, getting cells from the finishing colonies, etc. Again, when the virgins take their mating flight they have to run a gauntlet of drones, and mating is made certain. In addition all virgins are mated to your own drones, eliminating any chances of mating to drones in nearby apiaries or with drones from colonies in trees.

Hive-seat to accommodate the necessary equipment.

I do not believe queens go so far to mate as is generally supposed. I believe that Nature intended that the virgin should mate with drones from her own colony, for you will notice that bees never kill off their drones when they have a virgin in their hive. In their natural state, where bees are in trees, in many cases they are from half a mile to two miles from their nearest neighbor. What chance would a virgin have in mating with drones at such distances? Usually in the afternoon the young bees, the virgin queen and the drones all come out for exercise, and while circling within one or two hundred feet of the hive the virgin mates. Nature has put a check on injurious inbreeding, in that the drone that mates with the queen immediately dies, and if there should be an after-swarm the accompanying virgin would mate with a different drone.

The comb-box one of the handiest conveniences.

Formerly, when forming nuclei, we used to place a number on a wheelbarrow and take them to the hives, fill them and then set them on their stands. I prefer now to place them on their stands and take frames of brood to them. Right here let me mention two articles of yard furniture quite necessary to the comfort of the queen-breeder-the hive-seat and the comb-box. One should have a hive-seat large enough to accommodate the necessary equipment. In addition I certainly recommend the comb-box, one of the handiest conveniences about any apiary. This box is made of half-inch lumber and is large enough to hold seven frames. It has a bee-tight cover and can be used to store empty combs or frames of honey and to carry frames of brood and bees to form nuclei or put into the finishing colonies.

Into each nucleus, as it rests upon the stand where it is to be for the summer, we put a narrow division board feeder, which holds about one quart of syrup when full. We have found these much better than the two quart Mason jar with perforated lid, which we first used, for, while these jars give good results during a robbing season the robbers are apt to become a nuisance to the nuclei. This nucleus feeder measures 3/8 inch, inside measurement, is made out of ¼-inch material, and is the regulation size of a division-board. This feeder should be filled the day before a cell is given, and when this plan is followed there is practically no loss of cells.

Two days before time to form nuclei and introduce our first cells, the colonies from which the frames of brood are to be taken should be fed liberally. Any feeder will do, but I prefer to fill their bottom-board feeders at night for two nights. This is necessary to put the bees in condition to accept cells upon their arrival in the nuclei.

It is now the fifteenth, the time for introducing the cells from our first grafting. The comb-boxes are taken to the colonies that have been thus fed, and frames containing brood and some honey with the adhering bees are put into them. All frames should have considerable capped brood. Into each nucleus put one frame with its adhering bees, fill the feeder with one quart of thick syrup, and close the entrance.

Then go to one of the finishing colonies, take out the bar of ripe cells, and with a sharp, thin knife cut between the cells in case these have been built together, web-footed fashion. Then run the knife under the base of one, lift it off, put it very gently into the box in the hive seat, being careful not to invert or jar it in any way, and remove all the cells in like manner. Now go to the first nucleus hive to receive a cell. Take one from the box, lift out the frame and gently press the cell into the combs just above the brood. Replace the comb so the cell will come next to the partition which separates into the two nuclei, and thus be in a warmer place. Do the same with the other compartment of the nucleus hive, and replace the cover. Continue thus until your supply of cells is exhausted. Close the entrance tight so no bee can get out. In the bottom of the nucleus hive at the back end is a hole about one inch in diameter in the bottom-board. The hole is covered with coarse screen cloth, and left open all the time. This is quite necessary for, when the bees are confined, it affords them ventilation. In hot weather when robbers are troublesome, this enables one to close the entrance when the bees will get abundant ventilation through this opening.

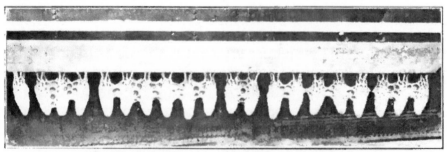

Built together web-footed fashion.

Leave the bees in these nuclei confined two nights, but liberate them after dark on the third night. By this time their old home is forgotten, and most of them will remain in this new location. A few of the old ones will probably go back, but not enough will desert to do any damage.

By following this method carefully, practically all cells should be accepted. If any are not, it is important to see that the feeder is filled when giving the second cell; for no matter how much nectar is in the fields, too few old bees are in the nucleus to gather it, and, when the young bees emerge, they will require lots of feed. If not abundantly fed at this time, they will tear down nearly all cells.

The day following cell introduction the virgin should emerge, and all is well. The following Saturday, when introducing other cells, look through the first and if any nucleus fails to have a virgin queen, it should be given another cell.

Number of Swarm Boxes Needed.

I have found one swarm box to the hundred nuclei sufficient. We have described the process up to the time of the emerging of the virgin queen, presuming we are using one swarm box. If the beekeeper grafts oftener than twice a week, more nuclei per swarm box can be run. If several hundred nuclei are maintained, it is advisable to have an extra swarm box to be used in "catching up" after a period of unfavorable weather, bad luck, mismanagement or accidents, which are liable to overtake the queen-breeder. I run six hundred nuclei and have eight swarm boxes. Very seldom are the eight all pressed into service. Usually but five are needed.

It is advisable to have always an abundance of cells, for, by the method described, they are easy to produce. To throw away fifty cells is much more economical than to be short five cells when needed. Not infrequently at the end of the season, I have a half bushel of cells that have been discarded. While it may look like waste to throw away hundreds of fine cells, it is the best way out. In the past I have sometimes tried to save them by using cages, etc.; but it is false economy. If the weather has been unfavorable for mating and you have a lot of ripe cells, while the virgins have not mated so as to be out of the way, just "junk" your cells and forget it. If the weather has delayed the mating three days you are set back just three days and you can't help it, unless you are a Joshua and can stop the sun while you catch up with queen orders; but most of us lack talent in that line. No, you cannot gain time by manipulated the clock, either. We have tried that. One of the ways in saving cells is to go through all the nuclei and cull out any queens that are not up to standard, giving them a cell; and should any virgin be missing, replace her with a cell also. In that manner part of the cells may be saved.

Another profitable way to use surplus cells is in re-queening, as described in Chapter XXXI, where a double brood-chamber is used. If the single brood nest is in service, the old queen should be removed, and the cell given immediately. During a heavy honey flow the cell will be accepted without ceremony; but if no honey is coming in, a heavy feed puts the colony in condition to accept the cell perfectly. If one has extra hive bodies or nuclei, these may be brought into use and the cells saved. These hive bodies of nuclei may later be united with the colony to be requeened, after first hunting out and disposing of the old queen as described.

By carefully following the methods we employ for cell-building as we have tried to describe, fine large cells are produce at little cost, so that cells are very cheap, and a goodly supply for all purposes is constantly at hand.

Chapter XX. Misfortunes of the Queen-breeder.

From the grafting of a larva to the taking out of a fine laying queen is a long step. Many things may happen to that larva to prevent her "coronation." Even if the queen-breeder has done the very best he knows, there will be accidents and blunders. Things may be going along nicely when suddenly the weather turns cool, the rain pours down and the wind blows. Now, a little of this does not hurt much, but let it keep up for several days and it hurts a whole lot. Cells are torn down in finishing colonies and in nuclei. Virgins fail to mate and come up missing. Grafting time is on you and you are not expecting this "duck-hunting" weather to continue so you do not feed. The result is that when you graft, there is a very poor acceptance, and you go sulking around with the grouches, saying to yourself, "looks like I am just about out of the queen business."

Now if you get in a muss like this and you happen to be a commercial queen-breeder, you will begin to hear murmurings from the "office force." She will want to know why it is that, although the output of queens is falling off, the correspondence is rapidly increasing. She says, "Here are a lot of letters asking why you have not sent those queens as you promised; and here is one who says he is going to report you to the bee journals if you don't send those queens by return mail!" Well, it serves you right if he does, for not having a "tornado clause" in your contract; so you turn you back on the "office force" and tell her something like this, "Oh, give them some of that threadbare, moth-eaten dope about unfavorable weather conditions and so on!"

But the skies are clearest after a storm, so the clouds roll away and a soft warm south breeze comes up. In the morning you go out into the apiary, feeling that this world is not, after all, such a punk place in which to live. A meadow lark sits on a metal hive cover washing its feet in the heavy dew. As it points it's bill up to the sky and begins to sing, you wonder why it was a day ago you were in the dumps. Bees come scampering in with their baskets bulging with pollen. Occasionally they lose some of it at the entrance of the hive where it gets water-soaked and makes a mess in the doorway. The day is clear, the sun is shining, the bees are bringing in nectar, and queen-rearing goes on at high tension again.

Now what are we going to do when bad weather hits us? As our beloved James Whitcomb Riley says,

> "It ain't no use to grumble and complain,
> It's just as cheap and easy to rejoice;
> When God sorts out the weather and sends rain,
> Why, rain's my choice!"

Nothing we can do will make the conditions as favorable as they are during good honey-gathering weather; but we can do a few things that will materially lessen the loss. When it begins to rain or turn cold, a good policy is to feed the cell-finishing colonies as well as the colonies that are to furnish bees for the swarm boxes, and give a little to the colony containing your breeding queen. If the weather is very cold, contract the entrances considerably, and give the bees the feed rather warm. If the weather turns warm immediately the feeding was unnecessary, but no harm has been done.

Another incident that often "plays hob" with the queen-breeder's hopes is a queen-cell overlooked above the excluder, when a virgin has emerged and tears down all the cells in that hive. One cannot guard too carefully against this. Every few days examine the frames in the upper story of the finishing colony for queen-cells. Shake off all the bees from each frame and look closely. Sometimes a very small cell will escape notice if a careful search is not made, and the virgin emerging from it will play havoc with all the cells. Such cases as the above are when the extra swarm boxes must be brought into play in order to make up as soon as possible for the cells lost. But if care is taken there will be a natural surplus of cells, so that it takes an unusual setback to require more cells than you normally have ready.

Another and perhaps the greatest problem with which the queen-breeder has to contend is that of robber bees. When no nectar is coming in, these rascals make life miserable for the beekeeper and interfere with queen production seriously. If a nucleus becomes weak they rob it out and, stimulated by their victory, hunt all over the yard for others of insufficient strength to resist their attacks. When a hive is opened these robbers fairly swarm into it. They soon learn to follow the beekeeper around so as to be on hand instantly when he opens a hive. Even if a strong colony is left open too long, it is robbed out. Colonies from which bees are taken to fill swarm boxes will be robbed if not carefully protected; and when the queenless, broodless, method is used, the colony is powerless against them.

In our yard when robbers are exceptionally bad we remove the brood and queen from the cell-starting colony late in the evening and graft early the next morning. In that we "put one over" on the robbers.

Chapter XXI. Records for the Nucleus Hives

If one has even as small a number as one hundred hives, some system of keeping tab on the condition within is necessary. Some keep a book containing the record. I tried this, but after it got pretty well stuck up with propolis, it failed to function and was indeed a closed book. Keeping the record on the hive with a pencil is better. But too much time is taken to place the record there and too much time is required to read it, since it is necessary to get up close to the hive in order to see it. A record to be satisfactory must be so made as to be seen readily at a distance. I tried various schemes until the present "block system," which has given such entire satisfaction, was worked out.

Four conditions within the nucleus hive must be known. They are Queenless, Cell, Virgin and Laying Queen. To indicate these conditions we use a block of wood 2 ½ x 2 ½ x 1-¼ inches. It is painted white on top and red on the bottom; while the sides are painted black, red, white and blue, reading to the right. The block is laid on the hive with the white side up. When the nucleus is queenless, the black side is to the front; when a cell is given it is turned a one-quarter turn making it red; when the virgin emerges it is turned to white, and when she begins to lay another quarter turn brings it to blue. These colors may be seen at a glance from a distance and the condition known without close inspection. These colors are suggestive. Black for queenlessness, conditions are dark within. Red when the cell is given and danger for the cell that is being introduced. When the block is turned to white, the presence of a virgin is denoted, white being symbolic of purity. The laying queen is what you are working for and denotes a "blue ribbon".

Sometimes there may be other conditions in the hive that need attention, such as a lack of stores or bees.

When this is the case, the block is turned the red side up, which shows that it needs attention. The reading on the side of the black may remain the same. This record is very simple and effective, and has proved very satisfactory. Nearly every one has his own "system" in such matters.

Now we come to the matter of dates, which is equally simple but may not be quite so easy to explain. After a laying queen is taken out two weeks will pass before another can be taken out even if all goes well, which condition does *not* happen in every case. Consequently we divide the hive into four imaginary parts for in the two weeks' period there will be need of four operations-four introductions, four removals of laying queens, etc. As you stand looking at the side of the nucleus hive, the first position is at the left end, the second position is just left of the center of the hive, the third position is just to the right of center and the fourth position is at the right end. Let us suppose we are going to introduce cells on Tuesday. This is at the first of the week, so we put in the cell, place the block at the extreme left end of the nucleus and turn the red side toward us. The next time we are introducing cells we examine this, and if the cell has produced its queen we give the block a quarter turn to white, but leave it in the same position on the hive.

When the block is turned to white, it needs no more attention until the time comes to remove the laying queen. When we introduce cells the next Saturday we go to those not having cells, supply them and place the blocks just to the left of the center since it is the last introduction of the week. Then turn the blocks to red. Do this in a similar manner on the following Tuesday and Saturday, when the blocks will be placed at the left of center and at the extreme right of the nucleus hive for these days respectively. Now on the following Monday the queens that have emerged from the first cells we put in should be laying and should be removed and the blocks turned to black. On Tuesday we can give this nucleus another cell, or a cell may be given immediately upon removing the laying queen provided the feeder was filled with syrup or honey on the day previous.

If every cell we put in resulted in a laying queen, the position of these blocks would never be changed on the nuclei and they would only be turned to different colors. But if we are introducing on Saturday, let us say, and we examine one that is read on the previous Tuesday position and we find that the virgin is missing, we should have to move the block down to the Saturday position. All complications that might arise are taken care of with this system. Now suppose we come on Monday to remove a queen and she has laid only half a dozen eggs, not enough so that we can properly judge her merit. If we left the block in that position turned to white, we should be unable to tell whether she belonged to the lot that were ready to lay or the lot that were due to lay two weeks later. So in that case we turn the block to blue and leave it until the next shipping day and examine again. However, suppose the virgin when the shipping day comes is not laying, but looks as though she might in a short time, you would not turn it to blue for she is not laying, and if you left it at whit it would not be examined till two weeks later. In that case we turn the block one-eight of a turn, which brings it midway between virgin and laying, and thus it will be looked after when the next shipping date arrives. Now suppose we look for a virgin and cannot find her. Maybe she is out for a flight, as it is well known that, as soon as they are able to fly, they spend much of their time flying out for exercise before mating. You do not want to turn the block to white, for if the queen is lost, that nucleus will remain queenless for too long a period and you will lose the use of it until the shipping date arrives when you will see it is queenless. In such cases the block is given a one-eight turn backward which brings it half way between cell and virgin. The next time you are introducing cells you will examine it, and if you can still find no queen, a cell is given. If the queen is there, it is turned to white if a virgin, or blue if she is laying.

These blocks are made at the planning mill. They are of hard wood so they are heavy and will remain in place. At first, I had some misgivings as to whether they would stay in position. I wondered if the wind might blow them off. Soon after putting them there a terrific wind and rain storm came along that threatened to blow everything away. Not a single block was moved in the least. I never had a bit of trouble except once, and that lasted only a short time. I would find the blocks moved or turned around while some were on the ground. Was it possible the wind did do it after all? The children do not play up there, for discretion forbids. One morning the blocks were disarranged again. It had been a very still night with no wind. Now we have an owls nest in our oak trees every year, and that owl and I are not on the best of terms, for when he can catch me out in the apiary after dark, especially if I am bareheaded, he seems to take much pleasure in scratching me on the head. This owl is one of the smaller editions, about four inches in height, but golly! how he can scratch! I made up my mind it was that consarned owl, and now that I had a double grudge I had a good case against him and I could shoot him, so I prepared for Mr. Owl. But one day I happened to look out and I saw what the trouble really was. A blue jay was hopping from one block to another in the most jubilant spirits. Sometimes he would kick a block off, and sometimes he would spin one around. Now I do not want to shoot any of our birds that build their nests in our woods, but this blue jay was throwing the system all out of kilter. Why, he had some marked queenless when they had a laying queen, and he had one marked a laying queen when it had laying workers! So I had to ask myself this question, "Am I engaged in bee or blue jay culture?" Decidedly I was in the bee business, so I took my shotgun and as the guilty bird started to fly away I shot him. There were two nests of blue jays in the trees, but strange to say there never was another block disturbed. The owl continues to scratch my head heedless of the bricks and clubs that are shied his way; but since he in no way interferes with queen-rearing we get along pretty well.

Position of blocks on nuclei.

The first nucleus contains a virgin queen that emerged from a cell placed in it the first of the week. The second nucleus in the same hive shows the block in the queenless position. The queen has just been removed on Friday. The third nucleus illustrates the block turned to white, indicating a virgin that should be laying the first of next week. The fourth nucleus indicates that no queen could be found when looking for laying queens; but it was though the queen might be flying out, so the block is turned between cell and virgin. Study this picture and see if you can tell from the blocks what day of the week it is supposed to be and where we are now working. The position of the red and blue could not be shown. These hives are too close for mating queens, and are placed in this manner merely to illustrate the block system.

Colonies packed in quadruple cases furnish brood for forming nuclei early in the spring.

Too much emphasis can not be placed upon the importance of keeping the nuclei in the best possible condition. They must be strong in bees with abundance of brood and rich in honey if the best success is to be achieved in incubating the cells and mating the queens.

Honey in Nuclei.

If plenty of honey is provided in the nuclei there will be no absconding; but, if not, the bees will abscond about the time the queen should begin to lay. You should bear in mind that a nucleus in the proper condition is a valuable asset and will pay you big dividends; but if it is allowed to run out of brood, in it gets weak in bees or has not sufficient honey, every thing will go wrong. Cells will be torn down, virgins will be killed, and bees will abscond. Take good care of your nucleus. It is the goose that is laying the golden eggs. Do not starve it or it will cease to lay. Frequently frames in nuclei may be exchanged with advantage to both. In case one has an abundance of honey and no brood, and another has an abundance of bees and brood, these combs should be exchanged and both will be put in better condition.

Laying Workers in Nuclei.

If, for some reason, several cells that are given to a nucleus are torn down, all brood will have time to emerge and laying workers will result. To cure them of this bad habit, take away their comb and give them another with unsealed brood and all adhering bees. They will usually accept a queen-cell, and the trouble is over. However, as is the case in many other lines of human activity, prevention is better than cure. Laying workers never develop while there is brood in the hive. So, whenever you find a nucleus that has no brood some should be given the bees at once. This puts them back into normal condition, and they will accept a cell and no laying workers will develop.

To resume our program, let us consider that it is now Monday, the 28th of the month. If all has gone well, the queens reared from our first grafting will be ready to be taken out from the nuclei. The cells were introduced on the 15th. The queens emerged on the 16th, the virgins should mate on the 23rd or 24th and begin laying on the 25th or 26th. By the 28th they should be laying enough so you are able to judge as to whether or not they have all the appearances of a good queen, and yet they have not laid enough so that injury will result on account of their removal from the nuclei.

Laying Queens and How Injured.

These queens are now ready for shipment or to be introduced to colonies at any time you are ready. It is a well-established fact that, when a queen laying to full capacity is removed from the colony and placed in a mailing cage, she seldom makes good at the head of a colony as far as prolificness is concerned. When it is necessary to remove a queen in the height of her egg-laying, as for instance, if one wishes to ship a breeding queen that is in a strong colony, it is best to place her in a nucleus for a few days. In this case she can reduce the number of eggs she lays, gradually becomes much smaller in size and therefore she can stand the trip better. If the queen is extra large in size and laying to full capacity, I have found a splendid "reducing exercise" as follows: Take out two frames of brood with bees and the queen, and place them in a nucleus hive on a new stand. In two days' time, move the nucleus to a new location, and on the next day move it again. The bees that fly out do not return. The nucleus is thus weakened in bees and is not getting any honey or pollen; so the queen rapidly curtails egg-laying. To remove her when she is laying in a limited way does not injure the queen in the least, for this is in perfect harmony with bee nature. For instance in the case of a swarm, after the swarm has gone out, a few eggs will be found which the queen laid within a very short time before her departure from the hive. Consequently, if you can get the queen down to laying only a few eggs before removing her from the nucleus, she will not be injured in the least.

All features of queen-rearing are fascinating; but I enjoy looking for laying queens more than any other feature of queen-rearing. As some who read this are interested in commercial queen-rearing, and others who produce honey may at times find it advisable to ship queens to other yards, I shall describe the process of mailing queens.

For very short shipments the common three compartment Benton mailing cage will do. However, I believe even in short shipments the use of the large cage that has six compartments is advisable. This cage is good for any shipment in the United States and Canada. For export, I prefer a cage nearly square with nine compartments, each of the same diameter as in the regular Benton cage. These compartments should be about one half deeper however. The secret of good shipments is to have the cage of sufficient size so that it will accommodate plenty of nurse bees and still leave abundance of room for them. There should be space for every bee to have a footing on the bottom and top of the cage and still leave room enough so there can be a space all around the bee equal to it's own size. If the bees get too warm, they then all have plenty of room to fan and thus keep down the temperature. In the six-compartment cage, when the weather is hot, I find that about fifteen nurse bees give the best result. If the weather is cool double that number is satisfactory. If it is necessary to ship queens when the weather is cold or where the route is through a cold district, as for instance going over the mountains in early spring or fall, the use of the export cage is advisable. More bees can be put into the six-compartment size; but in case the bees pass through a warm district they would suffocate. The cage should be large enough so that the bees can spread out when it is hot, and yet large enough to hold sufficient bees to they may form a cluster in one or two compartments to keep warm in case the weather turns cold. Such a cage is ideal as far as size is concerned.

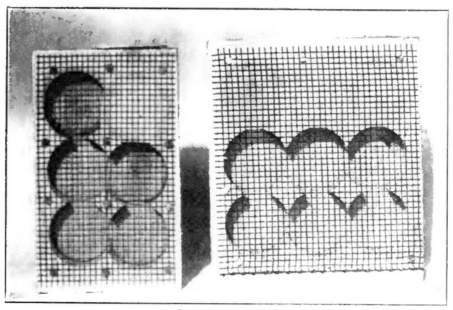

Queen cages.

Some years, the shippers of queens suffer a much heavier loss than in others. In the main, hot weather is the cause although the blame is usually laid on the queen candy. If the candy is made hard so the bees will not become daubed, it will give perfect results, at least for all ordinary shipments. Some have suggested that pure sugar candy lacks the nitrogenous food contained in pollen, and that this caused the loss in long shipments. Pollen could not be used in candy, as it would cause death by dysentery. Our friend, Allen Latham, queen-breeder of Norwichtown, Conn., has bee experimenting by mixing a little royal jelly with the candy as this would furnish the nitrogen and being predigested, would cause no intestinal disorder. We have been experimenting with this also, but have not gone far enough at this writing to come to any conclusion. However Mr. Latham's plan looks good.

We believe the principle cause of loss in hot weather is the lack of ventilation. Imagine a queen with her escorts, in a mailing cage with two little slits in the side of the cage packed in a mail sack, with all sorts of mail matter crowded all around the cage, and then perhaps with twenty-five or fifty more sacks piled on top of it and all out in the sun, with the thermometer reading one hundred in the shade. The wonder is that any survive. During very hot weather we use and recommend what we call "the hot weather cage." It is the regular mailing cage with a heavy fibre cover raised about one-eight of an inch with wooden cleats. This remedies much of the hot-weather troubles.

Hot weather cages.

Chapter XXIV. Queen Candy.

The other important feature in the perfect shipment of queens is the food during transit. Much trouble is experienced in shipping queens with candy not sufficiently hard. I have experimented quite extensively along this line by keeping a number of cages containing worker bees and different kinds of food in the attic, the basement and various other places under all sorts of conditions and temperatures, and found the chances of success greatly in favor of the harder candies. Not realizing how decidedly candy affects the bees, many make the mistake of using candy too soft. Watch the bees in the mailing cage. You will see them continually rubbing their sides against the candy. If it is a trifle too soft they gradually get it on their sides and into their breathing tubes. Then they fret, which makes a bad matter worse, and they die from worry and suffocation. Upon examination the dead bees do not appear candy-daubed at all, but that is what really caused death.

The commonest method of making cage candy is to take honey or invert sugar, put it in a vessel, set it on the stove and heat to a temperature of about 140 degrees. Then stir in as much powdered sugar as possible, put it on the bread board and knead in more powdered sugar until very, very stiff. You cannot get in too much sugar or knead it too thoroughly. Another difficulty is encountered in the use of this sort of candy made with powdered sugar. As is commonly known, powdered sugar contains from three to five per cent starch to keep it from becoming lumpy and hard. This does not make the best grade of candy, for starch causes dysentery to bees in long shipments, but for short shipments no harm is done. Powdered sugar without starch may be obtained by a special order, but it becomes so lumpy after a few days that fine, smooth candy cannot be made with it.

However, good candy can be made with either the sugar containing this small amount of starch or the pure powdered sugar. The pure sugar is to be preferred. During the war when powdered sugar could not be produced we made a fondant out of granulated sugar that gave excellent results; but it is a very exacting process to make it, so that we abandoned it. The candy that we consider the very best, however, is made from pure powdered sugar and homemade invert sugar.

As to the use of honey for cage candy, it is generally acknowledged that American foul brood has been spread in this manner. Even if the present laws are complied with, there is still danger. If for instance the shipment is accompanied with a certificate of health from an apiary, the candy may have been made of honey from some other apiary, disease infected; or, the diseased honey may have been produced some time before when the apiary was infected. The source and spread of bee disease are so mysterious and disastrous that we cannot be too careful of contagion, and while the party using such diseased honey would of course be ignorant of the fact, the damage is done nevertheless. Therefore not only should the breeder be particular but the purchaser should also exercise precaution and, for introducing the queen, he should never use the cage in which she is shipped. Let him always transfer the queen to a cage of his own and burn the shipping cage.

As to boiled honey, it is about the worst thing which can be used in making candy. Bees that have been fed on it for only one day frequently show distended abdomens, indicating they will soon die of dysentery.

For ordinary shipments, invert sugar has been used instead of honey. This may be purchased from dealers in beekeepers' supplies. With the process of making the very best grade of queen cage candy without the use of honey, there is no reason why honey should ever be used in making queen candy at all, even for extremely difficult shipments as is the case in export trade or in very hot dry weather. It has been considered that a fine quality of white honey is superior. However, the chemist tells us that honey and invert sugar are the same as far as the process of drying out is concerned. Now we know that commercial invert sugar is not so thick as honey, and that it does dry out more than honey. Pure water boils at a temperature of 212 degrees. If a heavier substance such as sugar is added, it requires a higher temperature to bring it to a boil. By this test one can easily prove that the commercial invert sugar contains more water than honey, which causes the candy made with it to dry out sooner than honey. The commercial invert sugar boils at a temperature of 245. After this fact was learned some invert sugar was made by using granulated sugar and tartaric acid, and it was boiled till it reached a temperature of 250 degrees, which is five degrees above the boiling point of honey. This gave a very thick, heavy syrup, so thick that it had to be warmed before it could be handled, which induced me to do some investigation along this line. The result of these experiments led me to make my own invert sugar, which is thicker than honey and therefore is superior to the best honey for queen candy. As was expected, candy made with this and powdered sugar does not dry out at all, no matter how hot and dry the weather is.

Receipt for Making Invert Sugar.

Put one-half pound of water in a kettle and place over the fire until it comes to a boil. Then add one pound of granulated sugar, either beet or cane sugar, they being exactly alike. To this add about ten grains of tartaric acid. As few will have pharmacists' scales to weight the acid, you can use about one-eighth of a teaspoonful or a small amount on the tip of a table knife. If trouble is experienced in getting the right amount, an empty 22-short rifle cartridge may be used as a measure. One of these holds 2 ½ grains, therefore four of these level-full will furnish the right amount. A little more or less will do no harm. Put in the acid and allow the syrup to boil slowly without stirring until a temperature of 250° is reached. Then pour it into a Mason jar. The cap should be kept on the jar as this syrup, being so thick, will absorb moisture if left exposed to the air.

A small amount on the tip of a table knife.

Making the Candy.

Take the bread board and cover it with powdered sugar two or three inches deep. Warm the invert sugar slightly, not above 120°, so the dough will work up easily. Pour a little invert syrup on the sugar and cover it with more sugar. Then work the dough up into a ball and knead the dough until it is so stiff one can scarcely pinch off a bit between the thumb and the finger. In fact, I have never been able to make this candy too stiff. Considerable kneading is required to get it just right. If the air is moist the candy becomes sticky, and more sugar should be worked into it before provisioning the cages. This candy will keep indefinitely. One winter we placed some above the hot-air register of the furnace, and in the spring it was still soft and the bees ate it readily. If worked long enough it will become quite dry, more like bread, but will not become hard. In that condition it will not gather moisture and will not run. As pure powdered sugar will become lumpy, if kept on hand too long, it is advisable to order twenty-five or fifty pounds and make it all up at once. When first made, if the weather is damp, it is well to keep the top covered with a little dry powdered sugar.

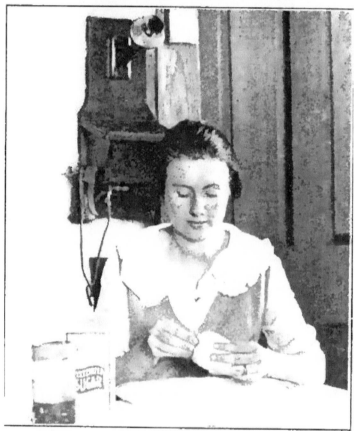

Pinch off a bit between the thumb and finger.

This makes it very convenient to pick up the nurse bees.

It is quite unnecessary to say anything about putting the queen candy into the compartments other than some suggestion as to the amount required. For shipments anywhere in the United States or Canada, one compartment full of candy is plenty if twenty-five or thirty bees are put into the cage. If fifty or more are placed in the cage or if the shipment is two thousand miles or over, two compartments should be filled. For export, it is well to fill three of the compartments in the export cage.

After provisional, the perforated tine that covers the hole in the cage should be swung around the tack placed in it so that it may be easily removed with the fingers. Having the cages provisioned, place them in the hive-seat on the left side. Open the nucleus hive and give the bees a very light puff of smoke from the smoker. Lift out the frame and find the queen. Pick her up with the right hand by taking the two wings between the finger and the thumb. With the left hand put the brood-frame a short distance down into the nucleus hive and tilt it over away from you so it will remain there. This makes it very convenient to pick up the nurse bees. Take up the mailing cage with the left hand and put the queen into the cage, closing the hole with the forefinger of the left hand. Then proceed to pick up worker bees by the wings and poke them into the cage, head first, closing the hole in the cage after each bee with the finger as before. When one gets accustomed to it, he can fill a cage in half a minute. Once in a while the beekeeper will get hold of a balky bee that insists on bracing with its front feed and refuses to go in. When the pressure is brought to bear with the finger, the bee runs out its sting, with the result that in time that finger becomes somewhat callous. A thimble might be used on the left forefinger, but this is awkward and un-sportsmanlike.

The Question of Cataleptic Queens.

In handling queens, many beekeepers have observed that once in a great while a queen suddenly becomes unconscious and sometimes dies. The reason assigned is that she took a cramp or had a cataleptic fit.

The author has observed this for the past twenty years. Some seasons it would occur but once; during others half a dozen times or more. In my mind there has always been some doubts whether or not the queen was cataleptic. One season, the loss was heavier than usual. One day two were lost. I say "lost" as the queens were always discarded after having a "fit," for previous experience had made me believe that they were permanently injured by having these "spells."

The day the two queens were lost, I observed how very similar was the action of the injured queen to the one that had been stung by another queen. There was a sudden collapse, then a slight quivering of the legs. In one case this lasted for over half an hour, when the queen slowly revived. In the other case, the queen quivered four about the same length of time and then died. It seemed certain to me that in some mysterious manner these queens were getting poison from a sting. Could it be that the poison on my fingers from worker-stings was causing the mischief? Investigation failed to substantiate this. I noticed that in one case the queen had taken hold of the top of her abdomen with a front foot, which might indicate that she had received a slight prick in the foot from her own sting. I therefore watched carefully and soon this belief was confirmed. The queen in taking hold of the tip of her abdomen exposed the sting. Then, in trying to get hold with the rest of her feet, she would strike right at the point of the sting. In this manner she undoubtedly received some of the poison. Since that time we have taken great care that a queen is not allowed to take hold of the tip of her abdomen, consequently no more queens have been afflicted with fainting spells.

There is considerable controversy on as to whether it injures a queen to clip her wings. This controversy has been on ever since the practice of clipping was started. Some claim it injures the queen, and some as stoutly maintain it does not. My experience leads me to agree with both factions. The following article, written by me and published in *"Gleanings,"* tells of my first attempt at clipping a queen.

AN AMUSING AND YET NOT UNCOMMON EXPERIENCE OF BEGINNERS.

The first colony of bees I got was in the ten-frame homemade hive, which I kept standing in the back yard the first year, not daring to go near it. In the fall, I went out one night and peeped under the cover, and was surprised to see that there was no honey. I supposed all a fellow had to do to get honey was get some bees and they would do the rest. Nothing succeeds like success, so they say. Not so here. Nothing makes me succeed like a failure, so I determined that next year those bees should make some honey or furnish a reasonable excuse.

I subscribed for Gleanings and got the ABC. Then the bee fever took hold of me in earnest. I studied the book night and day. I knew it all by heart. I got the "Facts About Bees," and learned it till I could recite it as easily as a minister can quote scripture. The argument in it was good. Everything in it was all worked out. How grateful I felt that everything had been learned for me, and all I had to do was to reap the benefits! I tried hard to be unassuming but inwardly I could not help feeling proud that I knew everything about bees.

Now the book said: "Catch the queen and clip her."

I did all this studying in the winter; and how I longed for spring to come that I might demonstrate what I already knew! How that winter persisted in staying with us, and how reluctantly did spring show her shining face! But at last, in the latter part of March, there came a beautiful, bright warm day-just the time for clipping the queen! I had never seen a queen, and my anxiety to view her majesty was something fierce. I had an assistant cover me with mosquito bar. I put on mittens and wrapped my wrists with rags. Then I fired up the smoker and prepared to go into action.

How I dreaded opening that hive! I felt a little pale, but my teeth were set and it was do or die. I was too big a coward to retreat while every one was watching. I must have been an aw-inspiring sight to those bees as I swooped down upon them dressed in armor, with the smoker spitting smoke and fire. I soon enveloped the hive in smoke, gave it a few jolts tore off the cover, then smoked again. Of course the bees cowed before such a vicious onslaught. Now, the books said, "Catch the queen and clip her." Clipping was the primary object of the expedition; but I saw where the books were right in saying "Catch the queen" before saying "and clip her." The only change in the wording of that I would make would be to precede that with "find the queen." I took out the frames carefully, and stood them around the hive in various places; but could not "catch the queen." I looked and looked. There were more bees in that hive than I had expected to see in ten hives. The separation of a mixture of the proverbial haystack and the needle would have been a cinch compared with the task in hand. I hunted all the afternoon, and had to give up on account of darkness. I was disgusted but not discouraged. This problem confronted me, "If I fail to find one queen in half a day, how long will it take to find several thousand queens?" (the number I expected to have in a year or two).

Nothing succeeds like a failure, and the next day I went after them with more zeal than ever. On lifting out the third frame my eyes rested on a bee the like of which I had never seen before. It was a long bee, and she walked with a more majestic treat over the comb and did not seem to be in such a rush as the rest of the bees. She was of a dark-brown color, and how handsome she looked! Verily this must be the queen! The queen of Sheba might have looked good to Solomon; but she was not arrayed like this one.

The next thing, "Catch the queen." I tried to make the catch, but she was not so easy. Just as I would close my fingers on her she was not there. At last I got hold of one wing but she buzzed around so that I let her drop.

If about one third of the wings is clipped off no harm can possibly come of it.

Again I got her by the wings and tried to transfer to the left hand, but her head did not stick out far enough for me to get a good hold, and she backed out and got away. Next time I shut down so hard that I was afraid I would kill her, and then let up so that she got away again. This time she dropped in the grass and I had a time to find her. The fourth time I held her between my left thumb and finger in a trembly fashion, much as a dog bites a rat, and probably the sensations to the rat and queen were similar.

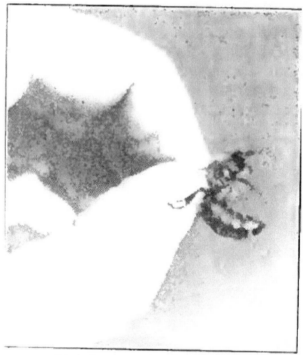

Pick her up by the wings.

I then got the shears. I forget whether they were a large pair of tailors' shears or the kind they use for shearing sheep. In my enthusiasm I had used them in prying frames apart, and they were more or less gummed up with propolis. I slid them under the wings and shut down. The wings bend over, but would not cut. I tried again and again until I either wore the wings apart or pulled them off. But I got them off and a leg with them. A little later I thought I would "shook" the bees into a new Danzenbaker hive, and I was astonished to find that they had a new yellow queen with wings of the regulation length and a full quota of legs.

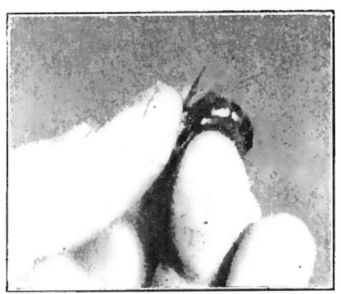

Leaving the wings projecting.

In the case described it undoubtedly was injurious to the queen to have her wings "cropped" as one correspondent puts it. However, if care is taken and clipping done properly it can be no more injurious to a queen to have her wings clipped than it would be in case the wing of a hen is clipped to keep her from flying over the fence and getting into the lettuce bed-it all depends on how it is done. If the wings are clipped too short, injury may result. In such case nerves and arteries may be severed. If about one-third of the wings is clipped off, no harm can possibly come of it. This bears out in practice, for in thousands of queens clipped in this manner, I never knew of one injured. The bees treat such queens in exactly the same way as though they had their entire wings and do not supersede them any sooner than the others.

Clip off about one-third.

In clipping queens' wings some prefer to hold her majesty by the legs, while some prefer to hold her by the thorax between the fingers and the thumb of the left hand. After trying both I prefer the latter method. To clip a queen, first pick her up by the wings with the forefinger and thumb of the right hand. Then pass her to the left hand, placing the thumb on the upper side of the thorax and the forefinger on the lower side. In this position her abdomen forms a curve over the end of the finger, leaving the wings projecting in such a manner that they may easily clipped. Then pass a blade of the scissors under the wings and clip off about one-third of the length. If preferred the wings on one side only may be clipped. It is not necessary to clip off much of the wings in order to prevent the queen from flying, as a laying queen has all she can do to fly when she has all of her wings in good order. The queen must never be clipped before she begins to lay, for of course, since she mates with the drone on the wing, she could not mate if her wings are clipped. As she never mates after she begins to lay, no harm is done by clipping her wings as described after eggs are laid. The wings of the queen do not grow out again after clipping, so one operation is sufficient for her lifetime.

Lest the beginner might not thoroughly understand the significance of the above, let it be understood that a queen must under no circumstance be clipped before she begins to lay, for it cannot be definitely ascertained that she has mated and fertilized until she begins to lay. As the queen mates only upon the wing, it can be understood that if her wings were clipped she could not fly and there-fore could not be fertilized. There are frequent cases where the amateur catches a swarm, and, if the queen is found, she is clipped to keep the swarm from absconding. One should be careful in such cases to be sure that the queen he clips is a laying queen, for, if the swarm should be an after-swarm with a virgin queen, clipping would render her incapable of mating. In such a case if the swarm were not given another queen it would be lost. In some cases the virgin would become a drone-layer and in others she would be killed by the bees who seem to think it was the queen's fault in not going out and mating. After the queen is killed by the bees, the colony soon runs to laying workers and is worthless.

Chapter XXVII. Introducing Queens.

Nearly thirty-five years ago, Mr. G.M. Doolittle wrote, "Perhaps there is no one subject connected with beekeeping that has received so much notice in our bee papers and elsewhere as has the introduction of queens." We find the condition in this respect very much the same today. Methods come and flourish for a time, and then quietly vanish. Many of these so called "new methods" were used and discarded before any of us were born. I have been guilty of making some startling discoveries only to find that they had been known many years ago and discarded because they were of no account. Other interesting facts I have come upon, only to realize after a certain period of time that they were advocated by others years ago. For instance, I proved to my own satisfaction that the presence of queen cells in a hive has little or no effect upon the bees' accepting a queen, provided no virgins emerged before the queen was laying. This is contrary to the general belief and I thought I stood alone in this, yet Mr. Doolittle says the same thing in "Scientific Queen Rearing." Lately the "Honey Method" of introducing queens has come forward and was thought by many to be something new, but Mr. Doolittle also describes that. Yes, and it is just as unreliable today as it was forty years ago. One year I used tobacco stems in my smoker to introduce queens, closing the entrance until the bees came out of their debauch. I thought I had something good as well as new, but we find that Alley used it many years ago. I think it is the best of any "smoke methods," but it is poor enough at that. We can all avoid traveling in a circle in this manner if we will only read all of the older publications in beekeeping. The old "masters" can teach us much. Many times we can find the very thing we are getting ready to "invent" and also the reason why it is "no good." All kinds of torture have been inflicted upon the innocent queen and bees during queen introduction. There are the "Starvation Method," the "Drowning Method," the "Honey Daub Method," "Peppermint," numerous "Smoke methods," "Chloroform," "Carbolic acid," etc. Yes, and away back in 1744 the English had the "Puff Ball Method." When one of these balls was "puffed" at the bees they became unconscious. Dr. Phillips tells me that they claimed it made the bees "forget," and when they came out of it they had forgotten all about the queen question and couldn't remember whether the queen they then possessed was the one they always had or not. No doubt many of the queens introduced in that manner also forgot to lay, and many would "forget" to live.

All of the above "heroic" methods remind me of the way the students of Anthropology tell us the cave man who lived a few hundred thousand years ago got his wife. Now the cave man was not at all sentimental in his make-up but rather intensely practical and conducted his business affairs in a drastic manner. He did not believe in all this serenading by moonlight, neither did he ever buy 30c gasoline to take his prospective bride joy-riding in his flivver, and he had no use for this swinging-on-the-gate stuff. Not he! When he wanted another wife to add to his heterogeneous collection, or if one was getting old and he thought she needed superseding, he picked up his club and went after another wife. As the people in those days were vegetarians, he knew where to look for one; so about supper time he would find her out in the alfalfa or sweet clover patch enjoying the evening meal. He walked up behind her, swatted her over the head with is club, and carried her back to his cave where she became his dutiful and obedient wife.

Now, he got a wife, sure enough and maybe it is a matter of sentiment, but don't you know I like the way we do it nowadays much the better! So likewise I like more sentimental methods of introducing queens. If the heroic methods just mentioned were sure in their results, we might overlook the rough treatment; but they are not. There is no method that will permit the taking away of a queen and immediately releasing another in the colony with anything like certainty, for it is entirely against bee nature and you cannot change bee nature. Let us try to understand and work in harmony with it. I am sure better results can be obtained.

Loss in Introducing with the Mailing Cages.

Probably the method used today in the majority of cases is that of introducing the queen in the same cage in which she was shipped through the mail. The loss of queens by this method has been frightful. Man who have had many years' experience as inspectors and are in position to know, have told me that they believe 50 per cent of queens are lost in this manner. An expert honey producer in California told me that he had kept track of his loss and found it to be one out of three in the introduction of all queens he brought through the mail. Various reports have come to me where six queens out of twelve are lost by using the common mailing cage as an introducing cage.

*(My friend M.T. Prichard, one of the best queen-breeders in the country, says that mailing cage is quite reliable for introducing provided the queen is kept cages 48 hours before the bees get at the candy to release her. He says if this precaution is observed the loss will be negligible. He recommends plugging the candy hole with a plug of beeswax for two, and in some hard cases, four days, at the end of which time the plug is removed and fresh candy put in.) Now if the bees would kill the queens outright the damage would not be so great, but frequently the queens are injured so they never make good; yet they remain at the head of the colony, possibly a year or more,

losing for the beekeeper the surplus that colony would have made had the queen been properly accepted.

It has often been a wonder to me how the beekeepers have been content with this heavy loss. If a stockman in buying cattle lost fifty per cent or even five per cent by having the cattle fight when he united the herd, he would look for better methods. The loss to the beekeeper is not so much in one lump, but the percentage is the same whether it is cattle or bees. In this connection we are reminded of the story told of a man who, in telling his friends how to teach their boys to swim, said: "All that is necessary to teach a boy to swim is just to catch the kid and pitch him right into the deep water. He will swim all right, for he has to. I know that method works, for I taught my eight boys to swim in just that very way, and lost only one out of eight!"

However, if this method of queen introduction must be used, the best way, in our opinion, to use this cage is to remove the queen from the colony to be requeened, take out a frame and set it away in the honey-house spread the brood frames apart, place the cage between the frames with the wire screen downward, and press the combs tightly against the cage to hold it in position. In this manner the bees cluster on the wire and get acquainted with the queen. It is well to tack a piece of tin over the candy for a day so the bees cannot release her too soon. When there is a honey flow on, this method will be successful, probably in four cases out of five. However, I believe this is the most unreliable method of any that I shall describe.

The Doolittle Cage.

The Doolittle cage is better. To make this, saw off two pieces from a broom handle, one five or six inches long and the other piece one inch long for the ends of the cage. Then construct a cylinder of wife-screen cloth the size of a brood handle to fit between the ends. The wire screen is tacked permanently to the short piece. A hole about three-eights of an inch in diameter, to be filled with candy for introducing, is bored through the long piece, which is withdrawn from the wire cylinder in order to put the queen into the cage. To introduce a queen she is first transferred from the mailing cage to this one. A frame of honey is taken out to make room and the cage is placed down between the brood-frames, which are pressed to-gether to hold it in place.

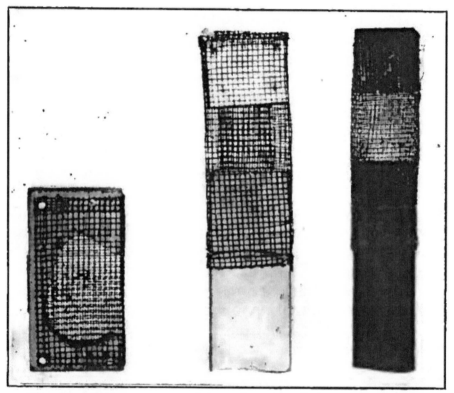

The Doolittle cage and two forms of the Miller cage.

The Miller Cage.

Dr. C.C. Miller constructed a cage similar in principle to the Doolittle; but he made his flat, using two wooden cleats so that he could shove it into the entrance or place it between frames without removing a comb to make room for it. A modification of this cage is shown with the queen excluding attachment. Either of these gives some better results than the mailing cage, but still there is considerable loss when conditions are not just right. Many of the Doolittle and the Miller cages are in use throughout the country. Some prefer one, some the other. With their use, the mailing cage is discarded, thus preventing the possible spread of American foul brood, as this disease has been scattered far and wide through diseased honey in the mailing cage.

The Push-in-the-Comb Cage.

Realizing the shortcomings of the above methods and reviewing the methods used in introducing queens, there seemed to be one style of cage that gave nearly perfect results, and that is the cage know as the "Push-in-the-comb cage." Mr. Doolittle used this cage and stated that not one queen in a hundred was balled when it was used. To make this cage take a piece of wire-screen cloth about six inches square and cut a notch out of each corner, so that when the edges are bent down it will form a bee-tight cage with one side open. The queen is placed on the comb, the cage is put over her and pushed down into the comb. In two days the bees usually burro under the cage and release the queen. If she is not out by that time, the cage is taken out and the queen released. This cage gives much better results than any of the others described, but it has some undesirable features. Sometimes it comes out of the comb or the bees gnaw away the comb, release the queen prematurely, ball and kill her. Moreover, since there is no convenient way of getting the queen into the cage, some are lost.

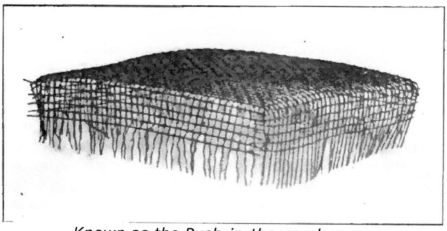

Known as the Push-in-the-comb cage.

Smith Introducing Cage.

Our Introducing Cage.

Realizing the heavy loss in queens, I have been experimenting for a number of years to perfect a cage that is sure in its working by using the principle of the Push-in-the-comb cage, and overcoming the objectionable features. I believe I have succeeded.

Our cage is made of half-inch material in a rectangular form. On the inside of the frame is nailed a strip of heavy tin so cut as to form pointed teeth. Wire screen is tacked on the top. Two holes are bored through the frame, one in the end and the other in the side. Corks fit snugly into these holes. On the inside of one hole is a piece of queen-excluding zinc, while the other hole is used to admit the queen.

And press it in tightly.

Now take the cage containing the queen and allow her to run up into the introducing cage.

To introduce a new queen go to the colony to be re-queened and remove the queen. Select an old comb that has had many brood reared in it. In such a comb the midrib is tough, for it is covered with old cocoons. If the colony to be requeened does not contain such a comb, get one from another hive. This old comb is important. Stick the cage into the comb, place it between the knees and press it in tightly. Now take the cage containing the queen and allow her to run up into the introducing cage. See that both corks are in place and set the comb back into the hive. It will be necessary to remove a comb of honey to make room for the cage. In two remove the excluder cork and allow the bees to get to the queen, and in two days more the cage may be removed and the queen liberated and all is well. In extreme cases where the robbers are bad and one had exceptionally cross hybrids or black bees, it is best to leave the cage in the hive for three days before removing the cork over the excluder and for three days more after that, but I have never found this necessary. Space should be left between the cage and the comb next to it so the bees can crawl over the wire screen and get acquainted with the queen. If queens that are to b introduced have been received through the mail, it is a good plan to burn the cages unless the candy in the cages contains no honey. American foul brood has been scattered all over the country in queen-cage candy.

As soon as the cork is removed over the excluder the queen increases egg-laying at a rapid rate, because the bees can get to her and feed her. She lays in all the empty cells, then goes back and lays in them again and again, sometimes almost filling them with eggs. She, therefore, is laying at full speed, and when released from the cage will fill the frames with eggs at an astonishing rate. I have frequently found eggs in three frames the day after the queen was released.

The queen lays more eggs the first day after being released than she would in two or three days where she could lay no eggs in the cage, and had to build up to egg-laying after being released. She has greatly increased in size, so by the time she is turned out she is a large laying queen, with the odor of the colony; the bees have been feeding her and have passed in and out of the cage so that the queen is as much the mother of that colony as though they had reared her themselves. When this cage is used as above described, I have yet to lose a single queen. Others have done equally well. In fact, I do not believe a queen would ever be killed by the bees when the cage is properly used.

And lays in them again and again.

Upon several occasions, I have had queens killed because I had overlooked some cell from which a queen emerged and killed the one that I was trying to introduce. Upon several occasions, virgins from nuclei got into the hive, were accepted, and killed the laying queen as soon as she came out of the cage. Once when I was making increase I shook the bees into a new hive at the time I released the queen. She came up missing. There were robbers who probably killed her. I have had the same thing happen to a queen that had been in the colony for a year, so I do not consider that the losses mentioned were in any way the fault of the method of introduction. To avoid the possibility of having a virgin in the colony, one should examine the cage at the time of releasing the queen, and if there is another queen in the hive the bees will be balling the cage. In such cases hunt up the other queen and remove her, leave the cage in a couple of days more and the queen will be accepted. I have saved several queens in this manner.

Reasons for Acceptance.

This is the "sentimental" method, and it in no way injures the queen. It is well known that an old and failing queen can be introduced to almost any colony as the bees pay no attention to her. From the fact that the bees know they have a queen while she is in this cage and that she can not get out and lay in the combs in a natural manner, I believe they consider they have a queen that needs superseding, which is another reason they accept her so readily. I have noticed that at times they build a piece of comb in the space left vacant by the removal of the frame, and on this comb they start numerous queen-cells, expecting the queen to lay in them. Sometimes a dozen or more cells will be started on a piece of comb not more than four inches long and two or three inches wide.

Introducing Queens to Laying Workers.

As a rule, it is not worth while to try to introduce a queen to laying workers from the fact that they will not readily accept the queen and, even if it is successful, these old bees are not capable of acting as nurses; so that if it does build up at all, the colony is very slow in doing so. However, the following method had been entirely successful in putting the laying workers back in the job in a satisfactory manner: Take a frame of emerging brood from another colony and use the cage on that as previously described. Set this in the center of the colony of the laying workers and introduce as before. The bees readily accept the queen, and the frame of emerging brood furnishes nurse bees enough to give them a start. If late in the season and the colony is weak, it is well to give it two or three frames of brood at the time of introducing the queen.

Emerging (or Hatching) Brood Method.

One method without the use of any of the above-described cages has been used for many years with almost perfect success. It is that of placing the queen on combs of emerging brood after having brushed off all the bees. The main objection to this method is the time and work it takes. However, if one has a very valuable queen and does not mind the work, it gives excellent results.

The procedure is as follows: Take four or five frames of brood and put them over a strong colony above a queen-excluder. In ten days all of the brood will be capped over. Now take off these frames of brood from the colony and be careful to brush off every bee. Place them in an empty hive and stop up the entrance with rags so that no bee can get out. Take this into the house, remove the perforated tin on the mailing cage containing the queen, set the cage down on the bottom-board and close the hive. The queen and bees will crawl out of the cage on to the combs. Keep this hive in the house so that the temperature is even in order that the bees may emerge and not become chilled or overheated. In five or six days set this hive on a stand, open the entrance just wide enough for one bee to get in and out at a time. Watch to see that robbers do not overpower it, as it will be a couple of weeks before the bees can put up much of a defensive fight. Do not allow your curiosity to get the better of your judgment and induce you to open the hive for several days after you put in the queen, for, if there are not many bees emerged, the queen is apt to fly out. I once lost a fine imported queen in that way. This method gives excellent results when carried out as above described.

Emerging brood may be taken directly from the hives instead of placing them above a queen-excluder, but in that case the unsealed brood crawl out and die and make a mess.

Unsatisfactory Modification.

Some have suggested a change in the method and recommend that, instead of putting the hive with brood into the house, it be set over a wire screen above a strong colony in order that the bees may get the heat from the colony below. In theory this is fine, but in practice it is a pronounced failure. Many of the queens will be found dead when this method is followed. It does not look reasonable that the bees sting the queen through the wire, but that may be the cause of the death of the queens. My assistants in nailing up queen cages frequently get their fingers stung through the wire screen. One day when the weather was rather cold, I had a mailing cage containing queen and bees. I placed my hand over the wire screen to see if it would warm them up. Immediately a bee planted it's sting in the center of my hand leaving it there. It is possible that the bees sting the queen through the screen. At any rate many have reported losing queens when introducing them in this manner, above a strong colony over a wire screen; but the former method of removing brood to the house is an excellent one.

A Common Cause of Failure in Queen Introduction.

Regardless of what method is employed in introducing queens, it is very essential that all combs be put back in the same position that they were before removing. If this is not done, many queens will be killed even after they have been accepted and have been laying. When the combs are not put back as they were originally found, this is what is apt to happen: When the new queen is released from the cage she takes a "swing around the circle" to see to it that all queen-cells that have been started are destroyed. Now, if a comb of honey has been inserted in the middle of the brood-nest, the queen does not realize that there is more brood over the other side of this comb; consequently she does not go over to that side to attend to the destruction of any cells that may have been started there. She seems to think that she has been all over the brood-nest, and settles down to egg-laying. A few days later a queen emerges on the other side of the comb of honey, and sooner or later the virgin and the laying queen meet and the laying queen is always killed. I lost many queens in this way before I discovered the cause. Many who buy queens through the mail have trouble from this cause. They will wonder why it is that although the queen was laying profusely, she was "superseded" so soon, for she was found in front of the hive dead and a virgin was discovered in the hive. She was not superseded; she was killed by the virgin that came from the other side of the hive.

Why Queens Die in the Mailing Cage.

When the common mailing cage is used as an introducing cage, in some instances all the workers and the
queen are found dead after the cage has been in the hive.
We have had many inquiries as to the cause of this loss. I
am satisfied that the bees of the colony sting the workers
and queen confined in the cage through the wire. Some
beekeepers hive tried to shorten the period of queenlessness of the colony by putting the mailing cage into the
hive before removing the old queen. Their theory is that
the bees will get acquainted with the queen, and she can,
therefore, be released as soon as the old queen is removed. In practice, however, this failed to work. All workers and frequently the queen are killed in the cage, or else
killed as soon as released from the cage. It was found
necessary to leave the cage in the hive from two to four
days after removing the old queen. Therefore, no time
was saved but many queens lost. In the push-in cage,
however, there is not this loss from the stinging through
the screen, probably because the queen naturally stays on
the comb and out of reach of the angry workers of the
colony.

There does not seem to be this loss when the round
Doolittle cage is used, possibly for the reason that the
queen stays near the edge nearest the comb. When a
Doolittle cage is used with a queen-excluder inside the
cage and the entire canal three or four inches long filled
with candy, a large percentage can be successfully introduced. This lacks, however, the one important feature,
that the queen can not lay while in the cage, and when
releases is not received as readily as a laying queen.

Chapter XXVIII. Disposing of Nuclei at Close of Season.

When the queen-rearing season is over, it is of course necessary to dispose of the nuclei. In warm climates, part may be kept over until spring if they are strong in bees and rich in stores. In case one has extra queens, this is an excellent plan since there is always a great demand for queens in the early spring before any can be reared. This demand is caused by the fact that many colonies come out queenless, and if queens can be procured the colonies would be saved.

I find it profitable to winter our nuclei even as far north as Vincennes. Two twin nuclei, with standard Jumbo size of frames, are placed in a case and packed in sawdust much after the plan used in packing full colonies. In case the nuclei are well supplied with honey and have enough bees to fill the hives nicely, they winter as well as stronger colonies that have more room. If the honey producer can winter a few queens in this manner, he will find that they come in handy the next spring in giving them to queenless colonies and replacing queens that are failing. In requeening such colonies, the whole nucleus (queen, combs and bees) is set over the one to be requeened and united by the newspaper method, for this nucleus will not be needed for queen-rearing since it will probably be a number of weeks before queen-rearing can be started.

However, a large percentage of nuclei must be disposed of as they queens in them will be needed for colonies. An excellent way to accomplish this is to gather a number of nuclei in one place in the center where the group formerly stood, then in the midst of them place a hive with two or three frames of honey. Next, remove the frames from the nucleus hives and brush off all of the bees in front of the hive. Some go in, and others go back to their old location. Finding their nucleus gone they circle around till they comb back to the hive where the bees are fanning when they join these, and as all begin to fan they call the bees in the air to them. After this more nuclei may be emptied and all the bees will at once go into the hive. It is a good plan to have a queen tarpon the entrance in order to catch any virgin that is apt to be overlooked. I usually put in about three pounds of bees. Then I go to a colony that is week and dump the bees in front of the hive. They immediately run into the entrance. In this way weak colonies can be brought up to the required strength. If one wishes, he can put a large number of bees together and give them a laying queen, forming a new colony. I believe it is better to strengthen weak colonies, for we can usually find a few of such.

Now the question arises, "will not the bees go back to their old location the next time they take a flight?" No, they do not, for when they set up fanning as they run into the hive they seem to put themselves into the condition of a new swarm and they will stay anywhere they are placed. I have never seen a single bee return to the old stand after it had once joined the new hive. Another peculiar thing in this connection is the fact that bees never fight against the colony with which they are united and never kill the queen. They seem so demoralized that they are willing to accept things as they find them. In uniting them in this manner it is well to make the colony extra strong, for many of these nucleus bees are old and will die off before spring. This plan of disposing of the nucleus bees has been of much value to me, for it has enabled me to build up weak colonies to good strength and winter them over, where otherwise it would have been necessary to unite these weak colonies to save them, thus reducing the number of colonies.

Packing Cases for Nuclei.

Nucleus packing case.

Many honey producers use extra hive bodies or extracting supers for nuclei; but some will undoubtedly find it more profitable to make or buy a special nucleus hive. In using an extra hive body it is necessary to have an extra bottom-board and cover, so there is little if anything saved in using them. A twin nucleus hive, with room for two frames and a division-board in each side, is a splendid equipment and is hard to beat. It is a source of great satisfaction to have on hand throughout the season a number of queens to use in case of emergency. This is especially true in the spring, for in a large apiary there are sure to be at lest a few colonies which come through the winter queenless. If the beekeeper has an extra queen wintered in a nucleus, this can be united with the queenless colony and it will build up ready for the honey flow in fine style.

In wintering the nuclei in cases if the opening in the winter case were directly in front of the nucleus entrance, the two entrances would be so close together as to cause drifting. We, therefore, make a tunnel one-half inch inside measurement, which is placed at right angles with the nucleus. In this manner the bees fly out at the side instead of in front. The tunnel is whittled off round at the end, and a round hole is bored in the winter case. That the bees may easily find the entrance, the end of this tunnel is painted black. For packing, sawdust is used, four inches on bottom, and six inches on top and sides. It is important to use sawdust which has been kept under cover for one summer which is perfectly dry. Green sawdust in our experience is no better than no packing at all.

After the nuclei are united, the combs must be taken care of to prevent their being destroyed by the wax worm. If the combs are allowed to freeze a few times, they are free from that pest until spring, or if they can be kept in a cool place during the winter they are safe. Do not, however, store them in a warm place such as a basement or attic without first fumigating them. A good way to fumigate is to put them into a close room and burn sulphur. This must be done several times, as the sulphur does not seem to destroy the eggs of the moth. Bisulphide of carbon is more effective, but it is very explosive and is dangerous to use in a room. A good method to employ in using it is to stack the hive bodies up outside, putting a thin super cover on both the top and bottom of the pile. The top hive body should be empty to make room for a dish of the bisulphide of carbon. Pour half a pint into the dish and let it remain till it evaporates. To make doubly certain, another application should be made in two weeks. It is a good plan to examine the combs at times.

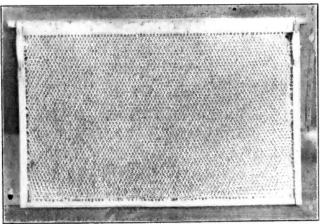

Comb made from Airco foundation drawn above an excluder.

Queen-Rearing for the Honey Producer.

Although I have now described the method of queen-rearing as practiced in a commercial way, the primary object of this book is to help the honey producer in rearing his own queens, for if a beekeeper expects to get the most out of his bees he must surely look after them carefully. Let us now adapt parts of the queen-rearing methods just described to the special use of the honey producer and recommend such changes as will best suit his needs. Let us consider that you are a honey producer operating two hundred colonies and upward. How far will the methods just described apply?

In the first place, I feel certain that the use of the swarm box, dipped cells and the manner of having cells finished above the queen-excluder are the very best for the honey producer to adopt. While a little more patience and study are required for its mastery, it certainly pays handsome returns for your time and labor invested. Many successful honey producers all over the world are using similar methods with the best of success. Many who use the grafting method and rear their own queens have informed me that with them it had simply meant the difference between success and failure. However, after the cells are completed and are ripe, they honey producer may branch off from the methods of commercial queen-breeder and adopt those best suited to his needs and circumstances.

Queen-Rearing from Commercial Cell Cups.

Mr. M.T. Pritchard, queen-breeder for the A.I. Root Company, Medina, Ohio, has probably reared more queens than any other man in the world. He uses the Root wooden cell cups into which are pressed the wax cups. He prefers the wooden bases because, he says, they protect the wax cups before and after they are completed by the bees; because they facilitate handling of the cells with their occupants; and last but not least, because he can easily separate the cells when they are finished by the bees. They enable him also to pick out the choicest and best cells from the cell starters and give them to the cell-finishers. These wooden cups are mounted on a bar by using a little pinch of wax. Mr. Pritchard prefers the queenless, broodless method for started the cells. He chooses larvae twelve hours old and endeavors to get the largest larvae of that age. He says, "the larger the better." He determines the age by keeping a record of the time the comb is given to the colony having the breeding queen. As soon as there are eggs in the comb it is an easy matter with him to determine the age of the larvae. He prefers larvae slightly curved but not coiled up.

Mr. Pritchard says there are times when he finds the cell-protectors a great convenience and the wooden cell cup fits the cell-protector better than the dipped cell. Mr. Prichard also believes it better to choose larvae near the center of the comb, as he considers the ones near the bottom of the comb not so well suited for grafting. The high quality of queens turned out by Mr. Pritchard is known the world over. A number of years ago the author visited Medina while queen-rearing was in full blast and learned a number of valuable features of queen-rearing from Mr. Prichard.

Requeening.

Some have preferred to introduce the ripe cell to the colony, while some prefer to introduce the laying queen. For the purpose of improving the stock as well as to replace worn-out queens I have requeened my entire apiary of several hundred colonies every year, and sometimes when I thought I had discovered a better breeder many were requeened two or three times in the year. I find the use of both methods of introduction, sometimes the cell and sometimes the laying queen, is very profitable. I believe it will pay all honey producers to have on hand a number of nuclei in order to have young queens to draw on when circumstances demand. In the main, however, since learning that it is a matter of feed that causes the bees to accept the queen-cells, I am using the cell method more and more.

The principle or law of the bees that a well-fed colony will not destroy cells should have a far-reaching effect with the honey producer, for we can now give ripe cells directly to strong colonies *immediately upon removing the old queen with almost unfailing certainty that they will be accepted.* We must remember that it is hungry bees that tear down cells; well-fed ones do not. When bees are hungry, they themselves tear down cells. When they are well fed they pass the task of cell destruction up to the queen. If she wants to swarm, she does not destroy them. If she does not wish to swarm, she attends to the destruction of the cells herself. True, the workers are doing most of the work; but if you catch the queen and remove her, the cell destruction ceases like magic if the colony has plenty of food. Therefore, if the colony is well fed and if you remove the queen, the cells are allowed to produce queens, and a strange cell will remain untouched the same as one which the bees themselves reared.

Therefore, if the honey producer raises a lot of nice large cells well supplied with royal jelly, he can requeen with very little labor and with small loss of cells. I strongly recommend that requeening be done after the honey flow is well along and the swarming season has passed or nearly passed. If colonies are given cells when they are preparing to swarm they will swarm anyway. They likewise swarm if a laying queen is introduced. I have found that, if a laying queen is introduced to a colony *before* it has any notion of swarming, it does not swarm that year; but, if preparing to swarm, it swarms with the new queen as readily as with the old one. The same thing occurs with a cell; if the cell is given before the colony has any notion of swarming, that colony is effectually prevented from doing so that season. However, except under unusually conditions, requeening with the cell method before swarming season is not advisable, for the absence of a laying queen at this time seriously affects the honey yield of the colony. In case one waits until the honey flow is well along, the colony may be requeened with little loss in strength, for the bees that would have been reared at that time would be too late to help in that harvest.

The method of giving the cell, which I have found entirely satisfactory, is to requeen during a heavy honey flow by simply removing the old queen and putting in a ripe cell. The bees seldom tear this down. However, to do all requeening during such favorable times is not possible. Just as the honey flow closes and the supers are off, you can remove the old queen in the evening and give the colony a heavy feed. Next morning give it a ripe cell, which the bees invariably accept. We must remember that in nature bees do not have cells except when they are receiving food in abundance, and we must duplicate these conditions if we hope to succeed.

After the cell is given, it is well, in a few days, to look to see if the queen has emerged. This may be determined by the appearance of the cell. If it has a small opening at the lower end, you may be sure the virgin is there. If the beekeeper has little time he can wait for ten days, and if eggs are present he may be sure the queen came from the cell given; but if there are other queen-cells started, it indicates that the cell given was destroyed. In such cases the frames should be removed and all bees shaken off in order that every one of these other cells that are started bay be destroyed. Queens coming from such cells would usually be inferior. As the colony has been without a queen for some time, it would be better to introduce a laying queen if you have some on hand, rather than give another cell, for this would leave it queenless so long that it would become greatly deleted in bees. Furthermore, if the second cell given should be destroyed, the colony would probably run into laying workers and be ruined. If you have no laying queen a cell should be given as previously described, feeding heavily as before. However, in this case it would be well to examine it one or two days after the virgin emerged from the cell, and in case this cell is destroyed give another immediately. We should bear in mind that cells are destroyed in this manner very rarely; but, in order to avoid ruining a colony from queenlessness, it should be carefully watched. However, if the work is properly done few cells will be destroyed.

Uniting Bees.

In the previous pages frequent mention has been made of uniting colonies. By far the easiest and safest method is what is known as the "newspaper method." The inventor of this system was Dr. Miller, and it is one of the many splendid things that he has given to beekeepers.

To unite two colonies, place two thicknesses of common newspaper over one colony on the brood-frames, set the other colony that is to be united on top and put on the cover. In a day or two the bees will gnaw away the paper and become acquainted. As they come in contact with each other, a few bees at a time, there is no fighting whatever. The bees in the upper hive body seem to realize that they have been moved, for they mark their new location and do not return to the old one. In very hot weather it is sometimes advisable to punch a hole in the paper with a common lead pencil to prevent suffocation.

Uniting Laying Worker Colonies.

If one has a colony containing laying workers to which he does not wish to introduce a queen, it can be united with any other colony by the newspaper method and the queen will not be injured in any way. The colonies should not be disturbed for a week or more, after which the workers will cease to lay. When this occurs, the hive body which originally contained the laying workers may be set on the bottom board of the present stand and the queen left with them, while the other hive body containing most of the brood may be moved to a new location and a queen introduced.

Making Increase.

There are many methods of making increase. If a large increase is desired, one should begin early in the season. For this, a good way is to proceed in the same manner as described in the first part of this book for forming nuclei, but it is better to give the new colony two frames of brood and bees instead of one. If done early in the season and a honey flow follows, a two-frame nucleus should build up into a good strong colony before winter. If no honey flow is on, they can be built up by feeding sugar syrup. The remainder of the hive is to be filled out with comb or full sheets of foundation. This method should be used only when a large number of colonies are to be made from a few. Where a limited increase is desired the method I have used for a number of years has given such perfect satisfaction that I recommend it in preference to any other.

It is as follows: Take the colony to be divided and set it temporarily to one side. On the stand it has occupied place an empty hive. Take out from the colony one frame of brood with the queen and place it in the empty hive. Then move the colony from which the queen and frame were taken to a new stand and introduce a queen to it or give it a ripe queen-cell, as previously explained. Fill out the hive that remains on the old stand with combs or foundation. The field bees from the colony just moved will return to their old location and build this colony up at a rapid rate. A queen is easily introduced to the moved colony, as the old bees that cause the trouble have gone back "home to mother's." Not enough bees will leave the moved colony, however, to injure the uncapped brood. If only sheets of foundation are used in the new colony it is better to give two frames of brood; but, if drawn combs are used, one frame is sufficient.

Mending Damaged Combs.

How often every beekeeper has dreamed of having perfect combs built clear to the bottom-bar, one hundred per cent worker-cells! All who have such combs please rise. Well, I shall remain seated with the rest. Even when by carefully wired full sheets of foundation we get fairly good combs, in time they get rounded off at the corners and later are drawn out into drone comb. Then mice get in and make holes in the combs. Wax moths do their work, and, as years go by, our combs become more and more filled with drone-cells.

Now, if we take a little pains we can have our combs continually improving and the drone-cells gradually diminishing. It is known that nuclei or weak colonies build worker comb only. Therefore, when we discover a damaged comb, let us put it into a nucleus for repair. If it contains drone comb, cut out the drone cells, and the bees in the nucleus will build worker comb in its stead.

Then in order to get the combs built to the bottom bar, draw the nails in that bar so they will be about ¾" of an inch lower. The bees will build comb to within one-half inch of it, the nails may be driven back into place thereby bringing the bottom-bar up snugly against the bottom of the comb, and you will have a perfect comb as the result.

If it is desired to mend combs when no honey is coming in, the bees will do excellent work at comb-mending by having their division-board feeder kept filled with sugar syrup or honey. Now it is unnecessary to have regular nucleus hives to mend combs, but any weak colony will do it if only two or three combs are given them at one time.

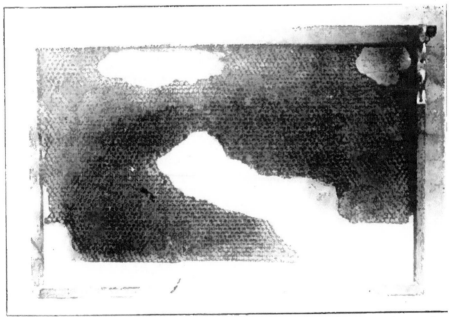

A "mouse-eaten" comb, with bottom bar lowered ready to be mended.

The same comb after mending.

It is the usual rule, where combs are mutilated or imperfect, to cut them out, melt them up and convert into wax. This, of course involved work and expense for foundation and perhaps new frames. Where frames are good and combs, except for holes (for all drone comb should be cut out), are otherwise good they can be repaired by the bees in the manner just described and save expense.

Chapter XXXI. Requeening Colonies About to Swarm.

What is said about giving a cell to a colony preparing to swarm holds good to introducing a laying queen; that is if a queen is introduced to a colony that is in the swarming notion, it will swarm just the same, the newly introduced queen going out with the bees. It is generally conceded that a colony headed by a young queen, one reared the current year, will not swarm.

Now, many amateurs, and some experienced beekeepers, too, have sought to cure the swarming fever by introducing a young queen. Failure is usually the result. A beeman, more progressive than his neighbor beekeeper, in order to get ahead of him, bought a queen from a breeder some distance away. He was very proud of the queen and rubbed it in on his neighbor, telling him he could not buy this queen for $10. It so happened that he introduced her to a colony preparing to swarm. He stopped his beekeeping neighbor and led him out to the hive to show him his fine new yellow queen. Before they got there, they met a large swarm of bees going to parts unknown. When they looked into the hive, there were a lot of queen-cells and few bees! A swarm and that new queen had taken to the woods. It was now his neighbor's turn. He said, "Say, I can now buy that queen for ten dollars can't I?"

However if a young laying queen is introduced to a colony before the colony has the swarming notion, my experience is that when the colony is run for comb honey it will very rarely swarm that season, and when run for extracted honey I have never had one swarm when requeened in this manner.

Hive Body Used as Nucleus.

A very popular method of requeening under certain circumstances is by using one of the brood-chambers where colonies are run two stories high. Certain localities have no early honey flow. The bees build up in the spring, go "over the peak of brood-rearing" and are on the down grade before the honey flow comes on. This is the case in many alfalfa districts.

A splendid method under such circumstances is the following: We shall suppose the colonies are in two-story brood-chambers. As the hives are getting well filled with bees and brood, remove one story, place it on a bottom-board, give it a cover, and put it close beside the hive body from which it was taken. To the queenless part give a rip queen-cell. This will in time build up to a strong colony. When the honey flow begins, kill the old queen and unite these two colonies by the newspaper method described further on. In this manner the colony has been requeened, and as there is a young queen in the hive, there is little danger of swarming.

By this system, the colony has the brood reared from two queens, and becomes a stronger colony than was possible to get from one queen. This same method can be used to advantage in the eastern states where there are a clover flow and a fall flow. In that case, the upper hive body is raised up and supers added between the two hive bodies during the clover flow. When most of the brood in the upper hive have emerged, it is set on a new stand at the side of the parent colony with enough bees to make a small colony. A cell is given in the regular way. This colony is allowed to build up until the fall flow opens, when it is united with the original colony after killing the old queen. By this method of requeening, the colony has been materially strengthened instead of wea-kened. In other words it has had *two laying queens* for some time instead of *no queen at all* as is the case where the old queen is removed and a cell given.

In case the double brood-chamber is used, one can be employed as a nucleus at a little outlay. All that is necessary is some extra bottom-boards and covers. When one makes a practice of introducing the queen-cell directly to the colony, I believe even then it is a splendid idea to have on hand a small number of nuclei. Laying queens can be kept in these to be used when needed. In working with the bees you occasionally run across a colony that needs a better queen, and if you have some laying queens in reserve in the nuclei, they can be drawn on when needed. Should a cell that you have given to a colony fail to provide a queen for it, a laying queen can be given thus preventing the weakening of the colony from lack of one over too great a period.

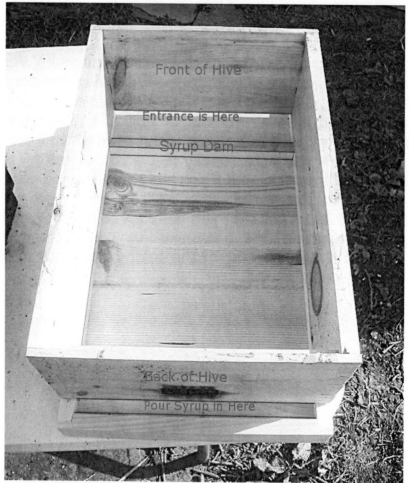

Transcriber's picture of a Bottom Board Feeder (the original picture was not clear)

If we use a large brood-chamber, we have to do little feeding, as a rule; but there are times when the honey crop fails, or we have been a little too enthusiastic in making increase, so that we find our colonies deficient in stores, with winter not far away. All beekeepers have been caught in such a predicament at one time or another. The feeder I prefer is made by nailing a strip across the bottom-boards in the deep side about two inches back of the entrance. In fact, when order the regular bottom-board I order an extra piece like the one used for the back cleat on the deep entrance. If this is not going to be used for some time, the bees will stop up all the cracks and make it water tight. If it is to be used at once, pour melted wax or paraffin along the cracks until it is tight. This feeder costs only three or four cents and is always there and never in the way. Moreover, it does not interfere in any way with the ventilation. If a driving rain comes up, this cleat keeps the rain from beating in. If you find a colony that needs feeding, all that is necessary is to slide the hive ahead about two inches on the bottom board and pour the syrup in by allowing it to flow against the back of the hive. It will thus spread out and run down into the bottom-board feeder. A funnel can be used if preferred. This bottom-board feeder holds about ten pounds of syrup. Of course, the hive must be level to prevent the syrup from running out. If a colony needs heavy feeding, the bees may be fed three or four evenings just at dusk, and before morning they will have all of the feed cleaned up away from robbers. This is a very good feeder for stimulating during queen-rearing also. If used every day for this purpose, it is best to have a thin board about two inches wide and the length of the bottom-board across the end as a sort of lid. Then, when feeding, this cover is raised and syrup poured in.

Preparing the feed for winter.

There are several ways of preparing syrup for the bees. Some give it warm, and some recommend cold syrup. I endeavor to feed early in the fall before the weather is too cold. A large tub or boiler is a good thing in which to mix the syrup. Pour in cold water till it is about one-third full. Then add granulated sugar and keep stirring until there is a saturated solution. After you have added so much sugar that it settles to the bottom about two inches thick and does not dissolve, no matter how much stirring you do, the syrup is right. Let it stand a few minutes. It will clear up and you have a nice, clear, smooth syrup. Now, when this is given to the bees they invert it and there is no danger of it's granulating in the hive. If you should follow this process and have the water warm, it would take up too much sugar so that when it became cold crystals would form and it would granulate in the combs. Another objection to the warm syrup is that, when the bees get in it, they become covered with sugar as soon as they dry off, and many are lost in that way. Of course, if warm syrup is given and you have the proportions exactly right, it is a good method but it requires much more work than the cold method just described. No measuring is necessary and water or sugar may be added at will and by stirring, the syrup can be made just right.

We should so conduct our apiary that feeding for winter is rarely necessary. Let us first assign to the bees the task of laying up honey for themselves. Then, after they have done this so that they have an abundance, we may consider that all they make above that amount rightfully belongs to us; but let us bet sure they have enough before we dip in. Sometimes in spite of care, however, the bees will need feeding. In case we have made a large increase or have had a late swarm or total failure of the honey flow, a number of well-filled combs of honey on hand for this purpose is a splendid policy. When a colony needs feeding nothing is better than combs of honey.

Chapter XXXIII. Requeening to Cure European Foul Brood.

It is now generally acknowledges that European foul brood is a disease of weak colonies and inferior or black bees. Many extensive honey producers testify to the fact that, if all colonies are headed by vigorous young, Italian queens and the colony kept strong, so far as European foul brood as a menace is concerned, they can forget it. During the winter of 1918-19 I was employed by the Government to do extension work in the Apiculture Department, my territory being California. During this time I had a rare opportunity to study bee disease, especially European foul brood. The conditions in California are favorable for the development of this disease. I found the statement made above concerning the cure of European foul brood to be correct. In many cases this disease threatened to put the beekeeper out of business until he began to rear queens from good Italian stock and requeen the entire yard. Then, in addition the colonies were given a large brood-nest, usually two nine-frame hive bodies and an abundance of stores. In such cases European foul brood ceased to be a menace. To be sure, it kept the beekeeper on the alert to see that no colony became weakened or had a poor queen. Now as we well know, strong colonies are the profitable ones, so it is evident that European foul brood makes a better beekeeper out of the one that has it. For this reason it is commonly known as a blessing in disguise. We admit that, when this disease first strikes an apiary and causes havoc before the beekeeper can form his defense, the incognito of the blessing is complete!

The method of rearing queens to eradicate this disease is the same as already given, but is better to introduce the ripe cell to the colony affected. Under no circumstances do the bees "clean house" as well as when they have a virgin queen. They clean and polish the cells to make ready for her to deposit eggs. Being without a laying queen they have very little brood to attend to, so they seem to devote all their energies to cleaning out the disease.

The important point in the cure of European foul brood seems to be to have a large number of bees in proportion to the brood. In any cure now being used this condition should be present. When we give a cell as stated, we reduce the amount of brood that the colony has, by removing the laying queen, thus making the cleaning out easier. When the queen is caged, the same condition is brought about. In some cases cures are effected by putting the brood in an upper story and confining the queen to the lower story with a queen-excluder. The same condition exists, for many of the bees leave the queen so that she slackens up on egg-laying, thus reducing the brood. Many have reported that, by putting a new swarm into a colony affected with European foul brood, an immediate cure was effected. This is the same condition as in the others, a large number of bees in proportion to the brood.

It is noticed that the first brood reared in the colony in the spring is not diseased. This same condition, plenty of bees to clean out the disease, is present. A little later in the season the disease is at its worst. This is due to the fact that the bees are rearing the maximum amount of brood in proportion to the bees, as the colony is rearing brood to the fullest capacity, and the old bees are rapidly dying off. If the colony survives and is built up strong, the disease disappears. This is due to the fact that the queen has reached her capacity in egg-production, so the number of bees in proportion to brood is greatly increased. However, if the colony is very weak, it is not a good plan to try to build it up. Better unite it with another colony. If European foul brood is very prevalent in an apiary, I would requeen the entire yard every year until the disease is stamped out. Then every colony that is not strong should be requeened, and if disease shows in any colony it should be requeened.

American Foul Brood.

Let it be understood that all which has been said about requeening to eliminate disease *does not* apply to American foul brood. As this book is a treatise on queen-rearing; we need not discuss American foul brood but only touch on one or two points. American foul brood cannot be cured by requeening. At the present writing, the "shaking treatment" is the only cure advocated. In Dr. E.F. Phillip's excellent book, "Beekeeping," (page 404) the disease and its treatment are described in detail. Many beekeepers wish to requeen colonies having American foul brood-not to cure the disease but to replace the queen on account of her age, for a good young queen is needed in the colony after losing all its brood and having to build up on foundation. The question is very frequently asked, "When is the best time to requeen a colony affected with American foul brood-before shaking or after?" I recommend that they be requeened after shaking; for, if a vigorous queen is introduced and allowed to lay heavily before shaking, she is liable to injured when this is done from the fact that her egg-laying is suddenly stopped, in the same manner as a queen laying heavily is injured by being placed in a mailing cage. Many have reported that queens that were very prolific before the disease was treated, were worthless after the colony had been shaken for its cure. Another reason is that it is not a good policy to open a colony affected with American foul brood any more than is absolutely necessary, on account of the danger of spreading the disease through robbing. After the colony has been treated and has several frames of brood, it is a good time to requeen. If the Push-in cage is used it will be necessary to get an old black comb from some other colony on which to use it, care being taken that it is from a healthy colony.

Chapter XXXIV. Finishing Cells in Queenless Colonies.

Sometimes, conditions are not right for building up colonies sufficiently strong to do the best work at finishing cells above an excluder in a queen-right colony. Nothing but failure will result in attempting to get good cells finished above an excluder if the colony is not *extra strong.* If these extra-strong colonies cannot be obtained, good results may be secured by giving the cells to a colony of medium strength made queenless. To use this method it is first necessary to get the cells accepted, as explained. Then take one bar of cells, go to the colony you wish to use as a finisher and remove the queen, at the same time giving a bar of cells. The bees will go right ahead and finish them in good shape.

The question comes up whether this colony should be queenless for a few hours before giving the bar of cells. It seems to make no difference, as the bees take right hold and go to work on the cells the moment their queen is removed. They seem to do this through their habit of completing a job that has been started. As the queen is not there to protest, the cells are finished. This colony may be used to build a second bar of cells, and even a third, but care must be taken to see that no cells of its own are allowed to furnish a queen.

Finishing Cells While Requeening.

A splendid plan for finishing is while requeening. Go to a colony that is to be requeened. Kill the queen, give the colony a bar of cells that have been started by the swarm box or queenless, broodless method, and at the same opening of the hive introduce a laying queen with the Push-in cage. In four days, by the time the queen is accepted, when you go to take out the introducing cage the cells will be sealed. They are taken out and given to some other queenless colony for incubation, or put over an excluder above a strong colony until ripe and ready to be given to nuclei. In this way you lose no time, for the cells are finished while you are introducing a queen. I find this very convenient late in the season after a honey flow when the colonies are not so strong as they are earlier in the season. As a rule, ten or twelve cells are enough for the bees to finish and do the best work. I never use this method exclusively but do so occasionally in conjunction with the method of finishing above an excluder.

Chapter XXXV. Cell Building During a Heavy Honey Flow.

All are agreed that the ideal time for cell-building is during a light honey flow. However, we have to take the weather and the honey flow as they come, and we seldom have a light honey flow for any great length of time. When the honey flow is beginning conditions are idea for a time; then, as it increases until it becomes a heavy flow, the kind that we all like for honey production, a number of elements interfere with queen-rearing, that much be overcome if we would succeed.

One of these is the scarcity of larvae of the grafting age, due to the fact that the bees are gather nectar so rapidly that the breeding queen is cramped for room to lay eggs. If the case is not too bad, an empty comb inserted in the middle of the brood-nest of the breeding queen will keep her on the job of egg-laying. If the honey flow is heavy, however, this will do little if any good, for the bees at once flood the new comb with nectar. It is their natural instinct, when an abundance of nectar is in the blossoms, to gather it and let everything else go, since other things, such as brood-rearing, cell-building and propolizing, can wait until the rush of harvest is over. Now this rush of nectar into the hive is caused by the field bees; so, if we wish to stop the supply in the colony containing our breeding queen, all that is necessary is to remove the field bees from the colony. This me be easily accomplished by moving the colony out to a new location. The workers, upon returning from the field, will go back to their old location, and if a hive is close to the former position of their own home, will enter without hindrance and put the fruits of their efforts into that hive instead of the one they formerly inhabited. The hive containing the breeding queen will, in consequence, receive very little nectar for several days. In the meanwhile, young bees will emerge, thereby making room for the queen to lay, and brood-rearing can go forward apace. If the honey flow continues, it may be necessary to move the hive several times to keep the excess of nectar from coming int.

Another difficulty, caused by a heavy flow, is found in the cell-finishing colony. Wax will be built all over the cells, sometimes completely covering them. In this case, since we want all the bees possible in the finishing colony, it is not advisable to move the hive as was done with that of the breeding queen. The remedy in this case is to remove all combs as fast as they are filled with honey, being careful not to take way any brood since that would weaken it. In the place of the combs removed, give empty combs. Foundation does not answer, for they will feel crowded for storage room and continue to build comb over the cells. If the bees still build comb around the cells, it is evident that they are yet crowded and an extra story of empty combs should be placed on top. Some extra lifting is necessitated to get at the cells but the results are well worth the work it involves.

Another difficulty arising from a heavy flow is found in the nuclei. This is not so serious as the former, and the only inconvenience is that the combs get crowed with honey and bulged at the tops so that they are removed with difficulty. If only one comb with a division-board feeder is used, the bees will get crowded and go over the division-board and build comb in the empty space. These small bits of comb when removed may be thrown into the solar wax-extractor and if much honey is placed in the newly built combs, it goes well with hot biscuits and is helped along with a glass of Jersey cream! At least, that rule holds good in this locality. IF much of this honey is found, cut it out, put it into tin buckets and tell the neighbors about it. It always moves off at a good price. This condition in the nuclei is remedied by giving empty combs or foundation. I usually prefer to leave one comb with them and give them a frame containing foundation. If the flow continues, the heavy comb is taken away and another frame with foundation is given.

Cell-Building at the Close of the Honey Flow.

By far the most difficult period of queen-rearing is at the close of the honey flow. The flow has restricted brood-rearing, and the fielders have worked themselves to death, so that the colonies are losing strength and the proportion of young bees is small-two serious conditions in queen-rearing. The feeders must be brought into use on the cell builders and on the hives containing bees for the swarm box. Search must be made through other colonies to find frames containing brood, which will be scarce for the reason that all have retrenched in brood-rearing. Some colonies have more brood than others. All combs not containing brood should be taken out and replaced with brood. If this is done after an early honey flow, such as white or sweet clover, cell building will go on nearly as well as at any other time.

After a fall flow, to keep the bees interested in the queen business is much more difficult. Finishing cells in a colony made queenless may be necessary. In case a few flowers, such as asters or goldenrod that the bees may work on, still bloom the queen-rearing season may be prolonged. If brood can be secured good cells can be built, but cold weather and scarcity of drones make queen-mating difficult.

Chapter XXXVI. The Quality of Queens.

It is a very simple matter to rear queens that are just "queens"; but to rear the very best, those that are long-lived and prolific, and to do this with uniformity under changing seasons and weather, require not only skill and experience but eternal vigilance which is the price of success. Bees observe the change in the nectar secretion much more readily than the beekeeper. From all appearances the honey flow is at its height, with no indication of slacking up as far as the beekeeper can see, but the worker bees may be seen astraddle of the drones, riding them to destruction. The workers see the slackening of the flow. As the only opportunity that the drones might have had is past and they will have outlived their usefulness before any more queens are reared, the bees seem to reason, "Why keep these drones around to eat up the profits?" so the poor fellows are driven out to starve. When the apiarist sees this condition he should put on the feeders, or a poor batch of cells will be the result.

Riding them to destruction.

It is the easiest matter imaginable to rear poor queens. I have seen queens in every stage of size and quality from a worker up to the very finest. Upon several occasions I observed, in nuclei, virgin queens, if we may call them queens, that were no larger than workers. The only way I could tell that they had a touch of queen make-up in their nature was the shape of their abdomen, which was more pointed than a worker's, and the color was little different, showing more yellow. The first one of these I found was a puzzler to me. I had taken out a laying queen and introduced a cell in the regular way. When I looked three days later the cell was torn down, and then it was that I discovered this pygmy queen. I left her to see what would become of her. She disappeared about mating time, as did the few others that were discovered later. Now this pygmy queen emerged from a regular worker-cell, as no sign of a queen-cell or an enlarged worker cell could be seen.

I believe the pygmy queen was reared in this way: When the laying queen was removed the bees began to feed some of the larvae profusely with royal jelly preparatory to rearing queens. When the queen-cell was given and the virgin emerged, these larvae that were given the extra feed or royal jelly were not destroyed but were left to go on and develop as workers; but, as they had received more of the queen food than a well regulated worker should, they took on a slight character of a queen. They have just enough queen nature in them to object to the presence of a queen-cell. Now why was this queen so small? Simply from the fact that she received such a scant supply of royal jelly.

I mention this occurrence to show that all grades of queens can be reared, with no distinct line between a worker and a queen. If the larvae are not supplied with food in abundance, inferior queens result. Then there are other grades of queens, a little larger than the pygmies, which emerge from a queen-cell, but they have been skimped in their food supply. These will be missing at mating time. Some are killed by the workers, that seem to realize the queen is worthless. Queens a little larger can be reared, and the percentage that is missing at mating time is large.

Those that do mate and lay are very inferior, laying very sparingly. They are usually superseded soon; but, if it is late in the season, they die in the winter and the colony comes out in the spring with laying workers. Then there are queens a grade higher that are fair layers. Next, there are good layer, and so on up the line until we get to the very best, that will keep a large brood-chamber packed full of bees and produce a colony that makes several hundred pounds of surplus honey above the average. These are the queens that bring you the profit and the ones you can raise if you play the game according to the rules laid down by the bees.

Chapter XXXVII. Drones.

All recognize the fact that the drone has as much to contribute toward keeping up the stock as the queen. Placing drone combs in colonies containing the best queens, has been advocated to rear drones. If this is done in a large way and drone comb given to a large number of colonies, good results may be obtained. I believe a much better plan, however, is to requeen the entire yard systematically and restrict the drones as much as possible by using full sheets of foundation. Enough drones are reared, and no colony will be injured, as would be the case if a lot of drone combs were allowed in a number of colonies.

Of course, if you are able to keep all drone comb out of the colonies some provision would have to be made for rearing drones, but I have net to see a colony that could not rear a few. Many say that, if they do have all pure drones, it would do little good as their neighbors have hybrids and blacks. There are many times when this would prevent pure mating of queens, but in many cases it would not. We are finding out that queens do not go so far to mate as was formerly supposed. A number have reported seeing queens mate within a very few feet of their hives. I have witnessed the same thing. If there are no bees nearer than a quarter of a mile, and if you have all pure drones in your own yard, my experience would prompt me to say that you will have very little mismating.

As is the case of American foul brood, a lot is laid to the neighbors that rightfully belongs to yourself. I have had several amusing experiences concerning the above. One man said he could not get rid of American foul brood because his bees got it from his neighbor a couple miles away; and when I called on that neighbor, he told me the same thing about the beekeeper to whom I first talked. IT was evident that both were spreading infection among their own bees. So it is in the case of drones; see that all drones in your own yard are pure and you will be surprised at the few mismatings which will occur. Of course, it is still more desirable if you can get the co-operation of your neighbor and interest him in better bees. Get him to requeen by rearing his own, or you can rear them and sell to him.

Normally, bees permit but one queen in a hive. Frequently, however, during supersedure the old queen remains for some time after her daughter is mated and laying. In such cases the old queen is so incapacitated that the bees and young queen do not seem to recognize her as a queen, and pay no attention to her. These old queens may be dropped into any queenless colony and are generally accepted. They may be utilized in this manner to carry a colony along until it can be requeened with a good queen.

Sometimes a freak case seems to violate all general rules. I have known two virgins to emerge in a hive and be the best of friends. They would mate and lay for quite a while; but sooner or later the bees decide this state of affairs is irregular and kill one of them. Sometimes two queens get into the same hive upon returning from their mating flight. They get along together for a while; but after a time the bees decide they cannot serve two masters. Once faction balls one queen, the other faction the other; thus both queens are so badly crippled that they have to be removed. I remember once when one queen had both wings nipped off close to her body and the other queen lost both legs on the same side. The bees were satisfied then and seemed to think that these two queens, since they both had been trimmed, were about equal to one good one.

A number of years ago there was considerable talk about the possibility of having several queens in one hive, and quite a good deal of experimenting along that line was done by many beekeepers. If it were possible to winter over fifty or one hundred queens in one hive and then have them to supply the demand for queens early the next spring, why, it was worth working for. Personally I was much interested. I used up all the old queens I had left from requeening and sacrificed several dozen good young ones. I learned some very interesting things but little of economic importance. I learned it was not difficult to introduce queens to each other, so they would be friends, yes, regular old cronies, always working together, and usually found on the same comb. The only discovery of importance was the fact that it is the *bees* that make plural queens in a hive impossible.

True, the queens usually fight, but they can be introduced to each other. However, the bees will swear allegiance to only one queen and declare war on all others. Once I put twelve queens in a mason jar to see what would happen. They began to fight, so I shook them around until they were tired. Five had been killed. The other seven called an armistice and apparently signed a treaty that was satisfactory to all. They probably laid the trouble to the dead ones entirely. They held out their tongues to each other and always clustered together. Now, I thought, my troubles were over, for I knew I could safely introduce them. I took six frames of emerging brood, brushed off all of the bees, put the frames into a hive and turned loose the seven queens. They staid together on one frame of brood, and as they emerged, they all began to lay. They certainly did furnish a generous supply of eggs, placing many in each cell. Now I reasons that, as the young bees emerged, they would never know how many queens they were supposed to have and would accept the seven all right.

Things went well for about two weeks, when this small colony began to assume the proportions of a real one. Then trouble developed very rapidly. Having arrived at the age of accountability, the bees decided there were too many queens. This multiplicity of stepmothers, was intolerable, so the killed one. In a few days they killed another, then another. I carefully examined them and found them balling another. For nearly a month this weeding process to get down to one queen continued, but they finally did it and saved the youngest and best.

In concluding this subject, I wish to give as my opinion that it is indeed useless to try to get the bees to tolerate more than one queen for any time long enough to do any real good to the colony. The above is given that it may save others costly experimenting. Of course, if one wishes to experiment merely for the fun of the thing, that is a different story.

In one experiment the stings of two virgins were clipped. These virgins could not fight but mercy, how they did wrestle! The bees stood aside merely interested bystanders until one of the virgins began to squeal, when the workers closed in and balled both queens and injured them so it was necessary to replace them.

As stated an old queen is readily accepted. Usually when a queen gets old and lays but few eggs, she is killed by the queen that supersedes her or dies a natural death. I observed one notable exception, however. I requeened a colony yearly for two years and then found an old queen in the hive that had been there for four years. I could tell her by the manner in which her wings were clipped. She had been in the colony, and a young queen was reared to supersede her. This queen was removed when one year old; and another a year later. As far as I could observe, this old queen had ceased laying altogether and was treated with absolute indifference by both bees and queens.

Chapter XXXIX. When to Requeen.

The question is asked many times, "When is the best time to requeen?" The answer depends upon the strength of the colony and upon the location. From the latitude of St. Louis south, when the colony is strong, late requeening is desirable. In fact, the later the better. In such a case no benefit will result to the colony requeened that season, for the queen would lay little if any before spring. Being strong, the colony would have plenty of bees to come through the winter in good condition. The queen would begin laying early in the spring and have the colony on the job for the first honey flow. If the colony is of medium strength requeening is preferable about August first, for then there is time to build up and go into winter quarters with young bees.

Further north, requeening should be done in August, for, on account of the long winter, it is desirable to have as many young bees as possible to carry the colony safely through, as young bees not only live longer but stand confinement better than old ones. In the north, if the colony is not strong it should be requeened in June or July, depending on its strength. In case one wishes to build up a weak colony for a fall flow, the earlier it is requeened in the season the better. Some of the colonies that have done the best work for me in a honey flow from sweet clover in June were requeened in November of the previous year. They were strong when they were re-queened.

In many parts of California, the colonies run down near the close of the season, winter poorly and build up on the first honey flow in the spring, thus failing to get a surplus from that flow. In such cases if the colony were requeened in August and given plenty of stores, the queen would build up the colony to good strength for winter. She would take a rest until January or February and then build the colony up strong, ready to give a sur-plus from the first flow.

Now from the fact that it has been advised to re-queen early in the season, many get the impression that it is not a good plan to requeen late in the season, even if the colony has a poor queen. This is not correct, for a good young queen is always better than an old one. Therefore, late in the season, if you discover any colonies that have inferior queens, by all means replace them with younger ones if they can be procured. If the colony is weak, it is probably better to kill the queen and unite it with another. If the colony is of good strength and a queen is introduced in October or November, she would be in the very best condition to build up that colony to great strength early the following spring.

How Often to Requeen.

A great difference of opinion exists as to how often to requeen. Some say requeen every year; some say, every second year; and some say, requeen only the colo-nies that have poor queens; while others say, let the bees do it themselves. Circumstances have to determine; but it is safe to say that there should be vastly more requeening than is practiced at present. I believe in most parts of California the honey crop could be doubled if each colony would be requeened every year with good Italian queens reared by the beekeeper himself, especially if a large brood-nest full of stores should be provided.

In many other parts of the United States, the same or similar conditions prevail. In the far north where the queen is idle such a long periods in the winter, possibly it does as well to requeen every two years as to do so every year where the honey season is longer and the queen kept on the job ten months. In this locality I have found it profitable to requeen every year, with an exceptional case where queens of unusual qualities are found which are desired for breeders. I believe a good rule that applies to all localities is to go over all colonies each year and replace every queen that is not the very best. In case a queen has a big brood-nest packed with brood, is large and has all indications of being a splendid, prolific queen, I would leave her another year. One can usually tell by the size and general appearance of a queen whether she is to be trusted to do the best work for another year or not. If there is any doubt about it, err on the safe side and requeen. Another point, that should not be lost sight of in requeening, is that one can continually improve his bees by breeding from the best. W should also keep in mind, if we requeen every year and keep all colonies strong, we shall not have to "cure" European foul brood, because we will prevent it.

Some have advocated that the best breeder is one that lives the longest. Now that depends on the mount of work she has done. I believe the best beekeeper is the man who can get the most eggs out of a queen in the shortest time. If managed properly, the queen should lay the bulk of her eggs the first year. In the north exceptional queens do well the second year; but, if a queen does even fair work the third year, it is evident she was not worked as she should have been during the two years previous.

Sometimes beekeepers believe they have discovered an exceptional queen from the fact that she lived four or five years. Possibly the queen *was* a good one; but I have some doubts about the beekeeper himself, for, if he had given her a chance, she could have done her work in one or two years instead of loafing along for four or five.

The question is frequently asked, "Does commercial queen-rearing pay?" So much depends on the person and locality that it is difficult to answer this question in a satisfactory manner. If one is in a reasonably good honey producing location, he can make much more money at honey production. Much more skill and experience are required to make a success of queen-rearing than to make a success of honey production. However, as we believe every honey producer should learn the queen-rearing business, some will prefer queen-rearing.

The main requirement for a successful commercial queen-breeder is a love for the business. If he has that, no obstacle is too great for him to overcome. The many difficulties in queen-rearing on account of failures in the honey flow and unfavorable weather conditions will cause all but the stoutest hearts to throw up the business in disgust. However, some prefer to engage in the queen business for the enjoyment and satisfaction they get out of it, even if the financial returns are not so great as in honey production. To attend to ordinary queen trade requires the handling of many details, much time must be given to bookkeeping and correspondence; therefore, additional help is required in these lines. On the other hand, the demand for queens of high quality is great and a good queen-breeder should experience no trouble in disposing of his output.

Queen-rearing involves no heavy work, but it is an exceedingly busy job while the season lasts. It is an ideal occupation for women or men who cannot do heavy work, but who are willing to be on the job, early and late during the queen-rearing season. It is not a "get rich quick" occupation and takes a number of years of careful study and experimenting before one can get things lined up and moving smoothly. Considerable expense is connected with

it for office help, shipping cages and other supplies, sta-
tionary, and for sugar with which to feed the bees in cell-
building and swarm-box colonies. The honey crop largely
has to be sacrificed, for, as one uses brood to form nuclei,
the colonies, thus weakened by the loss of such brood will
do well if they build up strong and make enough honey to
carry them over to the next season. This holds true in a
season when the honey flow is good. In a poor season
large quantities of sugar must be purchased to build up
the colonies and provide winter stores.

Locality plays an important part. Many localities
similar to ours have a succession of light honey flows
which are suitable to queen-rearing, but not heavy
enough to give a surplus. For instance, our season opens
with peach bloom followed by pear, apple, locust and a
little tulip tree and tupelo. Then come white clover and
alsike, which are very uncertain, and are frequently mixed
with a vile honeydew that ruins the honey for market.
Sweet clover is next, followed by a short dearth. Sometimes a little honey comes from unknown sources, probably from throroughwort, figwort or ironweed. Then a little
climbing milkweed or blue vine is followed by the heaviest
flow we have, heartsease. This flow rapidly tapers off into
aster and goldenrod. If one is in a locality where there is
but one heavy flow he will do much better at honey production, since queen-rearing during a dearth of pasture
for any length of time is both difficult and expensive.

Personally, queen rearing is so fascinating and en-
joyable that I would prefer to rear queens, even if the
financial returns were but one-half as much as from honey
production; for, after all, it is the pleasure we get out of
life, not the money, that counts. We should follow that
occupation which most enjoy.

After all, what is money for? To purchase enjoyment
in one form or another. Therefore, if we are getting the

enjoyment from our occupation direct, it is the same as money and we save the middleman's profit.

To the commercial honey producer, I can truthfully say I believe there is a bright future for him. More and more we see health experts call attention to the value of honey as a wholesome hygienic food for young and old, for the ill and well. Let us all co-operate to get honey into more homes as a regular diet, at a reasonable price. Especially should we practice better methods of honey production. First, increase our output per colony, and then increase the number of our colonies. With these improvements, there is no reason why beekeeping should not be profitable and, as Dr. Miller used to say, "Just think of the fun I have had."

About the Author

Jay Smith was an active writer in the bee journals during his day and wrote two of the most loved and used of the queen rearing books. Both are still being followed by many today. He is most famous for this book, *Queen Rearing Simplified*, but *Better Queens* is the culmination of his work in queen rearing.

Better Queens

by Jay Smith

Better Queens

Originally printed 1949 by Jay Smith

Reprint 2011 by X-Star Publishing Company

Transcriber's preface

I wanted this book available because I think Jay Smith was one of the great beekeepers and queen breeders of all time. There are many queen breeding books by scientists or small-scale breeders, but this is by a beekeeper who raised thousands of queens every year. I think that is much more applicable to practical queen rearing. It is also a method that does not require grafting, good for those of us who can't see well enough to graft, and does not require the purchase of special equipment, good for those of us lacking in the funds to buy one of the graftless systems on the market.

I was going to do *Queen Rearing Simplified* first, but it occurred to me that from Mr. Smith's point of view this book, not that one, is the culmination of his work on rearing the best possible queens.

If you wish to make comments please send them to bees@bushfarms.com

_____

X-Star Publishing Company

Founded 1961

Dedication

I dedicate this book to my wife, Ruth Reno Smith, who has faithfully stood by me through thick and thin and by her help and encouragement has made this book possible.

Acknowledgements

It would be difficult to give credit to all who have helped me in the study of bees, but I must mention a few of those glorious saints who have gone on before: W.Z. Hutchinson, Dr. C.C. Miller, Geo. S. Demuth, G.M. Doolittle, Henry Alley, C.P. Dadant, Geo. W. York; dear old Prof. L.M. Kelchner, my art teacher of more than fifty years ago, who made the cover design of this book, and Dan Urbaugh, who induced me to go into commercial queen business.

Among the Living, I wish to thank my friend and neighbor George E. Judd, president of the firm of Judd & Derweiler, printers of National Geographic Magazine, for the excellent job of printing this book and to thank Mrs. Thomas Linger, the artist who made the cartoons.

Foreword

A philosopher has written, "Without regard to friends or foes, I sketch your world exactly as it goes." Now to sketch anybody's world and do an exact job at it is more than I wish to tackle. Nevertheless, I like the sentiment of the above lines so I shall adopt them in a revised form as follows: Without regard to friends or foes I shall attempt to sketch your world approximately as I think it goes. In other words, I shall go right ahead eating my grapefruit, let the juice squirt where it may. Moreover, I am writing this in my own crude way. Should I attempt to write in an approved rhetorical manner it would seriously cramp my style.

Moreover, I am writing this in the first person for I am telling just what I do with the bees. It is easier to say "I do so and so" than to say "It should be done so and so." I have greatly enjoyed reading the writings of Dr.

C.C. Miller for he tells just what he did. In the introduction to his book *Forty Years Among the Bees* he writes, "Indeed I shall claim the privilege of putting in the pronoun of the first person as often as I please and if the printer runs out of big I's he can put in little i's." Incidentally, I wish to acknowledge the encouragement I received in my early beekeeping days from Dr. Miller who showed interest in my writings in the bee journals, as his card shows.

Marengo, Ill. 3/6/19

My good friend,

The thing I don't like about you is your not writing more for the bee journals. You're one of the choice group whose writings I especially enjoy.

Blessings on you.

C. C. Miller

I have always preferred to read authors who tell *how they do things* rather than those who tell *how things should be done*. Often there is a great difference.

A fine looking lot of cells, don't you think? Twelve in number. Count them. What! Only eleven you say? I will bet you forgot to count the dam-sel, Mrs. Huber F. Smith.

Table of Contents

My First Recollection of Bees

My interest in bees dates back many years. I was born on a farm a mile southwest of Tampico, Illinois, October 13th, 1871, and one of my first recollections of that early life was watching my father in his shop making beehives. These hives were about a foot square and two feet high. The supers were mere boxes with glass on one side. When these supers were full they were taken off and honey, super and all were sold to the grocer.

I believe the shape of these hives was better than the shape of the hives in use today. In those hives the bees could better regulate the temperature, build up quickly and fill the supers. I believe my father made more net profit from his bees than most beekeepers make today, for those hives were intensely practical. He made the hives out of cheap material on rainy days. Of course nowadays, with foul brood prevalent, such hives could not be used.

When I was about three years old I remember king birds or bee birds, as they were called, used to catch my father's bees. Now he was a crack shot with the rifle, shotgun or revolver. I remember one day a bee bird was in the top of an evergreen tree catching bees. My father dropped the bird at the first shot using a .22 caliber Smith and Wesson revolver, the distance being about twenty-five yards. The bird's crop contained nineteen bees as I remember.

Going West

In the spring of 1883 we moved to Dakota Territory. All my father's bees were sold at auction. We located near Sand Lake, some twenty-five miles northeast of Aberdeen. We took no bees with us but my uncle Perry Badgley took a number of colonies in box hives as he said he just could not get along without his bees. His bees all starved to death the coming summer as there was absolutely nothing for them to gather. Little did we dream that with the coming of sweet clover the Dakotas would be among the leading states in the production of honey.

Getting the Bee Fever

While quite a small boy in Dakota I read a book on farm topics, and among them was an article on bees. As I remember the article it was not very authentic in its statements, but not knowing the difference I was fascinated with it and resolved that when I grew up I would have some bees. This idea never left me and after my marriage in 1899 and our removal to Vincennes, Indiana, I procured my first bees.

At that time I was running a private Business College, and as part payment for tuition, one of my pupils let me have two colonies of bees. They were in old-fashioned Langstroth hives with porticoes in front. That was in the spring of 1900. Eagerly I watched them all summer expecting them to produce a lot of honey. I had dreams of hot cakes and *my* honey. In the fall I cautiously peeked into the super after dark expecting to find at least enough fine comb honey to last us all winter. Alas, there was nary a drop!

Next spring there was but one colony alive and I began to realize there was something about bees I did not know, so with characteristic stubbornness I set about to learn something about them. I resolved that those bees should make me some honey or furnish a reasonable excuse.

My First Swarm

This one colony built up rapidly and swarmed. It was Sunday and my wife was planning to go to church and had requested the pleasure of my company. I had shaved and was hoping something might happen to side-track the church-going when something did. Out came that swarm! Of course it had to be taken care of, so I got out of the church-going. The swarm clustered on a young peach tree and the tree was ruined.

Fortunately we had a little 3 ½ x 3 ½ Kodak and my wife took the picture shown.

My first swarm

Beginning the Study of Bees

I then began in earnest to study bees. I got a hundred Danzenbaker hives. This hive had close fitting frames and violated everything Langstroth taught, or rather tried to teach. When those end bars got glued together they were solid as a rock. I used a wrecking bar as a hive tool. Years later that wrecking bar was stolen. I suspected a neighbor of stealing it. He had just bought some modern hives with long top bars! I put about a thousand dollars into those hives which later I found to be impractical. I burned most of them up. True, the manufacturer later apologized for inflicting such a monstrosity on the public, so we got apologies-but no reparations!

In spite of these financial losses, poor seasons and foul brood, I managed to increase the number of my colonies. As I grew in experience I learned to go more to the bee for my information. I learned that should I ask a dozen men a question I might get a dozen different answers, while if I asked the bees a question I got just one answer, and that the correct one. No, they never gave me a five to four Supreme Court decision but their decision is always unanimous.

For Christmas 1902 my wife presented me with the bee book Langstroth on the Honey Bee, Revised by Dadant. This book set me on the right track and I still think it the best book ever published on bees.

Go to the Bee Thou Sluggard

Solomon said, "Go to the ant thou sluggard. Consider her ways and be wise." I have often wished that Solomon had been a beekeeper instead of Samson being one for if he had been he never would have gone to the ant unless he carried a can of cyanogas. Nevertheless Solomon noticed something that many of us miss, and that is that insects have an instinct we might call an automatic intelligence which often is greater than anything in the head of man. The bees used the principle of evaporation for cooling the same as is used in our electric refrigerators and they used it probably long before man had learned to hang from a limb by his tail! However, Solomon rather squared himself with us beekeepers when he invited us to eat honey and acknowledged it to be good. So I tried to study bees themselves and learn their ways. I learned I could not force them to go my way but that I must go their way as far as is practical.

Man is handicapped in that he must go to school and study in order to learn anything, while the bee is born with its college diploma in its hand. The bee's education is complete. They agree on all points while if you can find two men who agree on anything you are to be congratulated. The bees know all they need to know and you cannot teach them any more. I learned I must be the pupil and the bee the teacher.

I made a careful study of queen rearing and bought breeding queens from Doolittle and Alley and also imported some Italian queens. One queen I received from Mr. Doolittle was a wonder. Her bees were of a beautiful yellow color, very gentle and splendid workers. So pleased with her was I that I used her for the foundation of our stock. This decision I have never regretted.

Beginning Commercial Queen Rearing

One day in 1912 State bee inspector, Dan Urbaugh, came to inspect my bees. I remember he stood for some time in silence watching the flight of the bees. Then he turned to me and said, "Why don't you sell some queens?" Then he said, "Those are the finest bees I ever saw." I asked him how he could judge bees by merely looking at them. He replied, "I can tell by the looks of them and by the way they act. Here I am right among them and none offers to sting." Having had little experience with bees except my own I took it for granted all bees were like that. I told Dan that I would greatly enjoy rearing queens if I could sell them. I shall never forget how he looked at me and replied, "Sell them, why of course you can sell queens like these and you will not have to do much advertising either." I rather had my doubts but thought I would try it. I made some mating hives of various models and when I had a dozen queens on hand I wrote Mr. Urbaugh I had some queens ready. I made the remark to my wife that maybe Mr. Urbaugh was just talking to please me and really did not mean what he said. In a few days I received a check from him for six queens. I said to my wife, "I thought he was just kidding, but that check doesn't lie."

Thus at Mr. Urbaugh's request I became a full-fledged commercial queen breeder. I could have gone into some other business and no doubt made more money, but doubt if I would have gotten so much out of life. I have given all there is in me to the thought of better bees and especially better methods of producing them. During the years that have passed since 1900 I have spent many thousands of dollars in experimenting. When I found a new mating hive that suited me better than the ones I was using I would burn the old ones in the furnace. As Josh Billings put it, "Experience is a gude teacher but the tooishun kums purty hi."

Short Course in California

One of the outstanding experiences which helped and encouraged me in my work was during World War I. I worked as extension agent under Dr. E.F. Phillips who was then in charge of the bee work at Washington. Dr. Phillips, Geo. S. Demuth, Frank Pellett, M.H. Mendleson, Dr. A.P. Sturdevant and myself conducted a short course in bee-keeping covering most of the state of California. The flu was rampant and we wore flu masks. We also were re-quested to sneeze in our handkerchiefs in order not to spread the germs. However, I learned it took a profes-sional bad man to be quick enough on the draw to get the drop on a sneeze after he felt one coming on! I have often said I do not know whether or not I taught the California beekeepers anything worthwhile, but they taught me a great deal. One thing stands out in my mind most vividly and that is what a splendid lot of men and women it is who keep bees. I doubt very much if any other industry can show a better and more honorable set of people.

Queen Rearing Simplified

In 1923 my very good friend Geo. S. Demuth, asked me to write a book on queen rearing and *Queen Rearing Simplified* was the result. It was the best that I knew at the time but with the passing of more than twenty-five years we have made so many radical changes for the better when quality queens are desired that now we use practically nothing described in its pages. However, it has given me much satisfaction to receive letters from bee-keepers all over the world thanking me for the help given in that book. It is my belief that *Better Queens* will receive a greater degree of appreciation. If this is true I shall feel amply paid for all the effort and expense I have been to in preparing this volume. With the passing of the years there may be minor changes in our system of rearing queens direct from the egg but the main feature will always remain if quality of queens is desired. With our present system all cells are as large and as well supplied with bee milk as are the cells produced by the bees during swarming or supersedure in nature, and let no one tell you he can beat nature in rearing queens. All we claim is that we can equal nature and that is enough.

(Link to Queen Rearing Simplified:
http://www.bushfarms.com/beesqueenrearingsimpli fied.htm)

Shall the Beekeeper Rear His Own Queens?

My answer to that is emphatically *yes*, providing he can rear *the best of queens*. Maybe some member of the family, a boy or girl or yes, maybe Mother, may rear the queens and find it not only the most profitable work about the beeyard but they will get a world of pleasure in doing it. Many will make a larger profit from half the number of colonies if they rear their own queens. Therefore, instead of being more work to rear their own queens it really would be less work for the time saved in caring for double the number of colonies not to mention the larger amount of money invested. Let me give just one instance. In California I had the privilege of observing the systems of two beekeepers who worked in opposite lines. One had two thousand colonies. When a colony died from becoming queenless the beekeeper divided another colony. In one division was the queen. In the other, the bees were expected to rear a queen. Sometimes they did, but having only old combs the resultant queens were not the best. He had to hire a large crew of men to take care of all these colonies to see that they did not die out. In addition, the interest on the money invested in that large equipment was considerable and if you consider the depreciation, that was an additional expense.

In contrast to this I met a man in Ventura County who had but 250 colonies. This man and his son reared their own queens and requeened every year. He told me that year their colonies averaged 200 lbs. per colony and they sold the honey wholesale for 22¢ per pound. It was plain this man and his son made more clear money from the 250 colonies than the man who operated the two thousand.

How Bees Rear Queens in Nature

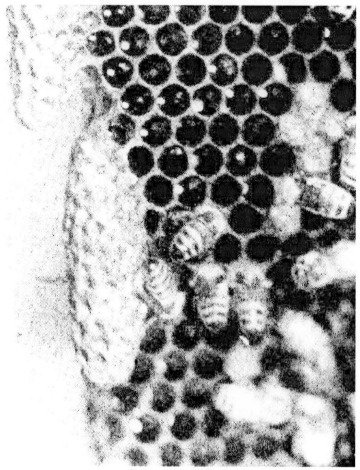

Swarm cells

As stated, let us now lay aside all man-made theories and go straight to the bees and see how they do it, for we must remember that if we are to rear the best of queens we must follow the bees in their work and duplicate their work insofar as it is practical. We must remember the proven fact that Mother Nature knows how to rear her own while man may use cheaper methods he never can improve on nature when *quality* of queens is considered.

When bees rear queens they do it in one of three ways-first, when preparing to swarm; second when they wish to supersede a failing queen; and third, when a colony becomes queenless for some reason such as an accident to the queen. In such a case they rear a queen by what we may term the "emergency" method.

Queens Reared During Swarming

When there is an abundance of nectar and pollen coming in from the fields and the colony is built up to good strength with plenty of brood in all stages in the hive and the bees become crowded, they may decide to swarm especially if the queen be an old one. As all colonies do not act alike some may swarm while others will not. The first step in preparing to swarm is the building of queen cells. The bees prefer to build the cells at the bottom of the combs or at the sides if there is room. These cells are pointed downward.

After the cells are nicely started the queen lays an egg in each. In about three days the eggs hatch and the larvae are surrounded with an abundance of bee milk. Sometimes the milk is placed in the cell before the egg hatches. The larva grows at an amazing rate. It has been stated that if a human baby grew as fast as a queen larva and weighed ten pounds at birth, in five and one half days it would *weigh twelve and a half tons*!

By the time the cells are sealed they contain nearly half an inch of dried bee milk. Now please consider the fact that if we wish to rear as good queens as the bees do when preparing to swarm we must duplicate their performance as nearly as possible.

In nature, when swarming, the bees seldom need more than one cell and never more than three or four, but they often build from one dozen to two dozen. Why this extravagance in cell-building I do not know but possibly the bees are secreting so much milk in their glands they want to get rid of it. This desire to get rid of this over-supply of milk is probably one of the causes of swarming. I have often prevented swarming by adding unsealed brood from other colonies. This gave the bees so much larvae to feed that they found an outlet for the excess of bee milk and the colony tore down the queen cells it had started.

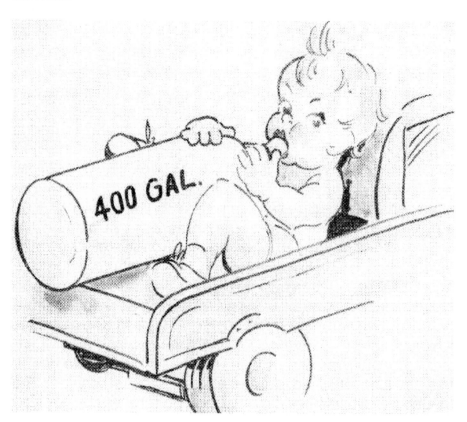

Queens Reared During Supersedure

In case the queen is getting old or, as is all too often the case when the grafting method is used, and the larva is underdeveloped, the bees may ask the queen to retire as they want to put a young vigorous queen in her place. In doing this the procedure is the same as in swarming but not so many cells are built. Usually from three to five. Conditions and locality may cause a variation in this matter.

The first virgin out will tear down the remaining cells unless the colony decides to put out an after-swarm. In due time the virgin flies out and mates with a drone and in two days she begins to lay. Sometimes either the bees or the new queen kill the old queen. At other times it appears that the old queen is sensitive or proud and thinks she is not properly appreciated and she with a few of her close friends simply up and pull out. This is called a supersedure swarm. Sometimes this swarm doubles up with another. I have known this to happen and the old queen was accepted and both queens carried on for some time. These old queens are readily accepted by the bees. I have dropped an old queen into a colony of laying workers where she was accepted and in time she reformed the laying workers and later the queen was superseded. The length of the life of a queen depends largely on the number of eggs she lays. In the North where the queen has a long rest period in winter she will live longer than she would here in Florida where she lays the year round. We find that for best results here in Florida the colony must be requeened every year. In the North if the queen is doing good work the second year she might be allowed to live the third year but as a rule two years is long enough to keep any queen. Any queen not doing good work should be removed regardless of her age.

Queen Alice

In Indiana we had a queen we named Alice which lived to the ripe old age of eight years and two months and did excellent work in her seventh year. There can be no doubt about the authenticity of this statement. We sold her to John Chapel of Oakland City, Indiana, and she was the only queen in his yard with wings clipped. This, however is a rare exception. At the time I was experimenting with artificial combs with wooden cells in which the queen laid.

Queens Reared by the Emergency Method

In case the bees are left alone and not interfered with by man, probably it is very seldom that this method is ever used. Where man interferes and opens the hive frequently the queen may be killed. In such a case, provided there are larvae of the proper age, the bees build queen cells over some of the larvae in the worker cells, such larvae originally intended for workers. It has often been observed that many of the queens reared by this method are not the best. It has been stated by a number of beekeepers who should know better (including myself) that the bees are in such a hurry to rear a queen that they choose larvae too old for best results. Later observation has shown the fallacy of this statement and has convinced me *that bees do the very best that can be done under existing circumstances.*

The inferior queens caused by using the emergency method is because the bees cannot tear down the tough cells in the old combs lined with cocoons. The result is that the bees fill the worker cells with bee milk floating the larvae out the opening of the cells, then they build a little queen cell pointing downward. The larvae cannot eat the bee milk back in the bottom of the cells with the result that they are not well fed. However, if the colony is strong in bees, are well fed and have new combs, they can rear the best of queens. And please note-- they will never make such a blunder as choosing larvae too old.

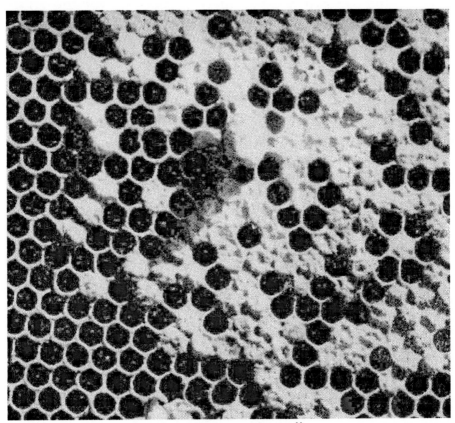

An emergency cell

Dr. Miller's Method

In reviewing the past it is interesting to note how many really good beekeepers discarded the grafting method after giving it a trial. A notable example is Dr. Miller. His method will produce just as good queens as can be produced. He gave a piece of new comb to the colony containing his breeding queen. As soon as the eggs hatched he trimmed the comb back to the larvae and gave it to a strong colony made queenless and broodless. The bees could easily tear down the new combs and build queen cells. For the person wishing to rear just a few queens his system is good. It would not do for the commercial queen breeder as too many cells are built together and it is difficult to cut out the cells without injuring them.

Another disadvantage to his system for cell building on a large scale is that the bees do not start as many cells with his method as they do with our system. After all it is little more work to mount the cells on bars as we do and we get equally good cells and more of them. Now please remember the *quantity* of bee milk the growing larvae receives is just as important as *quality* for this will be mentioned frequently in the following pages.

(Link to the Miller method: http://www.bushfarms.com/beesmillermethod.htm)

Artificial Queen Rearing-Its Development

It is not clear who first practiced grafting, or transferring larvae from certain cells to other cells. As far back as 1791, Francis Huber, the naturalist wrote: "In a hive deprived of its queen I caused to be placed some pieces of comb containing eggs and already hatched worms of the same kind. The same day the bees enlarged some of the worm cells and changed them into royal cells giving to them a thick bed of jelly. We then removed five of the worms placed in these cells and Burnens substituted for them five worker worms which we had witnessed hatching from the eggs forty-eight hours previously. Our bees did not appear to be aware of the exchange." (From *Huber's Observation on Bees* translated from the French by C.P. Dadant.) Therefore, the grafting method is more than 150 years old. Since that day many have placed larvae from superior queens into the cells of inferior queens.

(Link to Huber's book:
http://www.bushfarms.com/huber.htm)

G.M. Doolittle

The greatest impetus to the grafting method was given by G.M. Doolittle sometime in the 80's when he devised the method of dipping queen cells. Before that time the queen breeder had to depend on the few cells he could find in his hives. Some took soft wax and molded it around a stick making a crude cell, and these were offered for sale at a penny apiece. Mr. Doolittle, in his excellent book *Scientific queen-rearing*, states it came to him, "Why not dip the cells the same as my mother used to dip candles?" Soon after that Henry Alley brought out his system of rearing queens direct from the egg, the system on which ours is founded.

(Link to G.M. Doolittle's book:
http://www.bushfarms.com/beesdoolittle.htm)

The Grafting Method

The object of *Better Queens* is to be helpful to all who rear queens and not to criticize those who use the grafting system. After all, I am criticizing the method I taught in *Queen Rearing Simplified*, so it is perfectly legitimate to criticize oneself! Many who now are using the grafting system and who want to rear better queens will want the two systems compared. As most beekeepers know, by the grafting method we mean the method in which the larva from a worker cell is transferred to an artificial queen cell. We used that system for 33 years.

Not one of those years did we get the fine large cells which are necessary to produce full developed queens throughout the whole season. We found that when there was a light honey flow with plenty of pollen coming in, and if we kept the cell builders up to great strength, we could get a very high percentage of good queens. Even at its best we had to cull cells and virgins and frequently to discard laying queens that were not fully developed. Even then a few inferior queens would get by us which we had to replace. This never happens with our present system. We never have thrown away a cell for being too small, for all are alike. With the present system we have yet to see an undersized virgin. When using the grafting system, when there was no flow, it was well-nigh impossible to get good cells even though we fed sugar by the ton. Not one of those 33 years passed in which I did not long for a system with which I could produce those fine large cells which I had observed in colonies preparing to swarm, a system by which I could produce cells in quantities throughout the entire season.

The Desire for Better Cells Intensified

The wish for better cells was greatly enhanced while we were having a beekeeper's picnic at our place in Vincennes. I was asked to demonstrate queen rearing while using the grafting method and naturally wanted to make as good a showing as possible. As was our custom, I had built up several colonies to great strength by adding frames of brood and bees from other colonies. One colony of great strength swarmed just as the picnickers arrived. Our State Inspector, Yost, weighed the swarm and reported it weighed exactly 25 pounds. It looked as though as many more bees remained in the hive so they estimated that colony must contain at least *two hundred and fifty thousand bees* - many more of course, than could be the product of one queen. I was proud of the cells the colony produced, and for grafted cells they were fine ones. It so happened that this colony had built some swarming cells in the brood nest below the excluder. It has been stated that "Pride cometh before the fall" and I will say that my pride took an awful rumble when I saw those enormous swarming cells. I realized I did not know the first principle about raising really good cells. *Their cells were nearly twice as long as mine and had at least three times as much dried milk in them as did mine.* I realized as never before the shortcomings of the grafting system if *quality* of queens is desired. However with the inborn stubbornness for which I am noted, I resolved never to quit till I had a system in which I could duplicate the performance of the bees in building cells while swarming.

Fifteen Years of More Work

It took fifteen years of study and experimenting before I got the system perfected with which I can duplicate the work of bees building cells while swarming. At last we have it and we have made very few changes in it for several years, and I doubt if we ever will make any important change. At present, most queen breeders are using the system I taught in *Queen Rearing Simplified*. I feel confident most of them will adopt the new system we are teaching in *Better Queens*, for buyers of queens and package bees are beginning to demand better queens - queens that will not be superseded almost as soon as they begin to lay, thereby losing a crop of honey.

What the *American Bee Journal* Has to Say About Poor Queens

In October 1947 number of the *American Bee Journal* appeared this article:

"This is the season of disastrous queen failures. From every direction come reports of supersedure and queenlessness on an unprecedented scale. With an abundance of clover the bees have failed at a very critical time. The failure of the queens results in heavy loss to the colony. When egg laying ceases for ten days to two weeks in the spring much of the productive force is lost. Losses have been especially heavy with package bees. The cold and backward spring was unfavorable for replacement and in too many cases the colony has been lost entirely. Most certainly we need to know more about the reason for this serious condition which has cost the beekeepers a substantial portion of the possible 1947 Harvest."

Yes, We Need to Know More

So let me tell you. It is safe to say that 95% of the cases of premature supersedure are caused by *improper feeding of the young queen larva*. The grafting system simply will not produce the best of queens. A package shipper who uses queens reared direct from the egg reported to me that he did not have a single complaint about premature supersedure but, on the contrary, had scores of letters praising the performance of the queens and the splendid work the package bees did. Queens must be reared direct from the egg and the finishing colony must be fed *honey* for best results.

Experiments

Let us look at a few of the experiments I have carried on, many of which now look foolish. Mr. Eugene Pratt wrote a pamphlet entitled "Forcing the Breeding Queen to Lay in Artificial Queen Cells." Had I understood bee nature then as I do now I would never have tried it, for when he said "forcing" that would end it right there. I got the outfit and tried it. The wooden cups were corded up into a supposed comb and the queen was supposed to lay in it, and she did - about a dozen eggs then she went on a slow-down strike, then a sit-down strike, and finally a walk-out. The few cells I did get were very good but still were not flooded with the bee milk as I wished. So once again I sorrowfully returned to the grafting method.

Next I gave the bees only drone comb, as it is known the queen will lay worker eggs in drone cells if there is no worker comb present. The results were not quite so bad, still I could not get the bee milk into those cells in the quantities I knew were necessary to produce the best of queens. Then I had some wooden cells made smaller in size and square on the outside so they would cord up more like cells in a worker comb. It looked much like a comb of worker cells to me. Evidently it did not look that way to the bees or the queen for, after laying a few eggs, she went A. W. O. L. and had to be repatriated. The cells and the resultant queens were nothing to be proud of. I had not yet learned that *quantity* of bee milk was just as important as *quality*, and I just could not get the quantity of the bee milk into those cells. The reason was that it takes a large number of bees to provide the milk in abundance and it was not practical to use a large number in such experiments. Once more I reverted to type and went back to grafting.

The Alley System

Some may ask why I did not use the system used by Henry Alley in which the queen laid in the worker cells. I had tried his system and failed with it as many others have done. Mr. Alley recommended that we use combs in which bees had been reared as they were so much easier to handle, especially in hot weather. True, the combs are easier for *us* to handle but not so easy for the bees to handle, and as the bees have to do the work they should be consulted in the matter. I tried Mr. Alley's system using old combs. The bees did not accept the cells as readily as I wished and the resultant queens were no better than the ones reared from grafted cells. I learned later that the reason for the failure was that the bees could not tear down the cells and rebuild them into queen cells on account of the cocoons in the cells. They had to fill the worker cells with bee milk, floating the larvae out to the mouths of the cells much the same as when cells are built under the emergency system.

(Link to Alley's method:
http://www.bushfarms.com/beesalleymethod.htm)

Cells Built From New Combs

In the year 1934 I observed a *strong colony* in a hive containing only *new white combs*. They were hybrid bees and I had removed their queen intending to introduce a better queen. I had to ship out so many queens that I put off requeening this colony. We finally had a queen to introduce and on examining this colony I was astonished to see large well-developed queen cells such as I had seldom seen. Upon further examining this colony I found that they had torn down the comb and a number of worker cells and had built these fine cells over the very small larvae in the cells.

Then the thought came to me, why not try the Alley system using all new white combs? I went to my best breeding queen, removed all brood and placed a new white comb between two combs containing honey and pollen. The queen immediately began laying in it. After 24 hours this comb containing eggs was placed over a strong colony above a queen excluder. As soon as the eggs hatched I cut the comb in strips, mounted the strips on cell bars so when given to the bees the cells would point downward. Huber then removed the larvae in two cells leaving a larva in one till two bars were prepared.

On the Road to Success

These prepared cells were then given to well-fed bees confined in a starter hive. This was Saturday and I lay awake late in the night wondering if it would work. In fact I lay awake all through church the following Sunday pondering over this question. Would the bees accept the cells? If so, would they build them out sufficiently so the bees that were to finish them would recognize them as queen cells? Maybe they would tear them down or just finish them as workers. I had intended leaving the cells with the bees in the starter hive for the entire 24 hours but curiosity got the best of me, so right after church I looked at them.

Hurrah! Success at Last

When I examined those cells it was by far the most inspiring sight in my beekeeping life. Not only was every cell accepted and well developed but the larvae were fairly flooded with bee milk although they were so small as to be visible to only one having excellent eyes. They were much too small, even at this age, to be used for grafting. I cornered each member of the family and exclaimed (my wife says I yelled) "Here is where we discontinue grafting." No one can realize the joy and satisfaction that was mine as I realized at last I had reached the goal I had vainly tried to reach for more than a quarter of a century!

I realized there was yet much to be learned before I made the system practical so I could produce cells in abundance all through the season. However, I knew this could be done for the main object had been attained. I have tried many plans over a period of ten years or so now I see no chance for improvement.

The Difference Between a Queen and a Worker

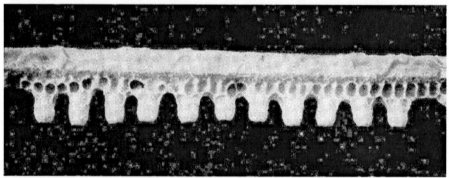

My first direct-from-the-egg cells

The difference between a worker and a queen is brought about by the different food each receives and the size of the cell it is raised in. Both are hatched from the same kind of egg, that is, a fertile egg. The drone is hatched from a nonfertile egg. Right here is an interesting phenomenon. Just how can the queen lay fertile eggs in the worker cells and nonfertile eggs in the drone cells? This she does with great regularity. A queen that has not mated with a drone will lay eggs but they will produce only drones. But to return to the question as to what makes a queen instead of a worker. Scientists who have done considerable work along this line tell us that the food of the worker larva and the queen larva is the same for approximately 48 hours. This food is bee milk secreted in the glands of the worker bee, such glands being located in the head of the worker. The discovery of these glands is attributed to Meckel, who discovered them in 1846. After about 48 hours the food of the worker larva is changed and it is fed honey and pollen. The queen larva is fed bee milk all through her larva life which is about five and one half days. Many have missed the all-important point and that is the queen larva receives *more of it*. And if we wish to produce the best of queens we must bear in mind that the *amount* of food the queen larva receives is fully as important as the kind of food it receives.

One of the Great Miracles of Nature

Let us see what a miraculous food this bee milk really is. The queen larva is given this food three and one half days longer than is the worker larva. This three and one half days' diet has made an entirely different bee out of it. Some have stated that there is nothing really miraculous about this. The queen being fed on more nourishing diet, develops into a fully developed female while the worker larva, being fed on less nourishing diet, is dwarfed in its growth. If this were true it would simplify the matter, but it is exceedingly far from the truth, for the worker bee is developed much more fully in a number of points. I believe I am the first to assert that the worker bee is as much a perfectly developed female as the queen but is *developed along different lines.*

The Worker Bee as Fully Developed a Female as the Queen

Let us look into this matter and note a number of features in which the worker is more fully developed than the queen. In the first place the worker has milk glands in its head in order that it may nurse the young larvae *and these glands are lacking in the queen*. If you ask most any mother I believe she will tell you that it is as much of a mother's job to nurse the baby as it is to bring it into the world. Yes, sir, when it comes to rearing bee babies the worker bees have the heavy end of it. It would be just as logical to say that the three and one half days' diet the queen received has dwarfed her so she cannot nurse her own babies.

No, neither worker nor queen is dwarfed, but they are entirely different bees both in physical make-up and temperament. The worker is more fully developed than the queen in a number of features. The worker has pollen baskets which are lacking in the queen. The worker has barbs on its sting while the queen has none. The sting of the queen is curved while that of the worker is straight. In all of this one can see the wonderful work of the Creator. Probably the reason for the queen's sting being curved is that when she is full of eggs it would be difficult to bring her battery to bear on her rival if her sting were straight. The worker bee has wax glands and a honey stomach, both lacking in the queen. The head of the worker bee is larger than the head of the queen, according to Snodgrass. There are several other slight differences between the worker and queen, all brought about by a change in diet for three and one half days.

Again the nature of the queen and the worker is entirely different. The queen will never sting a human being, while if you think the workers will not, you come with me. As stated, a queen will never sting anything but a rival queen. I might qualify that statement by saying a queen never stings anything but a queen, *or what she thinks is a queen*. I was stung by a queen once but I insist it was a case of mistaken identity, for she thought I was a queen. It happened thus: I had been requeening some colonies and in removing the old queens I killed them by pinching them between my thumb and finger. I had wiped my thumb and finger on my trouser leg. A virgin queen circled me a few times probably to adjust her bomb sights then made a pin-point landing on the spot where I had wiped my thumb and finger, and planted her sting in my leg. Yes, she thought I was a queen. While greatly appreciating the compliment, I would much prefer she would show her appreciation in a less militant manner.

Now as the queen and worker larva receive the same kind of food for the first two days, the reasoning has been that if larvae two days old are placed in artificial queen cells, perfect queens can be reared. As the attorney would say, "Your honor, I object." "While the larvae may be fed the same kind of food for the first two days they have not been fed the *amount* of food they would have been fed had they been reared in their own cells direct from the egg." "Objection sustained."

For Best Results Follow Nature

We should note that when bees are preparing to swarm or supersede their queen they fairly flood the young larvae just as soon as the eggs hatch and sometimes they put bee milk into the queen cells before the eggs hatch. This they do in such a lavish manner that after the bee has emerged there remains from a quarter to half an inch of dried milk. (I suppose we now should say "dehydrated milk.") This is the way Nature rears her queens and don't say Nature makes a mistake by putting in too much food. If this overabundance of milk were not necessary it would not be put there.

Therefore, if one does not have his cells built so there is a great abundance of dried milk left in the cells after the virgins have emerged, *he is not rearing the best of queens.*

Shortcomings of the Grafting Method

When Dr. Phillips was head of the bee culture laboratory in Washington, he wrote me that the word "grafting" was an improper term and that they did not graft but transferred. I wrote him it was refreshing to know there was one department in Washington that did not graft! In using the grafting method the larvae are left in the worker cells for two days where they are sparingly fed, for the bees are making workers of them. If much younger larvae are used they will perish, for they cannot stand such rough treatment. If you will examine the larvae two days old you will see very little bee milk around them. In fact, they are being "rationed." My experience has proved without a shadow of a doubt that such larvae have been starved in such a manner that they will never become fully developed queens no matter how lavishly they are fed after that. But wait-that is not the half of it. In grafting, you take the larva away from the starvation ration it has been getting and place it into an artificial queen cell, unless you have committed hara-kiri on it in the operation. Things are already bad enough, but putting it into that artificial queen cell!!-well, that is just about the blade of timothy that fractured the spine of the dromedary! We used to prime our cells with bee milk but, after careful examination, believe it was a detriment, for the first thing the bees do is to remove all the milk we had put in. Grafting in bare cells is better-or rather not so bad.

What Promised to be a Bright Idea Proved to be Very Dim

In order to get the cells filled with bee milk the same as they are when built during swarming, I allowed the larvae in the grafted cells to remain for two days till there was plenty of bee milk in them, then removed the larvae and put in young larvae. I hoped to get fine cells in this manner but the bees seemed to think otherwise. They accepted but a few of the cells and in some cases the larva was pushed over to one side of the cell and the bee milk all removed. In a few cases the bees accepted the cells but placed a little thin milk on top of the milk already in the cells. The few queens reared were no better than the ones reared by grafting in the regular way. This leads me to believe the dried bee milk is not suitable for larva food but rather is the crumbs left over from the feast. I believe the real food is the very thin milk that is fed to the larva. Then one may ask what is the advantage of having so much dried milk left in the cells. That merely indicates that the growing larva was fed in a lavish manner which is very necessary if quality of queens is desired. One queen reared as described above performed in a manner I never knew one to perform before or since. She laid drone eggs only but none of them died in the cells as is common with drone layers but all matured into perfect drones and that queen fairly filled that hive with beautiful drones. Scientists who have made a study of the subject tell us that bee milk is the same in all stages. I am inclined to doubt this. It may be changed in the moisture content only but I have observed that bee milk in the cell of a larva of one age will not be accepted by a larva of another age. When given, the bees immediately remove it and proceed to give the proper nourishment.

Many of our best beekeepers abandoned the grafting method after giving it a trial. I once wrote a paper, which was read at a meeting of the National Beekeepers Association, setting forth the shortcomings of the grafting method, although I was using it as the time. Dr. Miller wrote, "I say Amen to that."

Bees in All Stages of Development from Worker to Queen

Many seem to think there is a marked division between worker and queen. They seem to believe that the bee is either a perfect worker or a perfect queen. I believe I am the first to point out that bees can be reared in all stages from a worker to a queen and these stages are brought about by the *amount* of bee milk the growing larva receives. These different stages are not only brought about by the kind of food the larva receives but, don't forget, by the amount it receives. In our work we have observed bees in many stages as stated above. Upon a number of cases we have observed a bee like a worker but with a more pointed abdomen with a yellow color like that of a virgin. Such a bee remains in the colony and acts like a worker and in no way interferes with the queen cell that is given or with the virgin when it emerges. Whether or not such a bee goes to the fields or performs any useful work about the hive I have never been able to determine. Then there is a bee a few steps above that that is the meanest most contemptible and the most (censored) bee that ever existed. She is similar to the bee just described but a little larger and more like a queen but so small one might not notice her even tho she be in plain sight. She will tear down cells just as fast as you give them. What makes her such a pest is the fact that she occurs so seldom that you are not looking for her but are trying to find a regular virgin.

I do not know where such a bee ever flies out to mate, for when I finally get sight of her she never gets a chance.

A bee a few steps above the one described flies out to mate but never gets back. The bee above the last-mentioned flies out and mates, comes back, lays a few eggs and is superseded. All too many of such queens are sent out by breeders using the grafting method. A queen above the last may remain in the hive and lay sparingly, thereby losing a crop of honey and either dies in winter or is superseded early next spring. Then there are queens from this on up to the fully-developed queen that keeps the hive packed with worker bees which produce a bumper crop of honey. The first bee mentioned is reared in a worker cell while the next may be reared in a small queen cell that may often be taken for a drone cell.

The Mating Flight is Not for Weaklings

The mating flight of the queen is a very strenuous affair and it is quite evident that Nature made it that way in order to eliminate the unfit. When a dozen or a thousand drones take after the queen she flies in a zigzag manner so that only the strongest flying drones can overtake her. Upon one occasion when the queen and a horde of drones were in the air they made such a buzzing that Mrs. Smith asked if that was a swarm. It sounded like one. In mating, quite frequently both the queen and the drone fall to the ground. After the exhausting chase the queen will be unable to take wing again unless she be possessed with great vigor. While using the grafting method in rearing our queens we took it as a matter of course that a certain percent would be lost in mating. This loss we now know was caused by rearing weak queens. With our present system of rearing queens direct from the egg a virgin queen in the hive invariably results in a laying queen. In mating, the drone soon dies. It affects him much as it does the worker in losing its sting. The cut shows a worker bee removing the drone organs from the queen. This is a very rare photograph and was taken purely by accident.

Worker bee removing drone organs from the queen.

Sombre Side of the Romance of the Honey Bee

The honeymoon of the bee is short for no sooner does the queen become a bride than she is a widow, and no sooner does the drone become a groom than he passes away. There is one advantageous feature in this in that the bride is saved a trip to Reno. Possibly when the groom realized he is about to become the father of half a million children, and every blasted one of them females, it is more than he can endure. A queen that has never been to the altar will lay eggs but they will produce drones only. Hence the drone has no father but does have a grandfather on his mother's side. Neither does he have any sons but may have many thousands of grandsons. However, they never get to greet Grandpa for he has been gone months before they were born.

Some Complications in the Bees' Domestic Affairs

When a yellow queen mates with a black drone all her sons will be yellow like herself while her daughters will be partly black. Now the question arises, are these pure Italian drones and half-breed German bees full brothers and sisters? Can a drone and worker be full brother and sister? I wonder.

Food and Yet More Food Makes the Perfect Queen

The different grades of queens we have mentioned, from the poorest to the best, are brought about by the kind of food the larvae receive and what is of equal importance the *amount* of such food. If we wish to produce as good queens as the bees produce during swarming and supersedure, and don't let anyone try to make you believe he can do better, we must duplicate the bees' performance as nearly as possible. When building a swarming cell, after it is well started, the queen lays an egg in it. Just as soon as the egg hatches the bees fairly flood the cell with bee milk and they keep up this abundant feeding until the cell is capped over. From this it can easily be seen that the grafting method is entirely unsuited for rearing queens if quality queens are desired. Four or five pounds should be kept in the breeding hive. As this large force of bees have no brood to feed except the small patch in the new comb given *such larva receive exactly as good feeding as do the larvae in swarming or supersedure cells.* From what has been said it is evident that in order to produce fully developed queens that will keep a large brood-nest packed with bees the larvae must have an abundance of bee milk just as soon as they come from the eggs and *must have an abundant and constant feeding without interruption until the cells are sealed.*

Importance of Food for the Young

Let us look at the matter of food for the young. The newly born of any specie must have an abundance of food if proper development is obtained. Mature creatures may go on a fast for a period without injury and in many cases such a fast may be beneficial. Not so with the young. Many people grow old prematurely because they were given artificial food when infants instead of their mother's food. Many babies die because some quack doctor wants to improve on Nature and prescribes some new-fangled food rich in vitamins from A to Z. I know from sad experience. I believe if the truth were known there are many now behind prison bars who would not be there if they could have had their mother's milk for the first six months of their lives. For lack of proper nourishment their bodies and brains were not properly developed. I have observed the effects of food on the young of animals and fowls for many years and the more I observe these things the more I believe in proper and abundant feeding of the young.

Honey as Food for Man and Bees

It has been proved that the worst foods for man are white flour and sugar. Therefore, if sugar is unfit for human food, how much more unfit is it for food for bees?

Honey is the food God made for bees as bees produce more than they need it is evidently intended that we should have the surplus. Honey is rich in minerals and valuable enzymes which are totally lacking in sugar.

I have proved to my satisfaction *that the best and most vigorous queens cannot be reared when sugar is fed to the cell-building colony.* Only honey should be fed if you wish to rear those prolific long-lived queens. To prevent robbing when feeding honey a feeder should be placed on the back of the cell-building colony the same as that on the breeder hive.

It is well known that improper diet makes one susceptible to disease. Now is it not reasonable to believe that extensive feeding of sugar to bees makes them more susceptible to American Foul Brood and other bee disease? It is known that American Foul Brood is more prevalent in the north than in the south. Why? Is it not because more sugar is fed to bees in the north while here in the south the bees can gather nectar most of the year which makes feeding sugar syrup unnecessary?

We feed sugar syrup to the bees in the mating hives in order to get them to accept the queen cells but as the virgins are yet in the cells no harm is done. If the truth could be learned I believe it would be found that as soon as the virgin emerges from the cell it would go to the cells containing honey rather than to those containing sugar. The reason we do not feed honey in order to get cells accepted is that it would cause robbing, especially in the mating hives where the bees are few in numbers. Even at that we feed only when no nectar is coming in from the fields. Of course if disease is present honey should not be fed unless it is heated to destroy the disease germs. Heating would destroy the enzymes. Whether or not it would destroy the minerals I have no way of finding out.

A Homely Example

Let me give an example of the effects of an abundant feeding of the young. We bought a spotted Poland sow at a sale. She gave birth to two pigs only. We named them Theobold and Emogene. As these two pigs received a super-abundance of food they grew at such a rate as to make jack's bean stalk look like a mere sprout in comparison. Our county agent weighed them when they were six weeks of age and found that Emogene weighed 62 pounds while Theobold weighed 65. Later Emogene went to the butcher and he told me he never saw such a large hog. He did not have room to hoist it till he cut off its head. I sold Theobold to a farmer who said he had to butcher him as he was so big he could not be confined in any fence but would put his snout under the bottom wire, raise it up and the staples would come snapping out of the posts for many rods.

Now observe this: We had other pigs that had the same care after they were weaned but there were seven in the litter and they did not have the food while they were young as the two mentioned had, so the 7 were just ordinary pigs.

Importance of Food for the Queen Larva

The importance of food for the young applies especially to the queen larva as it grows at such an amazing speed. It grows as much in proportion to its size in one day as a calf does *in a year*. At this rate if we keep the young larva away from food while grafting for 20 minutes it is the equivalent to keeping a calf away from its mother a week. We keep a few registered Guernsey cattle and the calves are allowed to run with their mothers for two months. At the present price of milk each calf consumes a hundred dollars' worth of milk but the resultant animal is

well worth a hundred dollars more. Add to that the pleasure we get in rearing animals we are proud of!

The Case of the Two Virgins

From the looks of these two virgins one might get
the impression that one was wise and one was foolish
similar to those ten mentioned in that yarn Mathew spin in
which the ten stopped at a filling-station to get some oil.
The five foolish ones were so busy gossiping about that
bridegroom they hoped to see that they forgot all about
getting oil while the wise ones said, "Fill 'er up." But no.
The difference between these two virgins was simply a

matter of diet. The small one suffered from malnutrition during infancy.

Lament of the small virgin.

You may wonder what makes me so small
Why my bees got no honey at all.
It is sad to relate but the truth I must state
'Twas because I was grafted, that's all.
The large virgin has this to say:

Why I keep my hive well filled with brood
Is because I had plenty of food,
Raised direct from the egg, the only right way
That's why my results are so good.

An Amazing Spectacle

Having proved that perfect queens can be produced only by proper feeding of the queen larvae, I carried out an experiment to see how small a queen could be produced using too old a larva and one that had been reared in a worker cell and had been fed as a worker larva. This experiment astonished me beyond description.

The larva used was between three and four days old. A patch of such brood of about four square inches, counting both sides of the comb was given to some queenless bees. Now what did these bees do about this hopeless condition? First they chose two larvae which they attempted to make into queens. They fed them what food they had in their glands. Next, *they removed every single larva in the comb.* Why did they remove the larvae? I wish I knew the answer to that but my guess is they did it for one of three reasons and possibly for all three. First, they wanted to give all their milk to those would-be queens. Second, they wanted to use all the milk in those worker cells. Third, possibly they punctured the larvae and salvaged what milk they contained. In 24 hours they removed one of the queen larvae probably to use what little milk there was in the cell. They did their best but the cell never hatched. Therefore, when someone tells you that bees make such a mistake as using too large larvae when smaller ones are present, don't believe him. My observations have taught me that bees do the very best that can be done under existing circumstances.

So Let Us Get to Work

I can imagine someone saying, "all right, all right, you have us convinced, so get to work and tell us how to rear those super queens that will not be prematurely superseded and will produce those bumper crops." This I shall do. First, I shall take the reader with me through the method we use in rearing queens in a commercial way and then describe methods for the back-lotter who wishes to raise a dozen or maybe fifty in a season.

The Breeder Hive

The wood comb

The breeding queen hive is the crux of our system and has been developed after many years of experimenting. The difficulty has been to get the breeding queen to lay consistently in the new comb every day. At first we placed the new comb in which we wished the queen to lay between two frames containing honey and pollen. Sometimes she would lay in our new comb but all too often she would lay in the old combs and ignore our new comb. Next we made a hive with a compartment which contained the new comb only. It was only a partial success as the queen seemed not to like being caged in that way and often the queen would be lost, probably from worry or possibly the bees killed her for reasons they alone know. At last I hit upon our present system in which the queen lays in our new comb every day throughout the entire season and seems to enjoy it.

The cut shows the frame partly filled with wood, leaving a small comb five and one-half inches by nine and one half inches. These frames we shall call wood frames. It takes the queen but a short time to keep these combs filled with brood so she can devote her time to laying in our new comb. The standard size foundation will do very well but we prefer the size used in the Modified Dadant frames as it is deeper. A sheet is cut into four parts and fitted into standard frames. These frames of foundation may be drawn out in either the finishing colonies or in the front part of the breeding queen hive. When one of these new white combs is placed between the two combs partially filled with wood the queen at once begins to lay in it as the other two combs are filled with brood and the new comb is right between. This makes an inviting comb for the queen to lay in, which she does at once. A wooden partition separates the compartment containing the three combs from the rest of the hive. A cleat is nailed to the bottom of the hive below the partition. A space of an inch is left between the cleat at the bottom and the partition above. A zinc queen excluder is tacked over this opening. It is best to have the excluder at the bottom for if higher up the bees insist on filling the space with comb which they will not do when the excluder is placed near the bottom. When using the regular cover, the queen would sometimes get out of her partition due to the fact that wax or propolis would accumulate on top of the partition. To overcome this, the partition is extended three-fourths of an inch above the edge of the hive. We use two covers, one for the back part and one for the front. Then the regular telescope cover is placed on top. With this arrangement the queen never gets out of her stall. The cut will show how the hive is constructed better than a description.

Putting new comb into the breeder hive

We use a jumbo hive cut down to fit the standard frames. One may use the standard hive but it will be necessary to nail cleats around the bottom to make room under the frames. The entrance is about midway in front of the hive.

Keeping the Breeder Hive Stocked with Bees

If cells are started every day the bees in the breeder hive will rapidly decrease in numbers as the only emerging brood they will have is what comes from the combs in the wood frames, therefore, some way must be provided to keep up the strength of the colony. We open the starter hive in front of the breeder hive when we take the started cells out. Many young bees will take wing so when we remove the starter hive, these young bees will join the bees in the breeder hive. This is a great help but shortly they will need more bees in which case we shake the bees from not more than one comb from the starter hive. In this way their strength will be kept up. Sometimes in the past we dumped the bees from the starter hive in front of the breeder hive. This proved unsatisfactory as so many bees were apt to kill the breeder. When they did not the breeder was apt to stop laying for a time. Adding the smaller number of bees has never given us any trouble. We prefer the perforated zinc to any other excluder. Those wood and wire excluders are an abomination. I purchased a hundred of them years ago and now all I have to show for them is a bushel of wires and slats of wood. They cost more and are worth less. The wax worms will eat through the wood and the mice will eat holes in them. Some say the bees pass through the wire excluders better. I put that question up to the bees and I quote, "There is no difference. One is just as bad as the other". End of quote

The Feeder

This is an important part of the breeder hive. When no nectar is coming in the bees must be fed if we are to get that lavishly fed larvae we have pointed out is so necessary for best results. It is built on the back of the hive. A hole is bored in the back of the hive so the bees can get into the feeder. Slats are put in as illustrated, then melted wax is poured in to make the feeder water-proof and also to wax the slats tight to the bottom in order that no syrup or honey can get under these slats to sour. A cleat is nailed to the hive just above the feeder to keep out the rain. When feeding, the cover of the feeder is slid back and the honey or syrup poured in. As there is no bee disease within many miles of our apiary we feed off grades of honey. Honey is better food for the bees than sugar syrup but if disease is present sugar syrup should be fed. However, from years of observation I am confident the nurse bees cannot secrete milk with proper nourishment when sugar syrup is fed. I was surprised when I first put honey in the feeder. It was during a robbing season and I wanted to test it for I had a theory feeding with it would not cause robbing. It proved better than I expected for by looking at the bees at the entrance you could not tell they were being fed. With any other feeder I ever tried we had to feed at night or the robbers would rob out the colony. What causes robbing is that the bees get daubed with honey and then go out in front to get more air. The bees from the other hives pounce upon them and then proceed to clean up the whole hive. With our feeder few if any bees get daubed and those that do have to pass through a cluster of bees, then through the queen excluder and then another cluster of bees. By that time they are licked clean and do not have to go out for an airing.

Stocking the Breeder Hive

In case your breeding queen is in a fairly strong colony, the bees and brood are placed in the front compartment and one frame of brood and the queen is put in the back compartment. The two frames fitted with wood are placed in the back compartment with the frame of brood between them. Feed is then placed in the feeder. As soon as the queen has eggs and brood in the two outside combs, the frame of brood is removed and a new white comb is placed between the two combs. In case the colony is strong, most of the brood may remain in the front compartment, but later most of this must be removed to make room for the combs containing eggs. As stated, we should have several drawn combs ahead. When the brood in the front compartment is removed frames with sheets of foundation may be given. As the bees are being fed, they draw the foundation readily. The comb given should be left 24 hours when it will be well filled with eggs. It is then placed in the front compartment and another frame of new comb is given. This is repeated every day until there are three frames containing eggs in the front compartment. When the fourth comb of eggs is put in the front, the first one given will now contain tiny larvae just out of the egg and as this is practically all the brood that large numbers of bees have, the tiny larvae will be found to be fairly flooded with bee milk. In giving comb to the breeding queen, it should be trimmed back to the foundation provided the bees have built comb at the edges and bottom of the foundation. The queen prefers to lay in natural built combs rather than those built from foundation. I have used all kinds of foundation and find this to be true. The comb shown was not trimmed back so it can be seen that the queen laid in a sort of horseshoe shape.

The bees do not build cells with such regularity as the cells in foundation, therefore such comb is difficult to cut as many larvae will be destroyed. When the eggs are laid in cells drawn from foundation it is much better if the row of cells is uniform and easily cut in a straight line. Now let us examine the larvae in the cells to see if they are getting lavish feeding, the necessity of which we have so frequently stressed. The larvae are so small they can hardly be seen with the naked eye. They are flooded with bee milk fully as much as larvae in swarming cells. In these same cells they will remain while the bees go right on feeding them. Let us now prepare to start those larvae on the road to queendom.

Larvae ready for use in cell starting

The Starter Hive

The hive in which we start cells we call the starter hive. In the years that have passed since I began rearing queens in 1901 I have done a vast amount of experimenting. I used a starter hive when I first tried my hand at queen rearing. Then I would think I had a better way and would abandon it. Later I would come back to it, then leave it again for something I thought better. With some improvement in it we are now back to it to stay.

The starter hive

Many years ago Henry Alley used what he termed "the swarming box". He did not start cells in it but confined bees in it for some hours and at night dumped them into a hive with no brood or queen. He seemed to think it necessary to keep the bees queenless for some time to get them to accept the cells. Then the prepared cells were given to the bees in a hive containing two frames of honey and pollen. Mr. G.M. Doolittle said the bees would accept cells better if kept queenless for three days. In my experience I find the bees accept the cells just as readily after an hour or two as they do if left queenless for a longer period. We usually keep them confined in the starter hive from one to two hours which gives them plenty of time to clean up any honey that may have been spilled on them when shaking them into the starter hive. In later years Eugene Pratt gave the cells direct to the bees confined in a box and he called it "The swarm box". We prefer to call it "The Starter Hive" as it is used for starting cells and has no relation to a swarm. Our starter hive is made wide enough to hold five standard frames. The total depth is 17 inches. It has a wooden bottom with cleats at the sides making it suitable for feeding or water-ing bees. Three frames of honey and pollen are put in *but no brood*. In case no nectar is coming in, the bees that are to be used in stocking the starter hive should be fed three days before using them and this feeding should be continued each day until they are shaken into the starter hive.

Food and Water Necessary to the Bees in the Starter Hive

A matter that I believe most of us have overlooked is the necessity of food and water for the confined bees. All have observed how bees carry water in abundance when rearing brood. Now this water is even more neces-sary when rearing queens for the bees require more water when producing the milk in abundance. The best way to provide this is to fill the center comb with diluted honey. This not only provides the necessary water but the food it contains acts as a flow of nectar. In case some 50 cells are to be started from five to six pounds of bees should be used. In case from 75 to 100 cells are to be started from 8 to 10 pounds of bees should be put in. If larger number of bees are put in the starter hive should be kept in the shade or some cool place. The screens on the sides provide an abundance of ventilation.

Getting the Bees with Which to Stock the Starter Hive

In getting the bees for the starter hive we used to hunt up each queen before shaking the bees. This entailed a lot of work and sometimes when it was cloudy it was well-nigh impossible to find the queens. Again when robbers were troublesome, unless we found the queen quickly, we had to close the hive and go to another. In case we used the bees from three colonies, at the best it took at least half an hour and usually longer. With our present system the starter hive may be stocked with bees from three or four colonies *in five minutes*, and this too, rain or shine, robbers or no robbers.

The colonies are all in two stories with queen excluders between the two hives. The queens are confined below and the top story is kept will filled with brood. All that is necessary to do in filling the starter hive is simply to shake in the bees from the top story. The bees will be depleted somewhat, as some are used to keep up the strength of the breeder hive and the finishers, so unsealed brood should be given to these colonies from other colonies. We use the bees from each colony once in every four days. This is less work than it would be to use a larger number.

Preparing the Cells

Let us now consider the larvae in the breeder hive is ready for use. First some wax should be melted. An electric hot plate or a sterno stove with canned heat may be used.

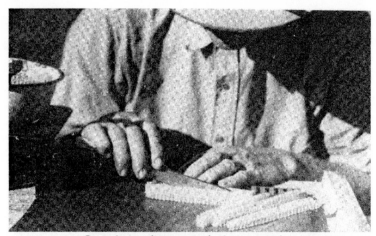

Cutting the comb into strips

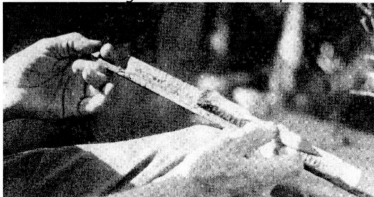

Painting the cells

The cooling box

Destroying the larvae

It will be necessary to destroy some of the larvae in the cells in order to make room for the queen cells. We have tried many methods but find the best way to destroy the larvae is to insert a spike in the cells and punch out the bottoms. An ice pick may be used if the point is broken off. We usually destroy two cells and leave one which gives plenty of room for the fine large queen cells. Sometimes when we are short of cells we destroy every other one. Some will be built close together and will be destroyed in cutting but one can get more cells in this way.

The comb from the breeder hive is now cut into strips one cell wide. With a small paint brush, wax is painted on the cell bar and the strips of cells placed on the waxed bar. Then with the brush wax is painted on the sides of the cell strips. The cells are then cooled in the wooden tank and more wax painted on till both sides of the strip of cells have enough wax to make the cells strong enough so they will not mash down when inserting the spike. Wax should be cool for painting the cells, nearly ready to solidify. We use a wooden tank four inches wide inside measurement, and half an inch longer than the cell bars. A cleat is tacked in one end under which the end of the cell bar is placed. At the other end of the tank is a nail driven through from the side extending about two inches into the box. As soon as the cells are waxed the bar is shoved under the wooden cleat at one end of the tank and the other end is placed under the nail at the other end. Then cold water is poured in the tank till it covers the melted wax. It is well to put ice cubes in the water for we want the cells to be firm so the larvae in the cells can be destroyed without mashing down the comb. While we prepare the cells as quickly as possible in order that they may be in care of the bees without delay, it has been astonishing to note what gluttons for punishment these well-fed larvae are. Frequently we have more cells than we have use for so instead of using the larvae, we set the comb aside for the larvae to die in order that we may use the comb again. Often after four days, when we give the comb to the bees in the breeder hive, some of the larvae will be rejuvenated or resuscitated, or whatever you wish to call it. An interesting feature is that although the larvae were away from the bees four days, they had not in-creased in size a single bit. This shows that larvae away from bees do not grow, so we should keep them away from the bees as short a time as possible.

Putting the Prepared Cells Into the Starter Hive

We use two bars in each frame, the space above and below the bars being filled out with wood. The starter hive is given a little jar on the ground so the bees will fall to the bottom and not fly out. The two frames of cells are set in place, the comb containing thin honey being in the middle. Cells will be built better in the middle of the combs as shown, for where the cells are near the top of the frame the bees are apt to build comb over them.

Bees in Perfect Condition for Starting Cells

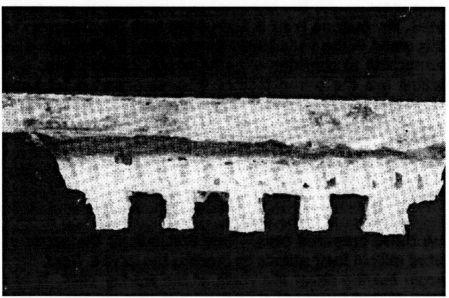

Cut #1 shows the cells as we prepared them especially for this picture. We bent the sides of the cells over to destroy the larvae in them leaving one cell and destroying two.

Let us now consider the bees in the starter hive. They have been nursing a horde of larvae in the colonies from which they were taken. They are secreting an abundance of bee milk. They are now away from their brood. They feel their queenlessness. The milk in their glands seems to cause them discomfiture. They want to rear queens and they want to get rid of that surplus of milk. When giving them the bars of cells they at once set up a joyful humming to spread the glad news that their wants are fulfilled. Often when I come with the prepared cells they set up a humming when the bars of cells are within a foot of the starter hive. Evidently they smell the larvae in the cells I am giving them. Since some who read this may not know, I will state that at the tip of the abdomen of the worker is a scent gland. By exposing this gland and fanning their wings, the bees spread the scent to the other bees. Mr. Eugene Pratt is accredited with the discovery of this gland in the year 1891. By the way Mr. Pratt gave us something to consider when he wrote, "Sugar, bees and brains are the prime factors in queen rearing." I have often stated that many of us fail because we use too large a proportion of bees and sugar!

Program of These Rejoicing Bees

Now what is the first thing the bees do upon receiving these prepared cells? They first unload the accumulated milk in their glands by feeding the larvae. *Next, they begin tearing down the walls of the worker cells.* It has been stated by many that the bees enlarge the worker cells. This they cannot do. They tear down the worker cells and build queen cells over the larvae. In this way they build the cells as they want them and do not have to put up with any man-made contraptions. The photographs will show how bees build their cells.

When starting cells from worker cells the bees first "raze" them and then "raise" them. This reminds me of an incident that happened years ago. I told my wife that a virgin had gotten into the hive where I had a lot of queen cells and this virgin was razing cells. She thought I said the virgin was "raising hell" so I let it go at that for I rather liked her version of the incident.

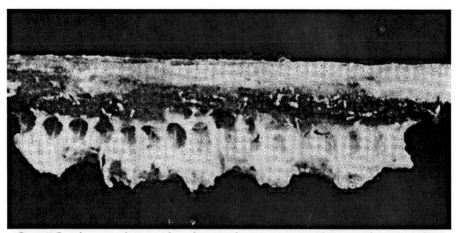

Cut#2 shows how the bees have torn down the worker cells and have started to rebuild. One cell at the right has already begun to take shape. This photo was taken only four hours after the cells were given to the bees. I had intended to take more pictures four hours later but darkness come on, so number three was taken the next morning and, as can be seen, the bees are rushing the work along at top speed. Note what a fine start they have and yet the larvae in these cells are no larger than those usually used for grafting.

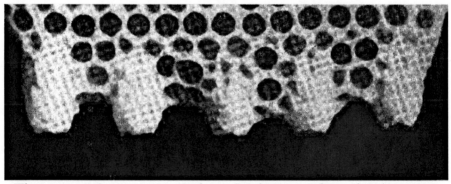

The next picture was taken 24 hours after the last one and now the bees have fully half an inch of bee milk stored in each cell.

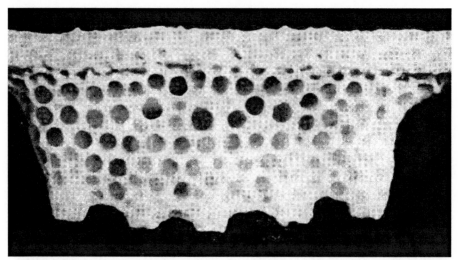

The fifth picture was taken 24 hours after the previous one and, as may be seen all the cells are capped over except the one at the right and it was capped over two hours later.

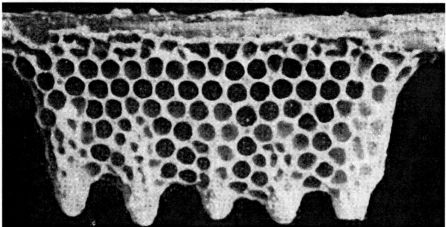

Finishing the Cells

While the bees are now busily engaged in remodeling the cells, we must prepare to have them finished, for they should not remain in the starter hive more than 24 hours for the best results. Again I shall tell what we do and then tell why we do it. The best colony for finishing cells *is a strong queenless colony*. The cut shows cells made by the grafting system illustrated in that excellent little book by Walter Kelly, *How to Grow Queens for Fifteen Cents*. Also is shown the use of wooden cell cups.

Thus far our system is foolproof as far as getting good cells started, for if we give them too many cells for best results, they will accept only the number which they can properly build. Right here is an interesting feature of bee behavior. While using the grafting method, the bees would accept all the cells we gave them, while in reality they were in no condition to properly care for one-fourth of that number. Why they accept more artificial queen cells than they can properly care for is due to a peculiar characteristic of the honey bee in that they will finish, or try to finish, any job that has been started, while had it been left to them in the first place they never would have started it. This trait may readily be observed if you place brood in a hive above a queen excluder with the queen below. The bees seldom start queen cells from this brood. But if cells be started and placed above the excluder the bees will finish them the best they can. Therefore, when artificial queen cells are given above an excluder, they evidently consider them as started cells so they do the best they can to finish them. When cells are given them as we do, they evidently do not consider them as started, therefore they accept only the number that they can complete in a perfect manner. Many times in Indiana we were trying to rear queens when conditions were not the best. Our customers were demanding their queens or their money back so we were trying to raise queens when conditions made it impossible to rear queens in quantities. Many of these cells, when the larvae were three days old, had little bee milk around the, barely enough to keep them alive. I knew from experience that only poor queens would result from such stunted cells and have taken my knife and shaved off hundreds of them and destroyed them. Had we been using our present system, the bees would not have accepted more than ten or a dozen in place of a hundred but would have finished them in perfect manner. To be extra conservative in the starter hive not more than half the number of cells they would accept. Therefore, as stated, thus far our system is foolproof as far as getting good cells started but from now on the beekeeper must take over, for as the cells now have been started the bees you give them to will try to finish them whether or not they can do a good job at it. Unless the bees that are to finish them are in the best of condition and strong in numbers you can still produce poor cells. The queenless colony that is to finish the cells must contain at least five pounds of bees and several combs of mostly capped brood. If no nectar is coming in, this colony must be fed liberally. Again, to be extra conservative, we have two colonies to finish the cells that have been started. If this is properly done you can produce cells every bit as good as those built by the bees during swarming. Bees must be added to keep up their strength. It is some help to them by giving them the cells with the bees that are with the cells. When more bees are needed the starter hive is dumped in front of the finishing hive.

Nurse Bees

Let us digress for a moment and discuss nurse bees. Some who have made a study of the subject tell us bees are good nurses at three or four days of age. Well if they have caught them in the act and really have the goods on them I will have to retract, but even so maintain that, if they are nurses at that tender age, they are might poor ones. I asked one professor, who claimed these young bees were good nurses, how he came to that conclusion. He said he had observed the young bees putting their tongues into cells where the young larvae were and he supposed they were feeding them. I asked him if he should see a boy crawl out of bed at ten o'clock at night, climb up into a chair and put his hand into the jelly closet he would deduct that the boy was putting the jelly there? Now, when these young bees put their tongues into the cells, do they give food or take it? In my experience I believe it is the latter. Acting up such conclusion I tried a plan in which I expected to get wonderful results. I shook bees from a dozen colonies in front of a hive in which I expected to rear cells. As I wished, the old bees went back to their own hives while the young ones went into the hive prepared for them. There were at least six pounds of those downy, fuzzy bees. They had been fed well so I gave them 50 grafted cells expecting to get wonderful results as I had been told these young bees were graduate nurses. Upon examining them next day not a single cell was accepted. All those little beggars did was just *eat*. That is what they are supposed to do, eat and develop and grow in order that they may be able to do any work about the hive and they can then do any work that is assigned to them such as nursing, gathering nectar and pollen, carry water or guard the entrance. I have found a bee worthless as far as doing any useful work about the hive until it is a week or more old. When ten days old it can do anything required.

Old Bees Good Nurses

Again some have told me that old bees cannot nurse as their milk glands dry up. I maintain such a statement erroneous. Many have observed that bees returning from the fields could not nurse. Of course not, for they were acting as fielders and not prepared to nurse. I will wager (a dime is my limit) that if these same bees were given several frames of unsealed brood, they would soon change from fielders and become nurses. Some of the brood might perish, for this was thrust upon them with no warning. But they would at once put on the white aprons and white caps with red crosses on them, take a hurried course in nursing, and in two or three days could show those younger bloods how to nurse. In a few days should you remove all the brood and give them grafted cells, they would do excellent work. Now just what are the requirements when taking this course in nursing? They devour pollen and honey in abundance and soon begin giving milk.

I have given a frame containing eggs only to a colony that had been queenless and broodless for some time. The bees removed many of the eggs as they were not nurses. Then they concentrated on a few and built a few very small queen cells. The eggs were given to them on too short notice for them to become nurses. Had they been given three day's notice they would have taken a course in nursing and would become competent nurses.

I have taken queenless and broodless colonies that had been in that condition so long that I was afraid laying workers would develop, and have introduced a laying queen. The presence of this laying queen served notice on them that they must take a course in nursing. This they did by eating pollen as has been stated and in three days, when the eggs began to hatch, they fairly flooded the cells with bee milk to such an extent that one might believe they were going to make queens out of the whole lot!

Probably in a normal colony the bees do the work best fitted to their age but, as stated, they can do any work required after they are ten days old. In the far North in the spring many bees would be six months old and all would be at least four months old. These old bees do a better job of nursing than the young ones that come latter, for European Foul Brood seldom attacks the first cycle of brood, while later, after the old bees are gone and the young nurses take over, European Foul Brood develops. This would prove that the old bees are better nurses.

Finishing the Cells-Continued

We use a single story to finish the cells unless there is a heavy honey flow on, in which case we add a second hive body but have the cells finished in the bottom hive. If we start four bars of cells in the starter hive it is best to give them to two finishing colonies. As the bees finish a job that has been started we must be very particular to see to it that they are in the best possible condition to finish the cells properly. In case we wish to start cells every day we have three sets of finishers, two hives in each set. In case one is rearing queens on a very extensive scale he can have four hives or more in each group and use two or more starter hives. After the cells have been in the starter hive overnight we lift out the frames of cells and give one to each of the finishers. The adhering bees are given to the finishers which helps keep up their strength. The frame of brood and cells are arranged as follows: At the side of the hive nearest to you is a frame of capped brood, then a frame of cells. Next day, group number two are given bars of cells, then next day group number three are given two bars of cells. Now, when you start the fourth batch of cells, the ones in the first group are sealed or nearly so and are given bars of cells, a frame of capped brood separating them from the first frames given. This is repeated till all finishers have three frames of cells each. When ready to give the next frames of cells, those first given now are what is known as "ripe cells." They must be taken out to make room for the next frames, and are ready to be introduced to mating hives. These cells will hatch the following day.

Three frames of cells

The photograph on page 60 shows the arrangement of frames of cells as they are painted white that they may be easily seen. From this on, when a frame of ripe cells is taken out another frame with started cells is given and this may be continued throughout the season. As the bees in the finishers become weak more bees must be added from the starter hive and capped brood given from other colonies. In case you do not need so many cells you can skip a day and the space in the finisher is filled by moving the combs over and should there be a honey flow an extra comb should be given to prevent the bees from building combs in the space.

Caution

Three or four days after giving brood you should look over the frames to remove any cells that may have been started. It is astonishing to note how many accidents can happen in those finishers no matter how careful one may be. We tack a zinc queen excluder to the bottom of the finishers to keep out virgins for if there is any place one can get it is sure to do so. These queenless bees frequently send out calls for a queen by fanning their wings so it is necessary to make the hives bee-tight. In one finisher I found a laying queen and removed her. A week later I found another in the same hive. Upon carefully examining the hive I found a small opening near the top of the hive where the cover did not fit closely and the queens returning from their mating flight were invited in. By making the hives bee-tight and by examining the brood carefully you will have no accidents, or rather I should say, few.

Reviewing the Past

Before giving the ripe cells to the bees in the mating hives, let us go back into the past and examine the various methods we have used in starting and finishing cells. I would prefer to omit this but some may think they have a better way and may try some of the plans I have tried and found wanting in one respect or another. I believe I have tried every plan that has any merit and many that have none. As stated, we used a starter hive in our first experiments in 1904. I used wooden cell cups and the grafting method. Just why anyone should use wooden cell cups is now a mystery to me. Possibly for the same reason I did-because I was taught to use them. Later we dipped the cells as Doolittle did and this saved a lot of time as we did not have to clean out the wooden cups and insert the cells into them. Later, I put the wooden cells into the furnace and they made a fine fire and kept the house warm for half an hour which was of more value than using them. I also bought a lot of queen nursery cages for the cells to hatch in and some of them did. Many did not and I raised a lot of half starved virgins that never amounted to anything.

Requirements of the Newly-Hatched Virgin

When the virgin emerges she is weak and frail like most newly-born creatures and needs the very best of care. She must be fed in the natural way by the bees until she is mature which is from 8 to ten days. To confine her in a cage and expect her to develop into a good queen is to expect the impossible. We found when a virgin was caged for three days she usually was delayed in mating three days so nothing was saved in time but much was lost in the quality of the queen. Some queen breeders have advocated keeping the virgin in the cage for five days. Such queens are not worth introducing. It would be better to allow the bees to supersede their queens in the natural way than to introduce such worthless queens. To be sure, there may be more money in it for the producer of such queens in case he could continue to find buyers but the buyers certainly would not make a profit in the deal. So my cages went into the furnace and did more good there than they did in hindering the development of the virgins. I wish to take just one more dig at another abomination and then we will get on to constructive work. That abomination is the wire queen cell protector. I understand that this bright (?) idea was conceived by two men about the same time and they came near developing a regular Kentucky feud as to who deserved the credit for inventing it. They might just as well have remained good friends for really there was nothing worth quarreling over. In using these cell protectors one is fighting bee nature and in such a fight the bee wins. The idea of the cell protector is that as bees tear down the cells at the side and not the ends, by protecting the sides the bees cannot tear them down so the virgins are allowed to emerge. Again we are trying to *force* the bees to accept cells they do not want for if they could get to them they would tear them down. So what are the bees going to do about it? They do just this about it. Figuratively speaking, they just sit down in their easy chairs and keep an eye on that cell and when that impudent baby dares to poke its head out of that cell they pounce upon it and tear it to bits. They simply will not tolerate having that unwanted baby planted on their door step. Once more we are attempting to force the bees to do something against their will.

The Remedy

What can we do about it? Do not attempt to force the bees to do your bidding but use appeasement. Feed the bees well and get them in that frame of mind such as you have after a big Thanksgiving dinner of roast turkey stuffed with oysters served with cranberry sauce and then topped off with hot apple pie with ice cream. After such a meal you "jest hain't mad at nobody." Well fed bees that have been queenless a day or two consider it a favor to have a ripe cell given to them. They accept it with thanks and at once begin to fasten it to the comb, air-condition it and eagerly await the birth of the baby, and when that happy event occurs they nurse it with loving care and give it a good start towards becoming a perfect queen. At any rate I had about a bushel of these protectors. What grieved me was that they were noncombustible. I often called them worse words than that. Since they would not burn I had to bury them, and as I remember there were no mourners at the grave and what I said would never qualify as a prayer.

Early Experience in Cell Production

I first used a starter hive with holes in the lid in which were placed flanged cups. A cushion was placed on top to keep the cells warm. The theory was that in giving the grafted cells to the bees as soon as they were grafted one could get better results. Lots of work and poor results were my reward. In order to get better results I put more bees into the starter hive. The bees would fret and if the weather were warm many would suffocate. Then the starter hive was abandoned and another plan tried. Two hives were set to the rear and a single hive placed where they formerly stood. Two frames of honey and pollen were placed in the empty hive. About half of the bees from the two hives were shaken into the hive that was to receive the cells. This gave very good results but I thought there must be a better way, for this system had two disadvantages. First as the colonies were deprived of their bees they ran down rapidly. The other objection was that often we wanted a larger number of cells than the two colonies were able to start and we had to prepare another set which made the labor excessive.

A Colony for Both Starting and Finishing

One season we used queenless colonies for both starting and finishing cells. Cells were given every four days. Before giving more cells, more bees to the amount of three or four pounds were shaken in from other colonies. These colonies did not accept the cells as readily as we wished but finished very well what they accepted. The bees never fanned and showed they were delighted to receive the new cells. The presence of other cells may have been a contributing cause. To sum it up, the results were not the best and the labor was excessive.

Three-Compartment Hive for Both Starting and Finishing

Another plan suggested itself. A three-compartment hive was made in which cells were both started and finished. The two outside compartments contained laying queens. These were separated from the central compartment by queen excluders. The bees from both sides flew from the central compartment. When giving cells, the openings between the compartments were closed and bees from both sides were shaken into the central compartment. Flight holes were then opened in the two outside compartments. Bees flying from these openings would join the bees in the central compartment. The plan promised much but proved to be about the flattest failure of anything I ever tried. They accepted very few cells and those they did accept were not finished well. So the hive was junked. I did not burn it in the furnace as I did the wooden cups-here in Florida we have a fireplace. I then realized what a stupid idea it was after all. Why have the two hives connected? Why not shake as many bees as you needed into the starter hive?

Starting and Finishing in the Same Hive Not Practical

In reviewing all the plans I have tried in the past I came to the conclusion that no plan is practical in which cells are both started and finished in the same hive. One can shake bees into a hive containing two frames of honey and pollen and allow them to both start and finish the cells. This may be a very good plan for the small beekeeper who wants but a few cells. However, this would not be practical for producing cells in quantities as it would take a large number of colonies and they would be greatly weakened.

Back to the Starter Hive

Once more I returned to the starter hive but built along different lines and it has given such perfect results that I am sure we will never use any other method for starting cells. When everything is considered it has more advantages and fewer disadvantages than any other system. Cut out of this starter hive has already been shown.

Queenless Colonies Best for Finishing Cells

As stated, we finish the started cells in queenless colonies. I finished cells above the excluder with a laying queen below for thirty-three years before I realized this is a very poor place to finish cells. Let us examine some of its shortcomings. In the first place the top hive is where the bees like to put their honey. During a honey flow they fairly bury these cells in honey. This smothers many of the larvae and the bees tear down the cells. Those not destroyed are smeared all over with honey making it a messy job in handling them. All of this could be overlooked if we were getting the kind of cells we wished, which we were not. The main objection is that since they have a laying queen below, they have about all the brood they can properly nurse, therefore they have neither the ability nor urge to build good cells.

Ability and Urge Necessary for Producing the Best Cells

Before going any further, let us look into this matter of "urge." Before bees can do the best at cell building they must have both ability and urge to do so. A weak, poorly-fed colony deprived of its queen would have the urge to build cells but would be lacking in ability. On the other hand, a strong colony headed by a young queen all in a three story hive packed with brood would have the ability to build good cells but would be lacking in urge to do so. To remove all their brood and queen would give them both ability and urge to build the very best of cells. We start cells in the starter hive because it gives us the very best results. The bees shaken into the starter hive have been nursing larvae, so by taking away these nurses and putting a large number of them into the starter hive they have ability and urge in the highest degree. After the cells are started, they may be given to a colony having less urge and all will be finished well provided the finishing colony be strong and well fed. When these nurses have been taken away from larvae they have been nursing they act as if in distress. Possibly the excess milk in the glands in their heads gives them a headache. The best headache remedy is to give them larvae to nurse. When the cells containing larvae are given to the bees confined in the starter hive, they at once begin fanning in order to spread the glad news and at once unload the milk accumulated in their glands.

Bees Having Access to Their Queen Lacking in Urge

As long as bees have access to their queen they will not do the best at cell building. When a frame of larvae is placed above an excluder with a queen below, the bees seldom start cells. With the laying queen below, the bees have about all the larvae they can care for, so they have little urge to build cells. It is plain then that finishing cells above the excluder with a queen below is not the best method when quality of queens is desired.

Another proof I have observed *thousands of times* is found in the hive containing the breeding queen. As before explained, the frame of eggs is placed on the opposite side of the excluder from the queen. I have yet to see a single queen started there even though there is plenty of larvae just the right age. However, should the queen be killed, or if I remove her to introduce another, thus making them queenless, the bees at once fairly cover this comb with queen cells. I wish to give just one more example where there is lack of urge to build cells when a queen is present, although in this case she did not even have access to the combs in the hive. My son Huber and I were requeening some colonies of hybrid bees we had acquired. One colony had a young laying queen and there was an abundance of brood in three stories. There was plenty of pollen and there was a light honey flow on. Hunt as we would we just could not find that black queen. As we had a queen in an introducing cage and as it was getting late, we had to do something about it. We knew we were using unorthodox methods but as an emergency existed, we thought the bees might overlook this little irregularity and accept the queen. We therefore set the hives aside, put a queen excluder on the bottom board, shook all the bees from the three hives on the ground, replaced the hives and put the introducing cage in the top hive. As the hive stand was high we doubted if the queen could find her way up to the entrance of the hive and if she did she could not get in.

We thought those bees up in the top story would be delighted to get our queen. We just *thought* that. The bees thought differently about the matter and killed our queen as soon as she was out of the cage. Now note this interesting fact-not a single queen cell had been started although it had been six days that queen had been below the excluder. We found this diabolical black queen on the bottom board and killed her and the bees accepted our next queen with thanks.

The Question of Mating Hives

The miniature mating hive

Before discussing mating hives I will tell a story and it is a true one. A number of years ago a man came to see me who was conductor on a passenger train. He had held that position for 25 years and had laid up some money. He said, "I think I have served the people long enough as I have worked hard as conductor. I have about decided to buy a little farm and about 500 colonies of bees and just take life easy and let the bees work for me. What do you think of it?" I replied, "As conductor you just think you've been working. Now if you want to know what work really is you get 500 colonies, do all the work yourself, then you will realize that as conductor you just *thought* you were working." He insisted there could not be much work caring for bees. All there was to it, you just put on the supers and when filled take them off and people would come running for the honey. He may have been reading an advertisement of a supply manufacturer that ran something like this, "Costs little to start. Practically all profit. No experience needed. Very little work," etc. What I am driving at is that most of us are like that conductor more or less and I hope less.

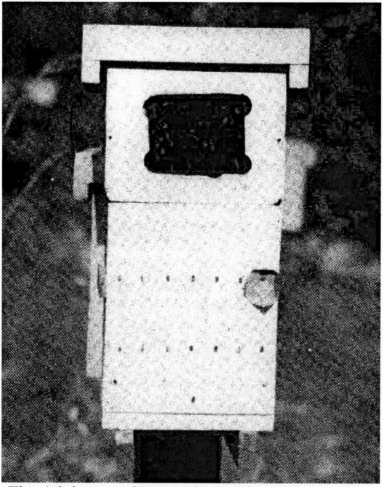

The Adolescent hive with ventilated box on top

Our five frame mating hive

That is the attitude many have in relation to mating hives. Many think that most any kind of box made of thin material will do. Nothing is farther from the truth, for the size and the way it is made is very important for best results. I have in the past, in an experimental way, made 16 different styles of mating hives. I have made them in sizes as small as the one shown in the cut, which used a single individual comb honey section about one inch square, and from that up to a two-story jumbo hive. In Indiana we used a mating hive that held three four-by-five comb honey sections with a division board feeder at the back. It did very well in Indiana but was a complete flop here in Florida as I rather expected it would be. Here is the general rule: The smaller the mating hive, the less number of bees it requires and the less feed it takes to operate it. However it requires much more labor to operate it as constant care must be taken to see that it does not run short of stores and to see that it does not become too weak in bees or that it does not become too crowded with bees or stores, for if any of these conditions are present, the bees will abscond.

The Adolescent Hive

The most successful small hive I ever used is what I term my adolescent hive. It is larger than the baby hive. It is made to hold three frames 5 x 8 inches. It is made of full three-fourths inch material and the cover is made of 2-inch material.

A small mating hive should not have a ventilator, as the air will blow in through it and cause the bees to realize their weakness and they will abscond. For an entrance a half-inch hole is made in the center of the hive. This hole is slanted upward to keep out the rain. If the entrance is placed at the bottom, the ants or robber bees will kill out the bees if they get weak.

When feeding, a battery filler can be used and a little food may be squirted into the entrance at night. A package box which is used for bringing bees from the outyard is made with wire screen sides as shown in the cut. Half a pound of bees will nicely stock this hive. A little feed is given, a ripe cell is placed between the combs, and the package box is placed on top. The entrance should be kept closed for 24 hours if the bees are taken from the outyard. If taken from the home yard the bees should be confined 48 hours, or many will return to their former colonies.

After confining them for the required time, the entrance is opened and the package box removed. To those who prefer a small hive this is by far the best of anything I ever tried. Even this hive requires close attention to see that it is kept in the right condition. We have wintered queens in such hives here in Florida but the results were rather uncertain so we have adopted a larger mating hive, one holding five standard frames. These hives are placed on concrete standards, two hives on each standard. The covers are made of one-and-one-half inch material. This cover keeps them warm in the winter and cool in the summer. They are heavy enough so no wind will blow them off. At first we formed them in spring. This weakened the colonies in the outyard to such an extent that they gathered no honey. Now we form the mating hives in the fall after the honey flow is over. We proceed as follows: As the bees in the outyard must be requeened, we take a package of one and one-half pounds of bees and the old queen. The bees and queen are put into the mating hive. As they have a queen it is not necessary to confine them if they are put in after dark. A young queen is introduced into the colony from which the package was taken. With a young queen, the colony soon makes up for the loss of the bees. After the old queen in the mating hive has been laying for a week or so she is removed and a ripe cell given. With this management, the colonies in the outyard are actually benefited. As we winter the colonies in the mating hives there will not be many to establish in the fall, only the ones that have run down for one reason or another. Some mating hive colonies may be too weak for wintering. In such cases a package with queen is put in the same as though the hive was empty. When there is no honey flow, our five-frame hive requires more feed than the smaller one. On the other hand the bees in our larger hive will gather more honey and therefore need less feed. While rearing queens we have but four frames in the hive as it is much more convenient to handle the frames. After queen rearing is over in the fall, an extra frame of honey is put in. A few may die out in the winter for one reason or another, but the rest will be so strong that they may be divided to make up the loss. When introducing cells, when no honey is coming in, we feed the night before giving the cells. Should any be short of stores, they are given an extra amount of feed. After these hives are nailed up we boil them in used crank-case oil. When dipped in cold oil, they will not take paint but when boiled in oil the oil soaks in and later they can be painted. We do not know how long such hives will last but as the oil penetrates the wood, and if we keep them well painted I believe they would last a hundred years. The entrance to this hive is in the center of the front. It is one and one half inches in diameter and the hole is slanted upward to keep out the rain.

Bees Abscond More in Florida

Why bees abscond more in Florida than in Indiana is not clear. It was stated they abscond because the ants bother them. We made the standards ant proof but it made no difference. Bees abscond here from the baby mating hives when they have virgin queens, something we never knew to occur in Indiana. Therefore, the bee-keeper must study conditions in his locality and act accordingly. As Joseph C. Lincoln would have one of his characters say, "If you got intellects, use 'em." My old friend, Allen Latham, uses a baby mating hive covered with tar paper and in his cool New England climate it gives him good results. Should his system be tried here in Florida, the bees would last till the sun got a good chance at the hives, when bees, comb and honey would come running out at the entrance. Our hives are placed on concrete standards so they come about waist high. The labor in handling them is much reduced with this arrangement. How anyone can tolerate having the hives low down on the ground is a mystery. We had ours that way in Indiana and my back still aches from it! Help will do a third more work with hives on standards.

Huber using the bee hopper

Introducing the Cells

In introducing cells no protectors are used but we simply press the cells lightly against the combs, placing them near brood as the bees will care for them better. A certain high brow told me cells do not hatch as only eggs can do that. Instead we should say the virgin emerges. From a grammatical standpoint he is correct but when I get ready to say anything I do not propose to let grammatical expressions or rhetorical technicalities hamper the thought I wish to express so, Mr. Highbrow, we shall continue to "hatch" our cells.

Stocking the Mating Hives

In stocking the mating hives it is best to take them from an outyard as they do not have to be confined so long. We do not weigh the bees but measure them instead. We have what we call our bee hopper. It is four and one half inches wide. The bottom is covered with galvanized iron. A scoop is made just wide enough to reach across the inside of the hopper. For the five frame hives we use about a pound and a half of bees.

Why Bees Tear Down Cells

Well-fed bees very seldom tear down cells. Just why I do not claim to know. I discovered this fact about thirty years ago. We were then using a mating hive holding two frames of honey. The bees tore down the cells just as fast as we gave them. I had been told that bees tear down cells because they are strange to them, similar to their reaction when introducing a strange queen. If that were true it seemed that little could be done about it. I sat down on a stump to think it over for most things can be worked out if we use the right formula. I remember that bees tear down cells worse at times than others. Why? It could not be because the cells were strange to the bees for they were always strange yet at times few were torn down while at others, as at present, nearly all were destroyed. Another reason was this: In introducing cells I often had some left over and would put them back into any hive I was using for cell building, often putting them into a hive that had not built them, yet in doing this many times, *there never was a single cell torn down.* That was proof that the bees did not tear them down because they were strange. Then why did the bees in the mating hives destroy the cells while the bees in the cell builders did not? Plainly it was because the bees in the cell-building colonies have been well fed while the bees in the mating hives had not. To test out my theory I fed the bees in the mating hives that were to receive the cells the next day. The more I thought of it the more certain I was that it would work. *But,* what would the bees think about it? Cells were given the next day and I was delighted to see a high percentage of them was accepted.

Well-Fed Bees Do Not Tear Down Cells

So my conclusion was that well-fed bees do not tear down cells. True, even during a heavy honey flow some cells will be destroyed but in such cases there may be a large amount of unsealed brood and few fielders so in reality the bees are not well fed. Liberal feeding will make acceptance sure. When there is a honey flow we do not feed but remove the laying queen one day and give a cell the next day and have very little loss. Often, when we had an abundance of ripe cells, we removed the layer and at the same time put in a ripe cell. When there was a honey flow we had a good acceptance but the results are rather uncertain so we do not recommend it.

How Long Should the Bees be Confined to the Mating Hive?

In case bees are taken from the outyard they should be confined to the mating hive one day and released after dark the day following. Some may ask why confine them at all when taken from the outyard? If liberated at once they do not seem to realize they have been moved and at once fly out without marking their hives with the result that many will gang up and go into one hive leaving the hive from which they came depleted in bees. This I have experienced in the past to my sorrow. At one time we took a hundred baby hives stocked with bees to an out-yard and at once liberated them. Immediately the air was full of bees and after circling about for some time they all tried to get into one baby hive and they covered it with bees about a foot thick! We surely had a time trying to distribute them. When confined for a day and released at night, they seem to forget their old home and behave as well-bred bees should. In case bees are taken from hives in the home yard close to where they are to be established they should be confined two days and released after dark. Most of them will stay but a few will be gifted with remarkable memory and will go back, but there will not be enough to materially weaken the bees in the mating hives.

Mating Hive Records

One must have some system of knowing what the conditions are within the mating hives. We have used a number of systems and out of all of them we have worked out a system that gives such perfect satisfaction that we never expect to make a change. We need to know the conditions within the mating hive for a period of two weeks, for by that time the virgin should have been mated and by that time she is a laying queen. Therefore, we have a row of copper tacks across one side of the hive, 14 in number, one for each day of the week for two weeks. These tacks are driven about halfway in. The first tack at the left represents Monday. By putting a mark over Thursday's and Sunday's tacks, we can see at a glance what days we are working on. Tags are made of galvanized iron two inches square. A square hole is made in the center of the tag. The corners are bent forward a little for easy handling. They are first dipped in yellow quick-drying enamel. After drying, each corner is dipped in enamel of the following colors. First black, then red then white, then blue reading to the left. Now let us consider the hive and it's signal relation to the queen. Let us say we introduced the first cell on Monday. The tag is placed on the tack at the extreme left and turned to red, indicating a cell has been given. We examine the hive Wednesday and if the virgin is there the tag is turned to white. Should the virgin not be there, the tag is turned back to black indicating the hive is queenless. When another cell is given, after the first one is not accepted, it is put in on Wednesday so the tag is moved to the third position and turned to red. This process is continued throughout the season. A picture of our mating hive has already been given. To indicate the condition within the hive we have blocks four inches long painted black so they may be seen at a distance. A number of these blocks are distributed among the hives, being placed on the stands between the hives. In case the hive can spare honey the block is placed lengthwise of the hive at the extreme left. If it needs honey it is placed crosswise in the same position. The block shown in the cut shows the hive can spare both brood and honey. If placed crosswise at the extreme right it indicates that the hive needs brood. If placed lengthwise on the same position it indicates the hive can spare brood.

Introducing Queens

Since the year 1900 I have spent much time in experimenting with introducing cages and hundreds of queens have been sacrificed in the interest of science or the lack of it as you will. I have tried all the fads such as daubing the queen with honey, sprinkling peppermint water on them, strangling the poor long-suffering queens with tobacco smoke and have tried many other such crack-pot methods with dire results. Some still just daub the queen with honey and drop her in. When you do that I suggest you say, "And may the Lord have mercy on your soul." Many queens "introduced" in that way will be killed outright but what is even worse, many will be injured and a queen may remain at the head of the colony some time before being superseded thereby losing a crop of honey. I have seen many queens that have been maltreated and then allowed to head a colony. One queen in particular was shiny with no hairs left and was completely devoid of clothing which might cause one to infer that she either was a member of a nudist colony or had lost in a game of strip-poker. In my years of experimenting I have come to the conclusion that the bees and the queen must become acquainted and no short cut can accomplish this. During a honey flow one may have a certain degree of success with any of these "quick" methods which will continue till disaster overtakes him. One party wrote us that he had never lost a queen when using the regular mailing cage so we could not interest him in any better method. Later he wrote that in introducing 12 queens he had lost five and wanted to know what was the matter. I wrote him that there was nothing the matter as that was perfectly normal. Probably he had formerly introduced his queens during a honey flow and the latter when there was no flow. Another party drove about 200 miles to our apiary to get five queens as he said he did not want them to go through the mail as they might get injured. As he seemed so careful I offered to lend him some of our introducing cages. He declined to accept our offer stating he just dropped the queens in honey and then put them right in among the bees. I told him I would not give a nickel a piece for queens introduced that way. Later he wrote that he lost two in introducing and wanted me to replace them. No, we did not replace them but did give him a lecture on introduction. I doubt if it took, for we never heard from him again. I have known queens that were not properly introduced to stay on the bottom board for several days before beginning to lay. Then they would lay for a short time and be superseded. I do not know what caused the injury as they did not appear to have been balled. The queen being a very nervous creature may have received a nervous shock which injured her. We must remember it is as much the attitude of the queen as it is of the bees that makes for perfect introduction. When the queen emerges from the regular mailing cage she is often frightened for she suddenly finds herself among strange bees. Often in her fright she goes taxiing over the combs squealing as she goes. As she does this the bees seem not to know just what is going on but realize there is some excitement and they want in on it so, as the queen goes by, a bee may grab a leg and others join in and form a ball around the queen either killing or injuring her. Sometimes they sting her but not often. Sometimes the queen's legs are broken. Such queens never amount to anything and the owner of such queens will report that he does not like that strain of bees for they just do not get honey.

How Soon May a Queen Be Released Among the Bees?

Our vast amount of experimenting along that line leads us to believe that when conditions are not favorable, as when there is no honey flow, the queens should not be liberated under *four days*. When conditions are favorable, as when there is a honey flow, three days will suffice. Even then, if released from an ordinary mailing cage, she may be injured. Now, how does our cage remedy all these defects? In our cage we use the principle that is acknowledged by all to be good. This principle is described in the bee literature of 60 years ago. That principle is allowing the bees to get into the cage with the queen instead of the queen getting out among the bees. As far as getting frightened is concerned, the case is reversed, for when a worker gets into the cage with the queen, it is the worker that is frightened. In my experiments, I have often observed the behavior of the first bee that gets into the cage with the queen. The first bee to get into the cage often would buzz around trying to get out. After a time it would flatten itself out and lay on the bottom of the cage, stretch out its tongue and offer food to the queen. She is too well bred to refuse so they become friends and band together for mutual protection. Soon another bee enters and it seems more frightened than the first for now there are two bees for it to contend with. However, they soon become acquainted and more bees come in and the queen begins to lay, the bees either eating or destroying the eggs. In another day the bees eat out the candy at the end of the partition and the queen goes out among the bees, not as a fugitive but as an accepted queen, and soon will have a large patch of comb filled with eggs.

Requirements of the Perfect Cage

First, the cage must be made of wood. A metal cage will burn the queen in the summer and chill her when the weather is cool. We sacrificed a number of queens by using a cage containing too much metal. We made a few cages of cork but the bees chewed up the cork in short order.

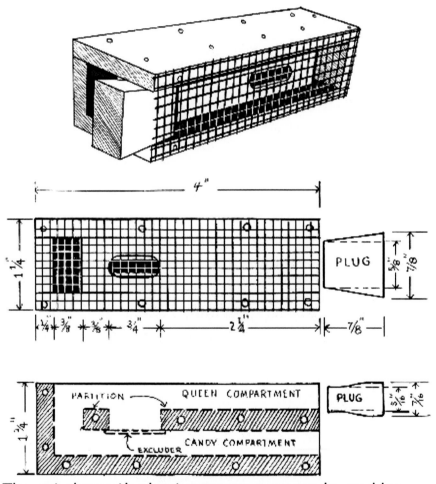

The cut shows the best cage we ever made, and by its use we have never lost a queen or had one injured due to the fault of the cage. We lose a few queens due to our own fault such as introducing a queen into a colony that is superseding its queen. We remove one queen, not knowing there are two, and in such a case our queen is killed as soon as it is out of the cage. Even at that our losses will not run over two percent and none whatever when everything is normal. Our cage has two compartments, one for the queen and the other for the candy. I have found that the bees eat just three-quarters of an inch of candy every 24 hours. This they do with surprising regularity.

The drawings of our introducing cage were furnished through the courtesy of *the Southern Beekeeper*. The drawings illustrate the cage better than photographs.

In the partition between the two compartments is a piece of zinc queen excluder. This is just two and one-fourth inches from the opening of the cage. The partition extends to within a quarter of an inch of the back of the cage. The compartment on the opposite side from the screen is filled with bee candy.

The queen is placed in the second compartment with no other bees. It takes the bees just three days to eat out the candy to the queen excluder. After the bees enter the compartment with the queen they eat from both ends of the candy, so they eat out the candy in one day and release the queen. When there is no flow, the candy compartment is filled completely so it is four days before the queen is released. When there is a flow, the candy compartment is filled to three-fourths of an inch of the opening, thus releasing the queen in three days. This cage is not patented, so anyone handy with tools can make it but it is best to have a sample.

The Push-in-the-Comb Cage

Years ago I brought out a cage with a wooden frame with metal teeth for pressing into the comb in order that the queen could lay in the comb before being released. When properly handled, this cage was sure in its results but too often it was not properly handled. It had to be used on old combs or it would fall from the comb and the queen would be killed. Then the hive had to be opened in order to remove the cork to let the bees into the cage and it was not suitable for outyard work. Again we could not use the automatic eat-out plan as the queen would remain on the comb and refuse to come out. Altogether, with the average beekeeper, it was not a complete success and we do not recommend it, nor do we now use it, as our present cage is just as sure in its results and the beginner can get just as good results as the experienced beekeeper. In using our cage, two combs may be spread apart and the cage placed between them. It may be placed in the super or on the bottom bar of the brood frame. The cage should never be placed on the bottom board as the ants may get in and kill the queen or she may be chilled.

Does Opening the Hive Cause the Bees to Kill Their Queen?

My answer to that is positively no! Often we open the hive as soon as the queen is out of the cage and in thousands of such cases we never had a queen killed or injured. A common belief is that the hive should not be opened until two weeks after the queen is introduced. In case the queen is not properly introduced, the bees will ball her whether or not the hive is opened.

The Home Yard Cage

We have what we call the home yard cage-the cut will show how it is made.

Often we have one or two queens to introduce in the home yard, and do not take the trouble to load a cage with candy, so we use our home yard cage. In the wooden side is a queen excluder. The queen is put into the cage and the cage is laid on the top bar or bottom bar with the queen excluder down so the bees cannot get in. In two or three days, depending on conditions as previously mentioned, the cage is turned over thus allowing the bees to enter. In one day more the queen is released. This is just as sure as the candy cage but requires more attention to opening the hive and turning the cage over at the right time.

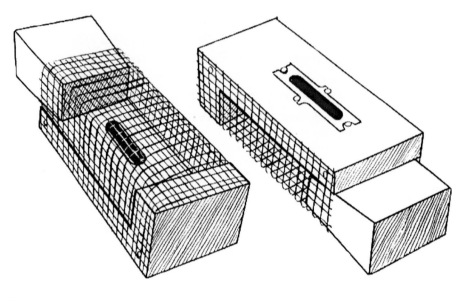

Bee Candy

Candy for the introducing cage or for mailing cages is made by mixing XXXX powdered sugar with honey. This should be kneaded as stiff as possible. If the honey is heated it will absorb more sugar. If one has bee disease, or is unable to get a health certificate before shipping, candy may be made of invert sugar and such candy is nearly as good as that made with honey. To make invert sugar, put half a pound of water in a kettle and let it come to a boil. Then add one pound of granulated sugar and stir in one-fourth teaspoon of tartaric acid and boil till a temperature of 248 degrees is reached. A candy thermometer is necessary for this. No certificate of health is required by the postal authorities when using invert sugar candy. All that is necessary is a statement that there is no honey in the cage. After the invert sugar is cooled it is used the same as honey in making bee candy.

Shipping Queens

Our experience has taught us that the reason so many queens die in the mail is lack of ventilation or because they are overheated. We have found that by wrapping the cages in cloth, they go through in good condition if they are not too long on the road. We also ship in mailing tubes which give even better ventilation. Shipments made with mailing tubes have gone to 43 foreign countries by air with almost no loss.

A Review

As we though it necessary to do a lot of explaining as we went along, we will now give the procedure step by step:

1. Stock the breeder hive as described, putting queen and one frame of brood and the two wood-comb frames in the back compartment, putting the frame of brood in the center.

2. Prepare as many colonies as are needed by putting them in two story hives with brood above the queen excluder and the queen below.

3. Cut Modified Dadant foundation in four pieces and put them in frames and get this foundation drawn in any colony suitable.

4. Take out the frame of brood from the back compartment of the breeder hive and put it on the other side of the partition, being sure the queen is in the back compartment. Then put in a frame of the drawn foundation. Leave it there for 24 hours, then put it on the other side of the partition. In case you wish to start cells every day, put in another frame of drawn foundation in the back compartment. Repeat this every day until the fourth comb is placed the other side of the partition, when the first is ready for use.

5. Put three frames of honey and pollen in the starter hive and put some thin syrup or diluted honey in the middle comb. Then shake in bees in the amount of five or six pounds from the hives previously prepared. Leave them three or four hours.

6. Take out frame of young larvae and cut into strips, mount them on bars and give them to the bees in the starter hive.

7. Next day remove the queen from two colonies to be used as cell finishers and give to each colony one frame of cells and dump the starter hive in front of the colonies from which the bees were taken.

8. When all finishing colonies have three frames of cells each, and the fourth is ready, remove the first given and introduce the cells to mating hives.

Starting Cells Every Third Day

In starting cells every third day, but two finishing colonies will be needed. In this case you put in a new comb in the breeder hive every third day and when it is placed the other side of the partition, an old comb or a frame of brood is placed in the back compartment between the two wood-combs. There are both advantages and disadvantages in starting cells every third day instead of every day. We are running more to the third-day plan. One advantage it has is that the drawn sheet of foundation may be left for the queen to lay in longer. Another is that as cells are introduced every third day, in introducing the cells, the ones introduced three days before can be examined. If using the third-day system, in order to keep the record straight, one should write down on a piece of paper the letters C S I, C S I in rotation, these letters being placed at the left hand of the paper, one below the other down the paper. C stands for comb and indicates that you gave a new comb to the queen in the breeder hive. S means that you start cells, and I means you introduce cells. Put the date and the day of the week after each letter. Of course in starting there will be no cells to start or introduce, but after some time the letters will have a meaning. In case one started cells regularly every third day there would be little use for the above memorandum as the condition of the cells in the finishing colonies would indicate what is to come next. However, in case you slipped one starting, all cells would look much alike and you might get things mixed up. At any rate I have done so and when I came to giving the started cells to the finishers I found the hive full of virgins!

Modifications for the Small Beekeeper

For the commercial queen breeder, or for the honey producer who wishes to rear 200 queens or more I strongly recommend the management and equipment that have been described. For the small beekeeper, or the back-lotter who wishes to rear a dozen or maybe 50 queens in a year, no special equipment will be necessary. Under such circumstances it is best to rear queens during a honey flow as conditions are more favorable and queen rearing is a simple matter compared to rearing them when there is no flow. When the honey flow is just beginning is the best time to start as there is an abundance of pollen coming in along with the nectar, and pollen is just as necessary as nectar.

For starting cells, first get the foundation drawn as described. Next set to the rear the hive containing your breeder, with the entrance in the opposite direction from the original position. In its place put another hive and in it put two frames solid with honey and pollen. Shake in half of the bees from the breeder hive and put the breeding queen into the prepared hive. Within 24 hours she will have a good patch of eggs in the new comb. The comb should be placed over an excluder on a strong colony. In case you do not wish to start more cells immediately, this hive with the two combs and the one with the eggs may be set over the hive formerly containing the breeder, setting this hive on its former stand. The two hives must be separated with an excluder, with the queen in the bottom hive. This colony must be well fed for we want the larvae that will soon come from the eggs to be well fed. As soon as the eggs hatch, turn a strong colony to the rear and in its place set a hive into which you put two frames of honey and pollen then shake out half of the bees from the hive you have moved into the hive with the two combs. In case the colony is not of sufficient strength, two colonies may be set to the rear and half of the bees from both colonies shaken in. Next cut the comb into strips as has been described and give a frame of two bars of prepared cells for this colony to both start and finish. In case one prefers to use a starter hive, one can be made by merely tacking a screen to the bottom of a hive. If this is done the hive should be set on blocks high enough to afford plenty of ventilation.

Mating Hives for the Small Beekeeper

In case increase is desired a hive may be moved to a new location and another hive set in its place. Into this hive put two or three frames of brood with adhering bees from the hive moved, being careful not to get the queen. It is well to wait two or three days after dividing this colony before giving a cell for bees will keep returning from the hive moved and are apt to destroy the cell given. As soon as the queen has mated and laying, the hive should be filled out with combs or frames of foundation or starters. In case feeding is necessary a good way to feed is to tip the hive back a little and under the front place a block. Then at night pour in a little syrup. Be careful not to put in more feed than the bees will consume during the night for if any syrup is left it will incite robbers as a bottom board feeder is the worst of any feeder to cause robbing.

Mating Queens in an Upper Story

In case one does not wish to make increase, a good way to mate queens is in an upper story of a two-story hive. A frame covered with wire screen should separate the two hives. Always use a wire screen and *never a queen excluder.* Some in the past have advocated using a queen excluder but in the great majority of cases the bees will kill the virgin as soon as she emerges. In the frame of the wire screen an opening of half an inch should be made for a flight hole. This entrance should be to the rear of the hive below or the queen in returning from her mating flight is apt to get into the hive below.

Hive with rear entrance

The entrance to the top hive should be closed for two days, otherwise too many bees will join the colony below. Syrup or honey should be put into one of the combs and a ripe cell given. After two days the entrance should be opened and the virgin will soon fly out and mate. When the honey flow permits, this is an excellent way to requeen the colony for we have *two laying queens* instead of having it queenless while introducing a queen or cell. The bees in the top hive will not suffocate from having the entrance closed for the screen will afford abundant ventilation. The best way to put syrup into the combs is with the use of a pepper box feeder. That is a mason jar with the cap full of holes. With this the syrup may be shaken into the combs. If you merely want to requeen, all that is necessary is to hunt up the old queen below and remove her and take out the screen. The bees will not fight as they have kept up their acquaintance through the screen. In case increase is desired, just move the bottom hive to a new location and set the top hive on the old stand. The lower hive should be moved as it is much stronger than the top hive. The bees for a time will try to get into the back of the hive but after a time they will make a chain around to the front door and all will be well.

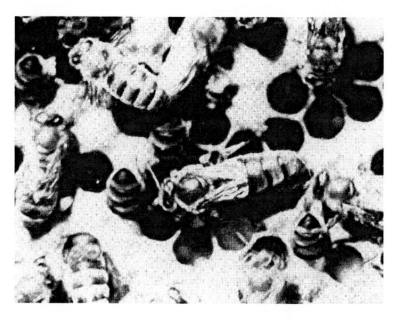

Requeening by Giving a Ripe Cell

As a rule we do not recommend giving a ripe cell to a full colony, as it weakens the colony by being deprived of a laying queen for nearly two weeks. Again, a strong colony does not accept a cell as readily as a weaker one, and if the cell is not accepted it will further weaken the colony. However, when the honey flow is over and more bees are not needed immediately, it is often desirable to requeen by giving a ripe cell. As it is difficult to find a virgin in a strong colony, one may give a cell and without looking for the virgin another cell is given three days later. In a week after the first cell is given, the colony may be examined and if there are no queen cells started, you may know a virgin is present. In case both cells are destroyed, or should the virgin be lost in mating, it is best to introduce a laying queen as the colony would become too weak if given another cell.

Clipping the Queen's Wings

The question has often been asked, "Does it injure the queen to clip her wings?" In case the wings are clipped too close to the body it may injure the queen as veins and nerves are cut. We clip the tip ends of both wings which prevents the queen from flying and in no way injures her. We do not advocate clipping one wing as the queen is apt to be injured. We used to clip one wing but too often just as we squeezed the scissors the queen would poke a leg between the blades with the result that she got an amputation along with the clipping. We prefer to have all the queens clipped in the outyard that we may know if any have been superseded.

Do Queen Cells Interfere with Introduction?

As far as we have been able to determine, the bees accept a queen when queen cells are present just as readily as when there are none. However, one must be sure the cells are not too far advanced. Cells that have just been capped or are about to be capped we pay no attention to but if they have the appearance of being well along and nearly ready to hatch, we go over the combs carefully and remove every cell. If this is not done a virgin might emerge before the laying queen is out of the cage and of course would kill the layer, for a fight between a virgin and a layer always results in the death of the layer. Even though the layer is out of the cage and laying, she might not get around to destroy the cells and a virgin would emerge and kill the layer for her lack of good judgment. A laying queen is often very reluctant to tear down cells as in nature she is very seldom called upon to do this, while a virgin is a pastmaster at this kind of sabotage for I have known a virgin to tear down over a hundred cells in 24 hours.

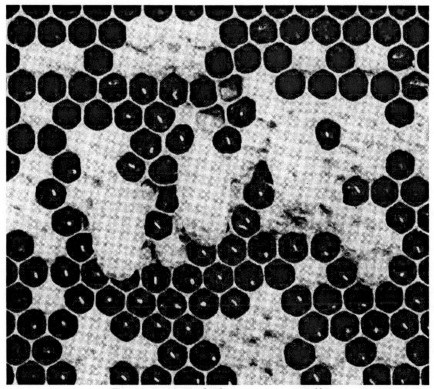

Eggs around the queen cells

The photo will show a case in point. A laying queen had been introduced and she seemed to be either a Jehovah's Witness or a Conscientious Objector for she refused to do battle with those cells. I photographed this and wanted to see the negative before destroying the cells and left them in till the negative was developed. Upon examining them I found a cell had hatched and a virgin was loose in the hive with my layer. She was too young to do battle but was in training and in a few hours more would have purged that layer. I killed the virgin but that layer had a close call. It would have served her right for being such a rank pacifist.

Breeding Queens

One must use his best judgment in selecting a breeding queen. Many have advocated breeding from the queen whose colony made the most honey. That would be a very good plan if one could be certain the larger yield was due to the better gather qualities of the bees. Many factors enter in that make for a better yield. For instance, when bees are out for a flight for the first time the air may be filled with them. In returning to the hive many of these bees are unable to locate their own hive among so many all set close together with the result that, when they alight at the entrance of any hive, they fan their wings, sending out the scent and thereby attracting in all the bees in the air. In this way several pounds of bees may be added to a colony which results in a better yield although the queen may be an inferior one.

The Colony that Gave a Bumper Crop

In Indiana we had an outyard laid out in the form of a triangle as that was the shape of the plot on which we had our bees. During the sweet clover flow one colony produced three supers of honey while the others averaged about two supers. In the fall that colony produced two supers of honey from smartweed and asters while the rest produced a little less than one super. Surely that colony that so far outdistanced the others must have a queen that would make an excellent breeder. I thought I would take a look at her but alas, when I opened the hive, I found it not only had no queen but was *fairly lousy with laying workers!* Just why then the big yield? This colony was located at the point of the triangle to the west and the fields of nectar lay to the west. It was evident that the bees in returning from the fields-maybe the ones out for their first load-stopped at the first hive they came to and kept it packed with bees.

Large Bees Better Gatherers

One of the most scientific experiments ever carried on in my opinion was conducted by Doctor Merrell at the Kansas experiment station. After some exact research work in finding that some bees actually gather more honey than others, upon examining them he found that they were *larger.* That is as we would expect. Which can haul the most goods in a day, a large semi-trailer truck or a pick-up? With this in view we have tried to choose for our breeders the ones whose workers are larger. We feel sure that if any progress is made in the future in producing bees that get more honey it must be by selecting the *largest drones*. Drones vary much more in size and shape and color than do either the workers or the queens so I believe we should proceed along this line.

Artificial Insemination

For several years we used the system developed by Dr. Lloyd Watson for inseminating queens. The system is a success and much credit should be given to Dr. Watson for perfecting it. However, in our case we could not afford to carry it on as it required most of the time of a skilled operator. My son Jay Alfred did this work and accomplished a great deal for the short time he had it in hand.

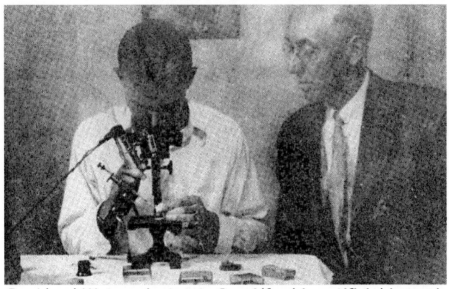

Dr. Lloyd Watson instructs Jay Alfred in artificial insemination

We are still interested in controlled mating, for no great advances can be made without it. Here in Florida we tried island mating one season, the island being located two miles from our apiary. To test the location to see that no drones were within mating distance, we took six mating hives with virgin queens but *no drones*. The idea was that if all six of these virgins failed to mate it would be proof that there were no drones within mating distance. In that case we would then weigh out our largest drones, take them there and get controlled mating. But it was not to be. Four out of six mated with the drones of our yard two miles away. We did not believe they would mate with drones at that distance, so we learned that much.

Many beekeepers claim they get pure mating as there are no bees other than their own within mating distance. I believe if they would try it as we did they would find they would get mating from drones far away. Nature is very solicitous about preserving the species and the queen and drones do court on the sly in a manner unknown to those who try to snoop into their private affairs. Possibly the antenna of the drone picks up the high-pitches sound of the queen's wings miles away. Who knows?

Bees and Chess

I love the game of chess possibly from my nature of tackling something I can't do. I am not much of a player but probably get more of a kick out of it than the masters do. I have often said that chess and bees are similar in a number of ways. First, the queen is the most important individual in both games. If one loses a queen in a colony of bees he may requeen from the larva in the lowly worker cell. If he loses his queen in a game of chess he may requeen from the lowly pawn, or rather *maybe* he can.

Again both games are so deep that the master minds of the past have not been able to delve to one-tenth of the depth of either game. It has been stated that if all the games of chess that can be played were numbered it would represent a string of figures a yard long. So also the bee problem is so deep it cannot be fathomed. And last, the similarity of both games lies in the fine class of good sports that plays both games. As Will Rogers said about congressmen, "they are the finest bunch of men money can buy." The friends we make at bee meetings and the friends we make by letter in all lands make life worth living. Let me quote from that prince of men, W.Z. Hutchinson, in his book *Advanced Beekeeping*: "fortunately, however, the perfection of man's happiness bears but little relation to the size of his fortune and many a man with the sound of bees over his head, finds happiness, deeper and sweeter than ever comes to the merchant prince with his cares and his thousands."

Amusing Happenings

Many amusing incidents occur while working with bees, two of which I wish to relate. These two incidents I wrote up for the journals some years ago. How often I have told them I do not know, but my wife says that if she had a dollar for every time she has heard me tell these stories she would endow a home for the feeble-minded. I do not believe there is anything personal in that remark.

The Tramp and the Bees

Linger

During the first world war I had some bees on a lot near the back door of a house where lived a dear old lady, Mrs. Mechlin by name. She was one of those kindhearted women who could easily be imposed on. I was teaching in the high school at the time and was out before breakfast working with the bees to get as much done as possible before school time. I was meditating over the fact that help was so scarce as so many of the boys had gone away to war. Apples were rotting on the ground because there was no one to pick them up. Sweet potatoes were rotting in the ground because there was no one to dig them, and other foodstuffs were in the same bad way. While thus thinking over these things, along came a big fat tramp who went to the back door of the widow Mechlin's house and put up the usual pitiful tale, "can you give a poor fellow a bite to eat? I haven't had a bite since yesterday." Neither had I, but I was working early and late trying to do my bit while that worthless no-account tramp was begging from that dear old lady. Mrs. Mechlin relied, "Why most certainly, my good man. You just sit right down in this chair and I will prepare you a good hot breakfast." I was furious. How I wished I could get him out among those bees. I bet I could put some pep into him. Why I could-Well, sir, believe it or not maybe my prayer was answered, maybe it was mental telepathy or maybe the stars were just right. I do not know what, but anyway kind Providence delivered him into my hands. He slowly arose, came out into the yard where I was and sat down on a hive not more than six feet from where I was working and right in front of the hive I was about to open. He asked, "What are those things?" I could hardly realize anyone could be so ignorant as not to know what bees are but evidently he did not. Thinks I, "Well, old boy, right here is where you are going to get your first lesson in bee culture." I reached down and took hold of the bottom of the hive, raised it about six inches above the stand and let it fall back with a bang. Well, sir, it was a great success. No soldiers ever responded to the command "charge" more nobly than did those bees. It looked as though all the bees in that hive had been shot out of a cannon and their aim was perfect. They covered the face of that tramp till I could hardly see it! He gave a screech, fell over backward, tipping over the hive he had been sitting on. He made a sort of barrel roll, got on all fours and started through the sweet-corn patch with a speed that would do credit to a trained athlete. The bees in the hive he turned over joined in the frolic but these did not sting him in the face. The tramp vanished forever from my sight. After a time Mrs. Mechlin came out with a large tray filled with delicious food, hot biscuits, hot coffee and cream and a big slice of beefsteak. She asked, "where has my man gone?" I replied, "I do not know where he has gone but if he keeps on the way he started he is there before this no matter where it is." Then I added "Furthermore, Mrs. Mechlin, I am confident that for some time in the future our journeyman will take his hand-outs standing." To think she had prepared that excellent delicious breakfast for that worthless tramp. And it *was* a delicious breakfast. I ate it.

The Professor and the Bees

One day a certain Dr. Hamilton called me up and informed me that he had just purchased some bees, five colonies in all, from a party living but two blocks from his home. He had moved the colonies home but reported that most of them had gone back to their former home and about a bushel of them were hanging on a post. He asked what he should do. I told him he had done too much already, but suggested that he move the colonies back, then after dark move them a couple of miles, then in a week or so bring them home. He thought that too much trouble and wanted to put a hive there for me to bring over a queen. This I did. The bees were furious and began stinging me before I got near them. I was slapping them right and left every time one lit on me sometimes before

it stung but usually just after. Here the Doctor inter-
rupted. "My friend, evidently you are not familiar with the
psychology of the honey bee. The bee is an extremely
intelligent insect and will never harm anyone unless it
thinks you are going to harm it. Any quick motion you
make causes the bee to believe you are going to harm it
and it therefore stings in self-defense. Didn't you know
that?" "No, I did not know that but I want to learn more
about bees so wish you would teach me. Please take this
smoker and drive the bees off from that post into the
hive." Very well, I shall take pleasure in teaching you the
fundamentals of bee psychology." He took the smoker and
calmly walked toward the bees. Being sure there was
going to be a scene I retired well out of range to watch
developments and, let me say, things surely developed at
a rapid pace. A bee stung him on the back of the neck but
he went calmly on. Another stung him on the ear, still his
serenity of disposition was undaunted. Then one gave him
one of those double-strength hot ones right on the end of
his nose. His placidity vanished instantly and he fairly
exploded. He dropped the smoker and brought his hand to
his nose with a vicious slap. In doing so he knocked off his
sailor hat displaying a shiny bald head. The bees fairly
covered his head and the professor went off on the run
and took haven in a small house that had a half moon cut
in the gable. Then I called to him, "Professor, don't you
know you should never fight the bees? Their psychology
will just not stand for it." As he rubbed the stings out of
his head he replied, "But such action on their part is
wholly incomprehensible and intolerable for I meant them
no harm whatsoever. I only mean them good and wished
to help them out of their difficulties and just see how
unappreciative they have been in attacking me with no
provocation whatsoever." "In other words, Professor, you
no longer consider the bees the embodiment of superb
intellectuality but just a diabolical pusillanimous palooka

and the idea that the bee has any intelligence whatsoever is all hogwash is that right?" The Professor declined comment. He never went near those bees again. One of the neighbors said the Professor refused to have anything more to do with them as they were so unappreciative of his efforts in their behalf.

Conclusion

Life to me has been such a paradox, so much happiness and joy yet so much misery and sorrow, that makes one wonder just what it is all about. I will close by quoting my favorite poem by Leigh Hunt, as it best gives my aim in life.

Abou Ben Adhem
Abou Ben Adhem (may his tribe increase)
Awoke one night from a deep dream of peace,
And saw within the moonlight in his room,
An angel writing in a book of gold:-
Exceeding peace had made Ben Adhem bold,
And to the presence in the room he said,
"What writest thou?" The vision raised its head,
And with a look made of all sweet accord,
Answered, "The names of those who love the Lord."
"And is mine one?" said Ben. "Nay not so."
Replied the angel. Abou spoke more low,
But cheerful still; and said "I pray thee then,
Write me as one that loves his fellow-men."
The angel wrote and vanished. The next night it came again, with a great awakening light,
And showed the names who love of God had blessed,-
And, lo Ben Adhem's name led all the rest!

THE END

Addendum: The Better Queens Method

(condensed from Better Queens By Jay Smith)

Transcriber's and Editors preface.

I found it difficult sometimes to figure out the next step from the entire book so I edited a condensed version of *Better Queens* because, while the full book is wonderful and educational and should be read before this condensed version, it chases a lot of "rabbits" and explains a lot of the reasons and explains a lot of the failures to help you avoid those. Assuming you have read the full text of *Better Queens*, and are convinced you want to follow his method and want it distilled down to just the method, here it is, still in Jay Smith's own words. When working through the method this makes a better "checklist". The pictures are available in the above text so refer to the full book for the pictures.

The Breeder Hive

The cut (see above) shows the frame partly filled with wood, leaving a small comb five and one-half inches by nine and one half inches. These frames we shall call wood frames. It takes the queen but a short time to keep these combs filled with brood so she can devote her time to laying in our new comb. The standard size foundation will do very well but we prefer the size used in the Modified Dadant frames as it is deeper. A sheet is cut into four parts and fitted into standard frames. These frames of foundation may be drawn out in either the finishing colonies or in the front part of the breeding queen hive. When one of these new white combs is placed between the two combs partially filled with wood the queen at once begins to lay in it as the other two combs are filled with brood and the new comb is right between. This makes an inviting comb for the queen to lay in, which she does at once. A wooden partition separates the compartment containing the three combs from the rest of the hive. A cleat is nailed to the bottom of the hive below the partition. A space of an inch is left between the cleat at the bottom and the partition above. A zinc queen excluder is tacked over this opening. It is best to have the excluder at the bottom for if higher up the bees insist on filling the space with comb which they will not do when the excluder is placed near the bottom. When using the regular cover, the queen would sometimes get out of her partition due to the fact that wax or propolis would accumulate on top of the partition. To overcome this the partition is extended three-fourths of an inch above the edge of the hive. We use two covers, one for the back part and one for the front. Then the regular telescope cover is placed on top. With this arrangement the queen never gets out of her stall. The cut will show how the hive is constructed better than a description.

We use a jumbo hive cut down to fit the standard frames. One may use the standard hive but it will be necessary to nail cleats around the bottom to make room under the frames. The entrance is about midway in front of the hive.

Keeping the Breeder Hive Stocked with Bees

If cells are started every day the bees in the breeder hive will rapidly decrease in numbers as the only emerging brood they will have is what comes from the combs in the wood frames, therefore, some way must be provided to keep up the strength of the colony. We open the starter hive in front of the breeder hive when we take the started cells out. Many young bees will take wing so when we remove the starter hive, these young bees will join the bees in the breeder hive. This is a great help but shortly they will need more bees in which case we shake the bees from not more than one comb from the starter hive. In this way their strength will be kept up. Sometimes in the past we dumped the bees from the starter hive in front of the breeder hive. This proved unsatisfactory as so many bees were apt to kill the breeder. When they did not the breeder was apt to stop laying for a time. Adding the smaller number of bees has never given us any trouble.

The Feeder

This is an important part of the breeder hive. When no nectar is coming in the bees must be fed if we are to get that lavishly fed larvae we have pointed out is so necessary for best results. It is built on the back of the hive. A hole is bored in the back of the hive so the bees can get into the feeder. Slats are put in as illustrated, then melted wax is poured in to make the feeder waterproof and also to wax the slats tight to the bottom in order that no syrup or honey can get under these slats to sour. A cleat is nailed to the hive just above the feeder to keep out the rain. When feeding, the cover of the feeder is slid back and the honey or syrup poured in. As there is no bee disease within many miles of our apiary we feed off grades of honey. Honey is better food for the bees than sugar syrup but if disease is present sugar syrup should be fed. However, from years of observation I am confident the nurse bees cannot secrete milk with proper nourishment when sugar syrup is fed. I was surprised when I first put honey in the feeder. It was during a robbing season and I wanted to test it for I had a theory feeding with it would not cause robbing. It proved better than I expected for by looking at the bees at the entrance you could not tell they were being fed. With any other feeder I ever tried we had to feed at night or the robbers would rob out the colony. What causes robbing is that the bees get daubed with honey and then go out in front to get more air. The bees from the other hives pounce upon them and then proceed to clean up the whole hive. With our feeder few if any bees get daubed and those that do have to pass through a cluster of bees, then through the queen excluder and then another cluster of bees. By that time they are licked clean and do not have to go out for an airing.

Stocking the Breeder Hive

In case your breeding queen is in a fairly strong colony, the bees and brood are placed in the front compartment and one frame of brood and the queen is put in the back compartment. The two frames fitted with wood are placed in the back compartment with the frame of brood between them. Feed is then placed in the feeder. As soon as the queen has eggs and brood in the two outside combs, the frame of brood is removed and a new white comb is placed between the two combs. In case the colony is strong, most of the brood may remain in the front compartment, but later most of this must be removed to make room for the combs containing eggs. As stated, we should have several drawn combs ahead. When the brood in the front compartment is removed frames with sheets of foundation may be given. As the bees are being fed, they draw the foundation readily. The comb given should be left 24 hours when it will be well filled with eggs. It is then placed in the front compartment and another frame of new comb is given. This is repeated every day until there are three frames containing eggs in the front compartment. When the fourth comb of eggs is put in the front, the first one given will now contain tiny larvae just out of the egg and as this is practically all the brood that large numbers of bees have, the tiny larvae will be found to be fairly flooded with bee milk. In giving comb to the breeding queen, it should be trimmed back to the foundation provided the bees have built comb at the edges and bottom of the foundation. The queen prefers to lay in natural built combs rather than those built from foundation. I have used all kinds of foundation and find this to be true. The comb shown was not trimmed back so it can be seen that the queen laid in a sort of horseshoe shape.

The bees do not build cells with such regularity as the cells in foundation, therefore such comb is difficult to cut as many larvae will be destroyed. When the eggs are laid in cells drawn from foundation it is much better if the row of cells is uniform and easily cut in a straight line. Now let us examine the larvae in the cells to see if they are getting lavish feeding, the necessity of which we have so frequently stressed. The larvae are so small they can hardly be seen with the naked eye. They are flooded with bee milk fully as much as larvae in swarming cells. In these same cells they will remain while the bees go right on feeding them. Let us now prepare to start those larvae on the road to queendom.

The starter hive

Many years ago Henry Alley used what he termed "the swarming box". He did not start cells in it but confined bees in it for some hours and at night dumped them into a hive with no brood or queen. He seemed to think it necessary to keep the bees queenless for some time to get them to accept the cells. Then the prepared cells were given to the bees in a hive containing two frames of honey and pollen. Mr. G.M. Doolittle said the bees would accept cells better if kept queenless for three days. In my experience I find the bees accept the cells just as readily after an hour or two as they do if left queenless for a longer period. We usually keep them confined in the starter hive from one to two hours which gives them plenty of time to clean up any honey that may have been spilled on them when shaking them into the starter hive. In later years Eugene Pratt gave the cells direct to the bees confined in a box and he called it "The swarm box". We prefer to call it "The Starter Hive" as it is used for starting cells and has no relation to a swarm. Our starter hive is made wide enough to hold five standard frames.

The total depth is 17 inches. It has a wooden bottom with cleats at the sides making it suitable for feeding or watering bees. Three frames of honey and pollen are put in *but no brood.* In case no nectar is coming in, the bees that are to be used in stocking the starter hive should be fed three days before using them and this feeding should be continued each day until they are shaken into the starter hive.

Food and Water Necessary to the Bees in the Starter Hive

A matter that I believe most of us have overlooked is the necessity of food and water for the confined bees. All have observed how bees carry water in abundance when rearing brood. Now this water is even more necessary when rearing queens for the bees require more water when producing the milk in abundance. The best way to provide this is to fill the center comb with diluted honey. This not only provides the necessary water but the food it contains acts as a flow of nectar. In case some 50 cells are to be started from five to six pounds of bees should be used. In case from 75 to 100 cells are to be started from 8 to 10 pounds of bees should be put in. If larger number of bees are put in the starter hive should be kept in the shade or some cool place. The screens on the sides provide an abundance of ventilation.

Getting the Bees with Which to Stock the Starter Hive

The colonies are all in two stories with queen excluders between the two hives. The queens are confined below and the top story is kept will filled with brood. All that is necessary to do in filling the starter hive is simply to shake in the bees from the top story. The bees will be depleted somewhat, as some are used to keep up the strength of the breeder hive and the finishers, so unsealed brood should be given to these colonies from other colonies. We use the bees from each colony once in every four days. This is less work than it would be to use a larger number.

Preparing the Cells

Let us now consider the larvae in the breeder hive is ready for use. First some wax should be melted. An electric hot plate or a sterno stove with canned heat may be used.

The comb from the breeder hive is now cut into strips one cell wide. With a small paint brush, wax is painted on the cell bar and the strips of cells placed on the waxed bar. Then with the brush wax is painted on the sides of the cell strips. The cells are then cooled in the wooden tank and more wax painted on till both sides of the strip of cells have enough wax to make the cells strong enough so they will not mash down when inserting the spike. Wax should be cool for painting the cells, nearly ready to solidify. We use a wooden tank four inches wide inside measurement, and half an inch longer than the cell bars. A cleat is tacked in one end under which the end of the cell bar is placed. At the other end of the tank is a nail driven through from the side extending about two inches into the box. As soon as the cells are waxed the bar is shoved under the wooden cleat at one end of the tank and the other end is placed under the nail at the other end. Then cold water is poured in the tank till it covers the melted wax. It is well to put ice cubes in the water for we want the cells to be firm so the larvae in the cells can be destroyed without mashing down the comb. While we prepare the cells as quickly as possible in order that they may be in care of the bees without delay, it has been astonishing to note what gluttons for punishment these well-fed larvae are. Frequently we have more cells that we have use for so instead of using the larvae, we set the comb aside for the larvae to die in order that we may use the comb again. Often after four days, when we give the comb to the bees in the breeder hive, some of the larvae will be rejuvenated or resuscitated, or whatever you wish to call it. An interesting feature is that although the larvae were away from the bees four days, they had not increased in size a single bit. This shows that larvae away from bees do not grow, so we should keep them away from the bees as short a time as possible.

It will be necessary to destroy some of the larvae in the cells in order to make room for the queen cells. We have tried many methods but find the best way to destroy the larvae is to insert a spike in the cells and punch out the bottoms. An ice pick may be used if the point is broken off. We usually destroy two cells and leave one which gives plenty of room for the fine large queen cells. Sometimes when we are short of cells we destroy every other one. Some will be built close together and will be destroyed in cutting but one can get more cells in this way.

Putting the Prepared Cells Into the Starter Hive

We use two bars in each frame, the space above and below the bars being filled out with wood. The starter hive is given a little jar on the ground so the bees will fall to the bottom and not fly out. The two frames of cells are set in place, the comb containing thin honey being in the middle. Cells will be built better in the middle of the combs as shown, for where the cells are near the top of the frame the bees are apt to build comb over them.

Program of These Rejoicing Bees

Now what is the first thing the bees do upon receiving these prepared cells? They first unload the accumulated milk in their glands by feeding the larvae. Next, *they begin tearing down the walls of the worker cells.* It has been stated by many that the bees enlarge the worker cells. This they can not do. They tear down the worker cells and build queen cells over the larvae. In this way they build the cells as they want them and do not have to put up with any man-made contraptions.

When starting cells from worker cells the bees first "raze" them and then "raise" them. This reminds me of an incident that happened years ago. I told my wife that a virgin had gotten into the hive where I had a lot of queen cells and this virgin was razing cells. She thought I said the virgin was "raising hell" so I let it go at that for I rather liked her version of the incident.

Finishing the Cells

While the bees are now busily engaged in remodeling the cells, we must prepare to have them finished, for they should not remain in the starter hive more than 24 hours for the best results. Again I shall tell what we do and then tell why we do it. The best colony for finishing cells *is a strong queenless colony.*

The queenless colony that is to finish the cells must contain at least five pounds of bees and several combs of mostly capped brood. If no nectar is coming in, this colony must be fed liberally. Again, to be extra conservative, we have two colonies to finish the cells that have been started. If this is properly done you can produce cells every bit as good as those built by the bees during swarming. Bees must be added to keep up their strength. It is some help to them by giving them the cells with the bees that are with the cells. When more bees are needed the starter hive is dumped in front of the finishing hive.

We use a single story to finish the cells unless there is a heavy honey flow on, in which case we add a second hive body but have the cells finished in the bottom hive. If we start four bars of cells in the starter hive it is best to give them to two finishing colonies. As the bees finish a job that has been started we must be very particular to see to it that they are in the best possible condition to finish the cells properly. In case we wish to start cells every day we have three sets of finishers, two hives in each set. In case one is rearing queens on a very extensive scale he can have four hives or more in each group and use two or more starter hives. After the cells have been in the starter hive overnight we lift out the frames of cells and give one to each of the finishers. The adhering bees are given to the finishers which helps keep up their strength. The frame of brood and cells are arranged as follows: At the side of the hive nearest to you is a frame of capped brood, then a frame of cells. Next day, group number two are given bars of cells, then next day group number three are given two bars of cells. Now, when you start the fourth batch of cells, the ones in the first group are sealed or nearly so and are given bars of cells, a frame of capped brood separating them from the first frames given. This is repeated till all finishers have three frames of cells each. When ready to give the next frames of cells, those first given now are what is known as "ripe cells." They must be taken out to make room for the next frames, and are ready to be introduced to mating hives. These cells will hatch the following day.

The photograph on page 60 shows the arrangement of frames of cells as they are painted white that they may be easily seen. From this on, when a frame of ripe cells is taken out another frame with started cells is given and this may be continued throughout the season. As the bees in the finishers become weak more bees must be added from the starter hive and capped brood given from other colonies. In case you do not need so many cells you can skip a day and the space in the finisher is filled by moving the combs over and should there be a honey flow an extra comb should be given to prevent the bees from building combs in the space.

Caution

Three or four days after giving brood you should look over the frames to remove any cells that may have been started. It is astonishing to not how many accidents can happen in those finishers no matter how careful one may be. We tack a zinc queen excluder to the bottom of the finishers to keep out virgins for if there is any place one can get it is sure to do so. By making the hives bee-tight and by examining the brood carefully you will have no accidents, or rather I should say, few.

The Adolescent Mating Hive

The most successful small hive I ever used is what I term my adolescent hive. It is larger than the baby hive. It is made to hold three frames 5 x 8 inches. It is made of full three-fourths inch material and the cover is made of 2-inch material.

A small mating hive should not have a ventilator, as the air will blow in through it and cause the bees to realize their weakness and they will abscond. For an entrance a half-inch hole is made in the center of the hive. This hole is slanted upward to keep out the rain. If the entrance is placed at the bottom, the ants or robber bees will kill out the bees if they get weak.

When feeding, a battery filler can be used and a lit-tle food may be squirted into the entrance at night. A package box which is used for bringing bees from the outyard is made with wire screen sides as shown in the cut. Half a pound of bees will nicely stock this hive. A little feed is given, a ripe cell is placed between the combs, and the package box is placed on top. The entrance should be kept closed for 24 hours if the bees are taken from the outyard. If taken from the home yard the bees should be confined 48 hours, or many will return to their former colonies.

After confining them for the required time, the en-trance is opened and the package box removed. To those who prefer a small hive this is by far the best of anything I ever tried. Even this hive requires close attention to see that it is kept in the right condition. We have wintered queens in such hives here in Florida but the results were rather uncertain so we have adopted a larger mating hive, one holding five standard frames. These hives are placed on concrete standards, two hives on each standard. The covers are made of one-and-one-half inch material. This cover keeps them warm in the winter and cool in the summer. They are heavy enough so no wind will blow them off. At first we formed them in spring. This weakened the colonies in the outyard to such an extent that they gathered no honey. Now we form the mating hives in the fall after the honey flow is over. We proceed as follows: As the bees in the outyard must be requeened, we take a package of one and one-half pounds of bees and the old queen. The bees and queen are put into the mating hive. As they have a queen it is not necessary to confine them if they are put in after dark. A young queen is introduced into the colony from which the package was taken. With a young queen, the colony soon makes up for the loss of the bees. After the old queen in the mating hive has been laying for a week or so she is removed and a ripe cell given. With this management, the colonies in the outyard are actually benefited. As we winter the colonies in the mating hives there will not be many to establish in the fall, only the ones that have run down for one reason or another. Some mating hive colonies may be too weak for wintering. In such cases a package with queen is put in the same as though the hive was empty. When there is no honey flow, our five-frame hive requires more feed than the smaller one. On the other hand the bees in our larger hive will gather more honey and therefore need less feed. While rearing queens we have but four frames in the hive as it is much more convenient to handle the frames. After queen rearing is over in the fall, an extra frame of honey is put in. A few may die out in the winter for one reason or another, but the rest will be so strong that they may be divided to make up the loss. When introducing cells, when no honey is coming in, we feed the night before giving the cells. Should any be short of stores, they are given an extra amount of feed. The entrance to this hive is in the center of the front. It is one and one half inches in diameter and the hole is slanted upward to keep out the rain.

Introducing the Cells

In introducing cells no protectors are used but we simply press the cells lightly against the combs, placing them near brood as the bees will care for them better.

Stocking the Mating Hives

In stocking the mating hives it is best to take them from an outyard as they do not have to be confined so long. We do not weigh the bees but measure them instead. We have what we call our bee hopper. It is four and one half inches wide. The bottom is covered with galvanized iron. A scoop is made just wide enough to reach across the inside of the hopper. For the five frame hives we use about a pound and a half of bees.

How Long Should the Bees be Confined to the Mating Hive?

In case bees are taken from the outyard they should be confined to the mating hive one day and released after dark the day following. In case bees are taken from hives in the home yard close to where they are to be established they should be confined two days and released after dark. Most of them will stay but a few will be gifted with remarkable memory and will go back, but there will not be enough to materially weaken the bees in the mating hives.

Mating Hive Records

One must have some system of knowing what the conditions are within the mating hives. We have used a number of systems and out of all of them we have worked out a system that gives such perfect satisfaction that we never expect to make a change. We need to know the conditions within the mating hive for a period of two weeks, for by that time the virgin should have been mated and by that time she is a laying queen. Therefore, we have a row of copper tacks across one side of the hive, 14 in number, one for each day of the week for two weeks. These tacks are driven about halfway in. The first tack at the left represents Monday. By putting a mark over Thursday's and Sunday's tacks, we can see at a glance what days we are working on. Tags are made of galvanized iron two inches square. A square hole is made in the center of the tag. The corners are bent forward a little for easy handling. They are first dipped in yellow quick-drying enamel. After drying, each corner is dipped in enamel of the following colors. First black, then red then white, then blue reading to the left. Now let us consider the hive and it's signal relation to the queen. Let us say we introduced the first cell on Monday. The tag is placed on the tack at the extreme left and turned to red, indicating a cell has been given. We examine the hive Wednesday and if the virgin is there the tag is turned to white. Should the virgin not be there, the tag is turned back to black indicating the hive is queenless. When another cell is given, after the first one is not accepted, it is put in on Wednesday so the tag is moved to the third position and turned to red. This process is continued throughout the season. A picture of our mating hive has already been given. To indicate the condition within the hive we have blocks four inches long painted black so they may be seen at a distance. A number of these blocks are distributed among the hives, being placed on the stands between the hives. In case the hive can spare honey the block is placed lengthwise of the hive at the extreme left. If it needs honey it is placed crosswise in the same position. The block shown in the cut shows the hive can spare both brood and honey. If placed crosswise at the extreme right it indicates that the hive needs brood. If placed lengthwise on the same position it indicates the hive can spare brood.

Bee Candy

Candy for the introducing cage or for mailing cages is made by mixing XXXX powdered sugar with honey. This should be kneaded as stiff as possible. If the honey is heated it will absorb more sugar. If one has bee disease, or is unable to get a health certificate before shipping, candy may be made of invert sugar and such candy is nearly as good as that made with honey. To make invert sugar, put half a pound of water in a kettle and let it come to a boil. Then add one pound of granulated sugar and stir in one-fourth teaspoon of tartaric acid and boil till a temperature of 248 degrees is reached. A candy thermometer is necessary for this. No certificate of health is required by the postal authorities when using invert sugar candy. All that is necessary is a statement that there is no honey in the cage. After the invert sugar is cooled it is used the same as honey in making bee candy.

Shipping Queens

Our experience has taught us that the reason so many queens die in the mail is lack of ventilation or because they are overheated. We have found that by wrapping the cages in cloth, they go through in good condition if they are not too long on the road. We also ship in mailing tubes which give even better ventilation. Shipments made with mailing tubes have gone to 43 foreign countries by air with almost no loss.

A Review

As we though it necessary to do a lot of explaining as we went along, we will now give the procedure step by step:

1. Stock the breeder hive as described, putting queen and one frame of brood and the two wood-comb frames in the back compartment, putting the frame of brood in the center.

2. Prepare as many colonies as are needed by putting them in two story hives with brood above the queen excluder and the queen below.

3. Cut Modified Dadant foundation in four pieces and put them in frames and get this foundation drawn in any colony suitable.

4. Take out the frame of brood from the back compartment of the breeder hive and put it on the other side of the partition, being sure the queen is in the back compartment. Then put in a frame of the drawn foundation. Leave it there for 24 hours, then put it on the other side of the partition. In case you wish to start cells every day, put in another frame of drawn foundation in the back compartment. Repeat this every day until the fourth comb is placed the other side of the partition, when the first is ready for use.

5. Put three frames of honey and pollen in the starter hive and put some thin syrup or diluted honey in the middle comb. Then shake in bees in the amount of five or six pounds from the hives previously prepared. Leave them three or four hours.

6. Take out frame of young larvae and cut into strips, mount them on bars and give them to the bees in the starter hive.

7. Next day remove the queen from two colonies to be used as cell finishers and give to each colony one frame of cells and dump the starter hive in front of the colonies from which the bees were taken.

8. When all finishing colonies have three frames of cells each, and the fourth is ready, remove the first given and introduce the cells to mating hives.

Starting Cells Every Third Day

In starting cells every third day, but two finishing colonies will be needed. In this case you put in a new comb in the breeder hive every third day and when it is placed the other side of the partition, an old comb or a frame of brood is placed in the back compartment between the two wood-combs. There are both advantages and disadvantages in starting cells every third day instead of every day. We are running more to the third-day plan. One advantage it has is that the drawn sheet of foundation may be left for the queen to lay in longer. Another is that as cells are introduced every third day, in introducing the cells, the ones introduced three days before can be examined. If using the third-day system, in order to keep the record straight, one should write down on a piece of paper the letters C S I, C S I in rotation, these letters being placed at the left hand of the paper, one below the other down the paper. C stands for comb and indicates that you gave a new comb to the queen in the breeder hive. S means that you start cells, and I means you introduce cells. Put the date and the day of the week after each letter. Of course in starting there will be no cells to start or introduce, but after some time the letters will have a meaning. In case one started cells regularly every third day there would be little use for the above memorandum as the condition of the cells in the finishing colonies would indicate what is to come next. However, in case you slipped one starting, all cells would look much alike and you might get things mixed up. At any rate I have done so and when I came to giving the started cells to the finishers I found the hive full of virgins!

Modifications for the Small Beekeeper

For the commercial queen breeder, or for the honey producer who wishes to rear 200 queens or more I strongly recommend the management and equipment that have been described. For the small beekeeper, or the back-lotter who wishes to rear a dozen or maybe 50 queens in a year, no special equipment will be necessary. Under such circumstances it is best to rear queens during a honey flow as conditions are more favorable and queen rearing is a simple matter compared to rearing them when there is no flow. When the honey flow is just beginning is the best time to start as there is an abundance of pollen coming in along with the nectar, and pollen is just as necessary as nectar.

For starting cells, first get the foundation drawn as described. Next set to the rear the hive containing your breeder, with the entrance in the opposite direction from the original position. In its place put another hive and in it put two frames solid with honey and pollen. Shake in half of the bees from the breeder hive and put the breeding queen into the prepared hive. Within 24 hours she will have a good patch of eggs in the new comb. The comb should be placed over an excluder on a strong colony. In case you do not wish to start more cells immediately, this hive with the two combs and the one with the eggs may be set over the hive formerly containing the breeder, setting this hive on its former stand. The two hives must be separated with an excluder, with the queen in the bottom hive. This colony must be well fed for we want the larvae that will soon come from the eggs to be well fed. As soon as the eggs hatch, turn a strong colony to the rear and in its place set a hive into which you put two frames of honey and pollen then shake out half of the bees from the hive you have moved into the hive with the two combs. In case the colony is not of sufficient strength, two colonies may be set to the rear and half of the bees from both colonies shaken in. Next cut the comb into strips as has been described and give a frame of two bars of prepared cells for this colony to both start and finish. In case one prefers to use a starter hive, one can be made by merely tacking a screen to the bottom of a hive. If this is done the hive should be set on blocks high enough to afford plenty of ventilation.

Mating Hives for the Small Beekeeper

In case increase is desired a hive may be moved to a new location and another hive set in its place. Into this hive put two or three frames of brood with adhering bees from the hive moved, being careful not to get the queen.

It is well to wait two or three days after dividing this colony before giving a cell for bees will keep returning from the hive moved and are apt to destroy the cell given. As soon as the queen has mated and laying, the hive should be filled out with combs or frames of foundation or starters. In case feeding is necessary a good way to feed is to tip the hive back a little and under the front place a block. Then at night pour in a little syrup. Be careful not to put in more feed than the bees will consume during the night for if any syrup is left it will incite robbers as a bottom board feeder is the worst of any feeder to cause robbing.

Mating Queens in an Upper Story

In case one does not wish to make increase, a good way to mate queens is in an upper story of a two-story hive. A frame covered with wire screen should separate the two hives. Always use a wire screen and *never a queen excluder.* Some in the past have advocated using a queen excluder but in the great majority of cases the bees will kill the virgin as soon as she emerges. In the frame of the wire screen an opening of half an inch should be made for a flight hole. This entrance should be to the rear of the hive below or the queen in returning from her mating flight is apt to get into the hive below.

The entrance to the top hive should be closed for two days, otherwise too many bees will join the colony below. Syrup or honey should be put into one of the combs and a ripe cell given. After two days the entrance should be opened and the virgin will soon fly out and mate. When the honey flow permits, this is an excellent way to requeen the colony for we have *two laying queens* instead of having it queenless while introducing a queen or cell. The bees in the top hive will not suffocate from having the entrance closed for the screen will afford abundant ventilation. The best way to put syrup into the combs is with the use of a pepper box feeder. That is a mason jar with the cap full of holes. With this the syrup may be shaken into the combs. If you merely want to requeen, all that is necessary is to hunt up the old queen below and remove her and take out the screen. The bees will not fight as they have kept up their acquaintance through the screen. In case increase is desired, just move the bottom hive to a new location and set the top hive on the old stand. The lower hive should be moved as it is much stronger than the top hive. The bees for a time will try to get into the back of the hive but after a time they will make a chain around to the front door and all will be well.

Requeening by Giving a Ripe Cell

As a rule we do not recommend giving a ripe cell to a full colony, as it weakens the colony by being deprived of a laying queen for nearly two weeks. Again, a strong colony does not accept a cell as readily as a weaker one, and if the cell is not accepted it will further weaken the colony. However, when the honey flow is over and more bees are not needed immediately, it is often desirable to requeen by giving a ripe cell. As it is difficult to find a virgin in a strong colony, one may give a cell and without looking for the virgin another cell is given three days later. In a week after the first cell is given, the colony may be examined and if there are no queen cells started, you may know a virgin is present. In case both cells are destroyed, or should the virgin be lost in mating, it is best to introduce a laying queen as the colony would become too weak if given another cell.

About the Author

Jay Smith was an active writer in the bee journals of his day and wrote two of the most loved and used of the queen rearing books still being used and followed by many today. He is most famous for his other book, Queen Rearing Simplified, but this book is the culmination of his work in queen rearing.

Lightning Source UK Ltd.
Milton Keynes UK
UKOW050618100713

213520UK00005B/24/P